The Second Rush

Mining and the Transformation of Australia

Published in 2016 by Connor Court Publishing Pty Ltd

Connor Court Publishing Pty Ltd
PO Box 7257
Redland Bay QLD 4165
sales@connorcourt.com
www.connorcourt.com
Phone 0497 900 685

ISBN: 978-1-925501-14-8

Cover design: Maria Giordano

Printed in Australia

List of shortened forms

AAEC	Australian Atomic Energy Commission
AAPC	Australian Aluminium Production Commission
ABS	Australian Bureau of Statistics
ACA	Australian Coal Association
AIDC	Australian Industry Development Corporation
AIS	Australian Iron and Steel
ALCAN	Aluminium Company of Canada Limited
ALCOA	Aluminum Company of America
ALP	Australian Labor Party
AMAX	American Metal Climax
AMIC	Australian Mining Industry Council
AMP	Australian Mutual Provident Society
ARCO	Atlantic Richfield Company
BA	British Aluminium Company
BCA	Business Council of Australia
BCPL	Bougainville Copper Proprietary Limited
BHAS	Broken Hill Associated Smelters
BHP	Broken Hill Proprietary Company Limited
BMR	Bureau of Mineral Resources
BP	British Petroleum
CES	Central Engineering Services
CHINALCO	Aluminium Corporation of China
c.i.f.	cost insurance freight
COMALCO	Commonwealth Aluminium Corporation Proprietary Limited
CRA	Conzinc Riotinto of Australia Limited
CSC	China Steel Corporation
CSIRO	Commonwealth Scientific and Industrial Research Organisation
CSR	Colonial Sugar Refining Company Limited
CZP	Consolidated Zinc Proprietary
CQCA	Central Queensland Coal Associates
DIDO	drive-in-drive-out
dwt	Dead Weight Tonnage
EEC	European Economic Community
ENI	*Ente Nazionale Idrocarburi*
EPDC	Electric Power Development Company of Japan
ERA	Energy Resources of Australia
E.S.and A.	English Scottish and Australian Bank
FIFO	fly-in-fly-out
GEMCO	Groote Eylandt Mining Company

GDP	Gross Domestic Product
GFC	Global Financial Crisis
FIRB	Foreign Investment Review Board
f.o.b.	free on board
GNP	Gross National Product
IEA	International Energy Agency
INCO	International Nickel Corporation
JCB	Joint Coal Board
JSM	Japanese Steel Mills
KACC	Kaiser Aluminum & Chemical Corporation
LNG	Liquefied Natural Gas
MGMA	Mount Goldsworthy Mining Associates
MINSEC	Mineral Securities Australia Limited
MITI	Ministry of Trade and Industry
MIM	Mount Isa Mines Limited
MKU	Mary Kathleen Uranium
MLC	Mutual Life and Citizens Assurance Company
NAA	National Archives of Australia
NABALCO	North Australian Bauxite and Alumina Company Limited
NGRP	New Guinea Resources Prospecting Company
NLA	National Library of Australia
NLC	Northern Land Council
NSW	New South Wales
OECD	Organisation for Economic Co-operation and Development
OPEC	Organization of the Petroleum Exporting Countries
POSCO	Pohang Iron and Steel Corporation
PMA	Petroleum and Minerals Authority
QAL	Queensland Alumina Limited
RTZ	Rio Tinto-Zinc Corporation
SEC	State Electricity Commission
SECWA	State Electricity Commission of Western Australia
TDM	Thiess Dampier Mitsui
TPM	Thiess Peabody Mitsui
UMAL	Utah Mining Australia Limited
USSR	Union of Soviet Socialist Republics
WA	Western Australia
WAPET	Western Australian Petroleum Limited
WMC	Western Mining Corporation
WTO	World Trade Organization

Table of Contents

Introduction

There have been two great minerals rushes in Australian history: the gold rush in the second half of the nineteenth century; and a wider minerals rush, which started in the 1950s and 1960s, the effects of which have continued to the present day. This is a history of Australia's second rush and the enormous changes it has wrought.

The discovery of gold in 1851, argues economic historian Ian McLean, "delivered a major shock to the predominantly pastoral economy of Australia and ushered in a dramatic episode in the country's prosperity".[1] Gold triggered the biggest economic disruption experienced in Australia to that time. Its "effects quickly permeated the social, political, and even cultural spheres, some so deeply that their impression has lasted to the present".[2] For David Hill:

> The gold era brought sweeping and lasting changes. It produced great wealth and ensured the financial viability of the precarious Australian colonies. It stimulated a dramatic increase to Australia's population, put an end to convict transportation, challenged the British class system, laid the foundations of Australian egalitarianism and played a key role in the establishment of Australia as a nation.[3]

The gold rush led to a rapid increase in immigration, a "rush" of peoples unhindered by controls across colonial borders or a national immigration policy. Nor were there any barriers to entry to the goldfields. To prospect for gold, an individual needed only a shovel, a pan, some food, a tent and the wherewithal to pay licence fees to the colonial authorities.[4]

In 1850 the non-Indigenous population of Australia's colonies numbered just 400,000. In the space of 10 years, the peak decade of the nineteenth-century gold rush, Australia's population trebled and the colonies were producing half of the world's gold. As a result, Australians suddenly found themselves the owners of a large proportion of a globally important strategic commodity, not unlike the inhabitants of the oil-rich Gulf States in the twentieth century. In the 10 years after 1851 Australia produced more than 40 per cent of the world's output of gold, which made up two-thirds of the exports of New South Wales and Victoria.[5] In 1861 gold's share of net national product was 15 per cent.[6] Gold was a tremendously important discovery for the Australian colonies. It was the basis of the monetary system of Great Britain, whose economy was the strongest in the world, and an international monetary system based on gold would develop in the late nineteenth century.[7]

The economic consequences of the mid-nineteenth-century gold rush were not confined to a few years. They lasted for about half a century. Production averaged three million fine ounces per annum in the 1850s. In the 1860s this had declined only to 2.5 million. The rate of production fell to about two million fine ounces per annum in the 1870s and then to 1.5 million in the 1880s. The peak gold-producing decade of the 1850s was thus followed by only a gradual decline in gold production over the three succeeding decades. Then, in the 1890s, another round of major discoveries in Western Australia boosted production back to three million fine ounces per year and helped to alleviate the effects of the 1890s Depression in the rest of Australia.[8] By 1903, when the Boer War had temporarily removed South Africa and the Transvaal from the world market in gold, Australia became the world's largest producer of gold again for a short time.[9] Ian McLean argues that it is appropriate to look at the entire half-century from 1851 to 1900 "as a single era of economic expansion, rapidly increasing population, and rising incomes" substantially influenced by the mining of gold in New South Wales, Victoria, Queensland and Western Australia.[10] No similar economic epoch would occur in Australia until the long period of economic growth from the end of World War II to 1973.[11]

Gold was the most important source of Australian export earnings between 1851 and 1870, making up more than half of the total. Throughout the four decades from 1851 to 1890, it remained a stable component of Australian exports.[12] Wool, which had dominated export receipts until

1850, took second place to gold in the 1850s and 1860s. Even when wool recovered its status as Australia's principal export commodity in the 1870s, gold still accounted for over 28 per cent of Australia's exports and for 16 per cent in the 1880s. Then in the 1890s, after the wool industry collapsed, exports of gold rose to equal those of wool, each making up 30 per cent of the total.[13] In the three decades after 1860, incomes continued to increase in Australia at a rate of 1.3 per cent per annum and the population rose from over one million to over three million. The growth in population enlarged the domestic market and workforce and stimulated industrial and agricultural development. In aggregate, the economy of the Australian colonies grew at an average rate of 4.8 per cent per annum between 1851 and 1869.[14]

Immigrants drawn to Australia by gold usually stayed. Despite the rising population, income per head was higher at the end of the 1850s than it had been at the beginning of the gold rush.[15] The 1850s saw the beginning of overseas borrowings by colonial governments when Victoria obtained a loan on the London market in 1858. Major political developments, too, occurred during the period of economic expansion in the second half of the nineteenth century. In the 1850s Victoria and Queensland separated from New South Wales, leading to the establishment, along with Tasmania, South Australia and Western Australia, of the six Australian States that would comprise the Commonwealth of Australia after 1901. Around the peak of the nineteenth century gold rush, five of the six Australasian colonies were granted "responsible government". Western Australia followed suit in 1890. After obtaining responsible government, the colonies gained independence in most matters except for foreign policy and defence, but they retained the Crown as head of state and their strong cultural ties to the mother country. But in the next half-century from 1900 to 1950, with only one or two exceptions, mining descended into the doldrums. From a peak of 10 per cent of gross domestic product (GDP) at the turn of the century, the mining sector's share fell to two per cent in the early 1960s. The resurgence of gold in Western Australia in the

1890s had seen total mineral exports average £13.1 million between 1898 and 1900. But from 1900 to World War I the production of gold declined sharply.

This decline was in part mitigated by the export of lead, zinc and silver from Broken Hill in New South Wales. In 1883 Charles Rasp, a boundary rider working for the Mount Gipps Station, thought that he had found tin in a crested outcrop known as Willyama in western New South Wales. When the samples were assayed, they turned out to contain not tin but silver and lead. Rasp and his associates then formed a syndicate to stake out the whole hill. By the end of the decade the syndicate had floated the Broken Hill Proprietary Company, whose mining of the silver-lead lode at Broken Hill had created a thriving metropolis of 20,000 residents. By 1888 Broken Hill was producing one third of the world's silver.[16]

Exports of coal, which had been discovered in New South Wales in the early nineteenth century, grew only fitfully after Federation.[17] On the eve of World War I, minerals accounted for 18.5 per cent of merchandise exports.[18] After the war and into the 1920s, the story of mining at an aggregate level was one of decline. Gold suffered because its price remained constant while inflation drove up wages and the cost of capital equipment. The prices for base metals also remained low. Between 1918–19 and 1929–30, the contribution of mining to Australia's GDP fell from 4.1 per cent to 2.4 per cent and the share of the mining sector in merchandise export income fell from 11.3 per cent to 7.5 per cent.[19]

During the 1920s a group of entrepreneurs known as the Collins House Group, which had come to own half the value of all metals produced in Australia, sought to compensate for low mineral prices by diversifying and buying shares in rival companies to secure global markets.[20] In 1915 the leaders of the Collins House Group, W.L. Baillieu and W.S. Robinson, purchased smelters at Port Pirie from the other great Australian firm, the Broken Hill Proprietary Company (BHP). Forming a new company called Broken Hill Associated Smelters, Collins House grew in time to acquire a power and influence equal to that of BHP. At the same time, BHP became a manufacturer of iron and steel for the Australian market.[21]

The Depression years of the 1930s were marked by a temporary resurgence in the fortunes of gold. In that decade, the price of gold almost doubled as a result of currency fluctuations, leading to a threefold increase in gold production and rapidly expanding exports. Gold exports reached £14.8 million and 12 per cent of merchandise exports in 1938–39 and played a major part in maintaining Australia's solvency during the Great Depression.[22] The other major part of Australia's mining sector – lead, silver and zinc – averaged only 5.2 per cent of merchandise exports in the late 1930s.[23] At the end of World War II, mining exports had fallen to £10.9 million or 7.2 per cent of merchandise exports. In 1951 minerals and fuels had fallen to make up little more than one per cent of total exports.[24]

Partly, this state of affairs reflected the phenomenon that most of the easily recoverable ores in Australia had been found and mined by the end of World War II.[25] With some exceptions – for example, silver, lead, zinc and copper at Mount Isa in 1923, scheelite at King Island in Tasmania in 1926, and uranium at Rum Jungle in the Northern Territory in 1949 – there were no major mineral discoveries in the half-century after Federation.[26] During this period, many of the profits from mining and managerial expertise were directed to secondary industry, making steel, smelting lead and refining zinc, but also manufacturing aircraft and ships and developing heavy engineering. Mining of black coal had begun in New South Wales early in the nineteenth century and had reached nearly two million tons in 1881, more than half of which was exported to other Australian colonies and to the United States, India and China.[27] Coal production increased from six million tons in 1900 to 12 million tons in 1913, but afterwards the Australian coal industry went into long-term decline as export markets turned to cheaper sources of supply or alternative products, such as oil. With one exception, Australia's annual production of coal did not regain its 1913 level until 1942. In the 1920s and 1930s, Australia's protective tariff dampened mining activity by raising costs of inputs used in the mining sector during a time when the prices of Australia's mining staples, gold and copper, were either stable or falling.[28] In the period from Federation in 1901 to the 1960s, Australia's economy turned inwards as the ratio of trade to Gross Domestic Product (GDP) declined. The protected manufacturing sector increased in importance from about 12 per cent to 28 per cent of GDP. During this period the government sector accounted for an increasing share of employment and economic output while the private sector was subject to an increasingly higher degree of regulation.[29]

Until 2013 there was only one comprehensive history of Australian mining, Geoffrey Blainey's classic, *The Rush That Never Ended.*[30] First published in 1963, this book built on earlier works on mining towns and company histories.[31] In the preface to the 1963 edition, Blainey explained that his history of Australian mining was of the precious and base metals mined in Australia. He touched on iron because "Australia's steel industry was nursed and nourished by a silver-lead mine", namely BHP's mine at Broken Hill. But he omitted coal because it "was less a speculative industry, was not an outback industry, had no metallurgical problems; its industrial tensions differed, its markets differed, and it was not such a dynamo of Australia's growth". [32] *The Rush that Never Ended*, as befitted such a seminal work in Australian historiography, went through five editions, the last of which was published in 2003, when the largest of Australia's modern minerals booms, inspired by burgeoning Chinese demand, was just beginning. Ten years later, at the end of the China boom, Malcolm Knox produced another one-volume history of mining in Australia, *Boom: The Underground History of Australia from Gold Rush to GFC*, that complemented Blainey's work.[33] Reacting to the dearth of books on mining, as opposed to other subjects like colonial history, bushranging and military history, Knox's book affirmed the centrality of mining in Australian history from the gold rushes onward.[34]

Blainey and Knox stress continuity: a mining rush that began with gold and never ended. The argument of *The Second Rush* is that there were two major rushes in Australian history, the first being the nineteenth- century gold rush, which began in 1851 and ended after Federation in 1901. For more than half a century thereafter mining in Australia went into eclipse until a second minerals rush began in about 1960 and, as this book will argue, continued at least until the end of the China boom in 2012. There are very few threads of continuity between these two rushes.

A comparison of Australia's long-term mining history with South Africa's is instructive. South African historian Jade Davenport has demonstrated how mining played a defining role in the course of South Africa's history over more than a century and a half. "Nowhere else in the world", she argues, "has a mineral revolution proved so influential in weaving the political, economic and social fabric of a society".[35] South Africa's trajectory was completely changed by the discovery of diamonds and gold in the nineteenth century as Australia's had been by the discovery of gold in 1851. But in the early 1930s, when mining, apart from gold, was languishing in Australia, mining industries were directly responsible for a staggering one quarter of South Africa's GDP.[36] As late as 1965 the Australian Committee of Economic Enquiry, chaired by Sir James Vernon and including Sir John Crawford as a member, observed that the mining industry in Australia, in contrast to that of South Africa, was "not a major sector of the economy".[37] The Vernon committee predicted that mining, then accounting for 10 per cent of exports, would increase only slowly until 1975. Reacting to criticism of the report's predictions about Australia's mineral economy, Crawford wrote in 1965 that he would "certainly deprecate flights of fancy which turn the Committee's $300 million for mineral exports in 1974–75 into a range of $600–800 million".[38] Australia's mineral exports in 1974 in fact reached $1.26 billion – four times as much as Vernon's Committee of Inquiry had originally predicted.[39]

As the Vernon committee was inquiring into the Australian economy, the second great minerals rush in Australian history was already beginning. It commenced in the 1950s and 1960s with discoveries of world-ranking deposits of iron ore, bauxite, coal, nickel, uranium and natural gas, some of them the largest in the world. In the space of a decade Australia was lifted to the front rank of the world's mineral exporters and its inward-looking economy became more exposed in the international marketplace.[40] Before the second minerals rush Australia's major export market was the United Kingdom and its exports were overwhelmingly pastoral and agricultural. By the 1970s the mining industries had become a major exporting sector, resulting in Japan replacing the United Kingdom as Australia's largest trading partner. By the mid-1980s mining had surpassed agriculture as the largest source of Australian goods and services exports and Northeast Asia had become the major destination for Australian exports. In the 2000s Chinese demand meant that minerals and fuels were making up more than half of all exports. The second rush was punctuated by three booms: the Japanese-inspired minerals boom from the early 1960s to 1973, the resources boom from about 1977 to 1982 and the China boom from 2002 to 2012.

The second rush featured minerals that had not played a prominent part in earlier Australian mining history or that had languished in the first half of the twentieth century. One such mineral was coal. The coal industry had developed in colonial times but had then gone into decline during the first half of the twentieth century when it became under-capitalised, poorly performing, crisis-ridden and domestically orientated. For precisely these reasons Blainey largely omitted the coal industry from the first edition of *The Rush that Never Ended*. Another mineral that had not played a prominent part in the first edition of Blainey's book was iron ore, which he had merely touched upon as laying a foundation for the Australian steel industry that developed in the first half of the twentieth century. From 1938 until 1960 the Australian government prohibited the export of Australia iron ore in the belief that reserves of this strategically critical resource were limited. But after 1960, with the relaxation of the ban, iron ore became an outwardly focused industry and one of Australia's most important export industries. By the 2000s it had replaced coal as Australia's top export-earner.

Another important new industry was based on bauxite. This mineral had been mined only in very small quantities in Australia until the 1960s when huge reserves of newly discovered payable bauxite were developed in Northern Queensland and the Northern Territory and a way was found to upgrade previously known deposits of bauxite in Western Australia. The mining of bauxite and its refining to produce alumina and then smelting to produce aluminium metal is a major component of the history of the second minerals rush. Blainey addressed one aspect of this industry in his history of Alcoa of Australia, published in 1997.[41]

The Second Rush concentrates mainly on the recovery and development of "hard minerals" but includes a discussion of petroleum and natural gas, which were both discovered in Australia in the 1950s and 1960s. Oil was found in Bass Strait in quantities that shifted Australia from total dependence on imported oil to 70 per cent self-sufficiency in the 1970s. Natural gas began to transform industry and society in the 1960s and 1970s and to be exported from the 1980s. *The Second Rush* covers commodities that had been mined in Australia from the nineteenth century, such as gold, which made a spectacular recovery in the 1980s, as well as copper, lead, zinc and manganese. It also covers others that were developed only in the twentieth century such as nickel, uranium, natural gas and oil. Central to *The Second Rush* are industries based on two metallic minerals, iron ore and bauxite, and one non-metallic mineral, coal.

The first chapter of the book discusses the transformation of the coal industry between 1955 and 1972. The post-World War II history of Australian coal is in part a story of how governments promoted modernisation and change by regulation. Abandoning the earlier policy of *laissez faire* towards the coal industry, federal and State governments set up the Joint Coal Board in the 1940s. The board modernised the New South Wales coal industry and encouraged the development of coal for export in the 1950s and 1960s. The other aspect of the post-World War II story of coal is that of discovery. Australian, American and Japanese companies were motivated to explore the Bowen Basin in Queensland to find coking coal deposits and develop them for the Japanese steel industry.

The second chapter examines the relaxation of the federal ban on the export of iron ore, which permitted the beginning of an Australian iron ore export industry. Historians have tended to view the economic policies of the Hawke Labor government in the 1980s as the crucial reforms that pushed Australia in the direction of greater openness. This book supports Ian McLean's argument that these policy shifts began much earlier than the 1980s with changes including those to Australia's coal and iron ore policies in the 1960s.[42]

The third chapter traces the evolution of Western Australia's Pilbara region, where the vast majority of Australian iron ore had been discovered, into a world-ranking iron ore province that provided the basis for what became Australia's dominant export sector in the 2000s. The development of the Pilbara was based on the high quality of the iron ore deposits, their proximity to Japan and the success of Australian and foreign-owned mining companies in borrowing capital based on long-term contracts with Japanese steel mills. By these means the WA government and Australian mining companies were able to construct a "state in miniature" in the Pilbara.[43]

The fourth chapter discusses the development of Australia's bauxite mining empire in Queensland, the Northern Territory and Western Australia in the period from the 1950s until 1972. The development of Australia's bauxite deposits produced a wave of industrialisation in Australia not seen since BHP's entry into steel making in 1915. This industrialisation occurred not only in Western Australia, Queensland and the Northern Territory, where the bauxite was mined, but also in New South Wales, Victoria and Tasmania where it was refined into alumina and smelted into aluminium.

The fifth chapter discusses the discoveries of oil, gas, nickel, uranium and manganese and examines the impact of the Japanese-inspired minerals boom on Australia in the late 1960s and early 1970s. This period saw the beginning of a significant change in Australia's export mix away from an overwhelming dependence on pastoral and agricultural products in the direction of mining. Minerals and fuels represented barely three per cent of Australian exports at the first half of the 1950s, were still only six per cent in the mid-1960s, but had risen to almost 20 per cent in the early 1970s. [44]

The nineteenth-century gold rush had been triggered by what Ian McLean describes as "an increase in the known endowment of gold (a supply-side shock) rather than an increase in demand leading to a rise in price (which at that time was fixed)".[45] The second minerals rush, beginning in the late 1950s and early 1960s, was driven by the dynamic of rapidly expanding demand, first from Japan, later from Northeast Asia more broadly, and later still from China, but also by changes in government policy on resource development and by discoveries of minerals that were a response to rising regional demand and changing government policies. These connections are complex and will be explored in some detail in *The Second Rush*.

In general, the countries comprising the West had pulled ahead of most other parts of the world in terms of production and wealth in the twentieth century. One of the most successful of these countries, the United Kingdom, was Australia's main export market from 1900 to the 1960s. Some developing countries, however, were able to make up the gap with the West. Beginning with incomes per head as little as a fifth or a quarter of those of Western countries, such modernising countries were required to achieve per capita growth rates of more than four per cent per annum if they were to close the gap with the West within about two generations (or 60 years). To do so they needed to assemble all the elements of an advanced economy – steel mills, cities, electricity plants, shipping lines and so on – at the same time.

One of the first countries to achieve this feat was Japan. Though already an advanced industrial company in the early twentieth century, Japan had been devastated by the effects of World War II. After the war, per capita income grew in Japan at an average rate of 5.9 per cent per year between

1950 and 1990, with the highest rate of eight per cent between 1953 and 1973. This latter period of high growth coincided with the minerals boom in Australia.[46] In the period after 1973, the Japanese economy continued to grow, but at a slower rate than in the 1960s and early 1970s. Income per head in Japan jumped from a fifth of US levels in the 1950s to almost 90 per cent in the 1990s, "a spectacular, indeed unprecedented catch-up".[47] By the 1990s, after Japan had caught up with the West in terms of capital per worker, education per worker and productivity, it could grow only as fast as other advanced countries: by a percentage point or two each year.

Chapter Six examines the mining industry in the Whitlam years, from 1972 to 1975 – in this twilight period straddling the end of the "minerals boom" and the start of the "resources boom" in the second half of the 1970s. The period of the Whitlam government coincided with the ending of the Japanese-inspired minerals boom, the first oil price shock and the onset of "stagflation", a term describing an economy simultaneously experiencing high inflation, high unemployment and lower growth rates of aggregate GDP. This was the time in which the Whitlam government introduced radical policies to unwind the increasing amount of foreign ownership of the mining industry. Although the Whitlam era coincided with a drying up of new foreign investment in mining, the already established and now leading minerals industries, coal and iron ore, nonetheless experienced strong growth. In these industries the second minerals rush gained momentum despite the malaise in Western countries triggered by the steep rise in the price of oil in the 1970s. Australia therefore enjoyed a measure of insulation from the worst effects of the 1970s downturn. As the chapter shows, however, mining played an important part in the politics of the Whitlam government, including its defeat in 1975 based on a campaign against unorthodox means of attempting to raise money for mining activities.

In the years from 1976 to 1983, Australians continued to experience "stagflation" and slower economic growth under a federal coalition government led by Malcolm Fraser. But they also experienced another resource boom, triggered by the world oil price increases engineered by the Organization of the Petroleum Exporting Countries (OPEC) cartel in 1973 and again in 1979. The rise in the price of oil triggered a boom in the energy resources sector, particularly steaming (thermal) coal, coal-fired electricity plants, the uranium industry and the aluminium and natural gas industries. The "resources boom", which is examined in Chapter Seven, was of shorter duration than the minerals boom and lasted from about 1977 to about 1982. During the resources boom, Australian governments tried to respond to growing unemployment and the long-term decline of Australia's manufacturing industries by stimulating the growth of outwardly orientated resources industries like the coal and aluminium industries. In effect State governments were attempting to push Australia along the path to greater openness, and some time before the economic reforms of the federal Labor government in the 1980s.

The resources boom was punctured by a serious international recession in 1982–83. It was one of the main causes of the defeat in March 1983 of the Fraser government, which had pinned its fortunes to the continuation of the resources boom. Just before the start of the resources boom in 1976, the mining sector accounted for 27 per cent of Australian exports, the rural sector for nearly 40 per cent and manufacturing 15.6 per cent. At the end of the resources boom in 1983, mining accounted for 32.6 per cent of goods and services exports, more than the rural sector's 31.5 per cent and considerably more than the shares of the manufacturing and services sectors at 17 per cent and 17.9 per cent respectively.[48]

The years from 1983 until the turn of the twenty-first century are described in Chapter Eight as a period "between booms". They were an interlude between the end of the resources boom in 1983 and the beginning of another resources boom in the first years of the twenty-first century known as the "China boom". In the 1980s inflation and unemployment remained stubbornly high. Aggregate rates of growth were higher than in the 1970s, (3.5 per cent compared with 3 per cent), and incomes rose at 1.9 per cent per annum compared to 1.3 per cent per annum in the previous decade. But, concludes McLean, "there was no return to the prosperity associated with the 1960s when the average increase had been 3.2 per cent".[49] Due to persistent conditions of oversupply, mining commodity prices remained under constant pressure in the 1980s and 1990s. To take one example, the nominal traded price of iron ore was almost exactly the same in 2000 as it had been in 1980.[50]

The recession years of 1982 and 1983 ushered in two decades of depressed prices for Australia's principal mineral commodity exports. The iron ore industry in the Pilbara experienced much tougher times in the 1980s and 1990s than in the 1960s and 1970s. Similarly, the coal industry faced a long period of hard bargaining on prices with its overseas buyers. Yet despite the tougher times for coal in the 1980s and 1990s, coal remained Australia's largest single export earner for most of those years, and production and export levels increased significantly. After lean years in the first half of the 1980s the aluminium industry made a strong recovery in the late 1980s and early 1990s, and it was joined by gold, which made an astonishing rise to become one of Australia's top five commodity exports in the late 1980s. Between booms, moreover, the mining sector as a whole frequently outperformed the rural sector in the 1980s as an export earner. During this period between booms, in individual sectors of the mineral economy – coal, aluminium and gold – the second rush continued. Also in this period one of Australia's largest mineral deposits, at Olympic Dam in South Australia, was developed.

Following a recession in 1990 and 1991 Australia experienced a "sustained period of growth and prosperity characterized by rising living standards, declining levels of unemployment and inflation, and only minor fluctuations in economic activity".[51] Though mining contributed to Australia's improving economic conditions in the 1990s, those years were not a period of resources boom, partly because of the slowing of Japanese economic growth (and therefore its demand for Australian resources) and partly because of the Asian economic crisis of 1997 and 1998, which particularly slowed the demand for Australian coal. But on the whole, the period between booms was one in which the mining sector was either the largest or second-largest export-earning sector and the mining lobbies had become among the most powerful in the country. In the 1980s the mining industry mounted a concerted and mainly successful campaign against national land rights for Aboriginal peoples but after the High Court's Mabo decision in 1992 it adjusted to a new regime based on negotiation and partnership with Indigenous peoples.

Ian McLean interprets the period from 1851 to 1900 as a single era of economic expansion, influenced by gold. *The Second Rush* makes a similar case for viewing the period from 1960 to 2012 as an epoch comparable with the golden half century in Australia between 1851 and 1900. In large part, this was the result of the transformation wrought by mining that reached its apogee in the "China boom" of the first part of the 2000s. Chapter Nine deals with the period after 2001 when China entered a new phase of growth that ushered in the most dramatic phase of the second rush in Australia – the so-called China boom from 2002 to 2012 – and resulted in the further transformation of the mining industry into Australia's overwhelmingly dominant export sector and iron ore into its single most important export. The industries that boomed during the period from 2002 to 2012 were those that had been developed in an earlier part of the second rush. The China boom of the first part of the 2000s developed into what was called a "commodities super cycle" through the massive Chinese fiscal stimulus in the wake of the Global Financial Crisis of 2007-08. But mining companies and governments both misunderstood the nature of the China boom. When the Chinese later returned to a strategy of more moderate and consumption-led growth, the mining boom in Australia ended and was followed by a period of plummeting commodity prices, lower revenues for governments and generally less prosperity for Australians. We are yet to see whether this is merely the end of a boom, or whether it represents a more fundamental shift: the end of the second rush.

Aerial view of Port Kembla and coal-loading wharf in foreground, 1964. NAA: A1200, L50013

Moura mine operated by Thiess Peabody Mitsui Coal Proprietary Limited, Central Queensland, 1966. NAA: A1200, L53373

1

The Transformation of the Black Coal Industry, 1955-1972

In the 1860s mining in Australia was dominated by gold. A century later three industries – coal, iron ore and bauxite – were in the vanguard of a mining renaissance. In the interwar period and into the 1950s the coal industry was inefficient, inwardly focused and riven by industrial disputes and unable to meet local demand. Coking coal supplied the Australian steel industry and steaming coal was used for generating power, town gas manufacture and rail transport. But mineworkers were handicapped by irregular employment and high incidences of accidents and respiratory diseases.[1] After the Great Depression of the 1930s, the coal industry partially recovered but remained essentially unprofitable and under-capitalised.[2] A 15-month industrial dispute in the 1930s decimated the coal export trade and a 10-week stoppage in 1940 hampered Australia's war effort. These strikes preceded an even more serious industrial dispute – the national coal strike of 1949. Coming at the end of a decade of chronic industrial strife, the 1949 strike interrupted gas and electricity generation and gravely accentuated the coal industry's difficulty in meeting domestic demand.[3]

The 1949 strike provoked a counter-attack on the mining unions and their communist leadership from the Chifley Labor government. This did not, however, stop the government from being defeated at the December 1949 general election.[4] Also symptomatic of the poor condition of the coal industry in the 1930s and 1940s was the absence of an export dimension. In the nineteenth century, Australia had sold coal to North and South

America, the Pacific Islands and Asian countries. Exports peaked in 1907 at 2.645 million tons.[5] But after 1921, coal exports steadily declined. By the late 1930s a mere 400,000 tons was exported. Between 1921 and 1968 exports of black coal to continental Europe were negligible and Pakistan was as far west as Australian coal was exported.[6]

In Australia up to the 1950s there was no national coal industry. Rather there were separate State-orientated coal industries supplying domestic coal-fed power stations, black coal for railways and some houses. Brown coal in Victoria and South Australia and poor quality black coal in Western Australia was uneconomic for export. Mining was generally underground in New South Wales and Queensland and the coal industries in those States were reliant on supplying coal for local steel and metallurgical industries. The two key Australian States mining black coal in the early post-war period were New South Wales and Queensland. In 1961 New South Wales produced 18 million tons, compared with Queensland's three million tons.[7] Harold Raggatt, a historian of mining, attributed New South Wales's predominance to the location of its coalfields near ports and industrial centres such as Wollongong and Newcastle.[8]

A powerful interpretation of twentieth-century Australian history views the 1980s as the crucial decade in which the Australian economy was opened up to international influence. According to this interpretation, the settlement reached after Federation in 1901, comprising protection for manufacturing industries and centralised wage fixation, broke down in the 1980s when the Hawke Labor government promoted greater integration of the domestic and world economies, including by floating the dollar, reducing tariff protection and introducing competition from foreign banks.[9] Revising this interpretation of Australian history, Ian McLean suggests that the policy shifts to greater openness began decades earlier than the 1980s. He points to reforms such as the 1957 Agreement on Commerce with Japan, the end of export bans on certain minerals and the phasing out of the White Australia Policy in the 1960s.[10] For McLean:

> Taken together, these nudged the economy in directions somewhat

> counter to those it had taken for much of the twentieth century. Thus the 1980s are best viewed as the period of most active and comprehensive policy change in this sense, but not its origin.[11]

In providing an account of Australia's second minerals rush, *The Second Rush* helps explain the origins of Australia's steps towards greater economic openness before the 1980s through the development of mining. In contrast to the process of deregulation in that decade, the history of the transformation of the coal industry is a case study of how comprehensive regulation transformed the NSW coal industry into a highly efficient industry in the space of two decades. In combination with the later development of black coal deposits discovered in Queensland, government involvement would lay the basis for the development of the coal sector as Australia's leading export industry from the late 1970s to the early twenty-first century.

The Revival of the NSW Coal Industry

The transformation of the coal industry began in 1946 when the Chifley Labor government set up a Royal Commission into the industry in New South Wales. Chaired by Supreme Court Justice Sir Colin Davidson, the royal commission identified four prerequisites for the stability and growth of the coal industry: preserving industrial discipline; maintaining confidence in the sanctity of agreements and the efficacy of the law; collecting and publishing essential facts and statistics; and fostering innovation.[12] The Chifley government accepted the recommendations of the royal commission and acted in concert with the NSW Labor government to remedy Australia's "coal problem". The symptoms of this "problem" included an inability to meet local demand, a history of industrial strife, unemployment and irregular employment, and a low rate of return on mining investment.[13]

Reacting to the catastrophic state of the coal industry in the 1930s and 1940s, mining unions exerted strong pressure on governments to nationalise it. This raised the question of which government had the authority to do so. The Australian Constitution gave no express power to the Commonwealth.[14] However, the States had sovereign power to acquire all the assets of the coal industry without compensation. But the Davidson Royal Commission report noted that:

> [n]o Government in the British Empire has yet attempted deliberately to rob citizens of their possessions by legislation and it cannot be presumed that Australian States will be the first to adopt such predatory tactics.[15]

Following the Davidson report's recommendations, the coal industry was not nationalised as the British coal industry was in the 1940s.[16] Instead of nationalisation, the path taken was comprehensive state control and regulation. The Commonwealth and NSW governments "intervened directly in coal in a way and on a scale unparalleled in Australian political and economic history".[17] In 1946 the federal and NSW parliaments passed Coal Industry Acts that established a Joint Coal Board (JCB). The acts were passed in the same form in both parliaments except that "the powers granted to the Board to control collieries and compulsorily to requisition and resume land, buildings, plant, machinery and equipment were contained only in the New South Wales Act".[18] The legislation enabled the federal government's financial powers to rehabilitate a vital industry and its powers over interstate commerce to be employed with the State's competency on regulation of mines, health and social welfare.[19] As the JCB stated in its first Annual Report:

> In broad concept the aim of the Coal Industry Acts is to regulate, assist and rehabilitate the coal industry within the framework of private ownership, with the proviso that where necessary the Board has the power to step in and control and operate coal mines and ancillary enterprises.[20]

The JCB consisted of a chairman and two other members charged with ensuring that the NSW coal industry produced coal in sufficient quantities to meet local and overseas demand. The board was also responsible for conserving, developing and working NSW coal resources to the best advantage of the public interest. Federal and State legislation empowered the board to ensure the economical use of coal, to maintain essential services and activities and to promote the welfare of workers engaged in the coal industry. Only if the cooperation of owners and miners was withheld was the JCB to consider owning and operating mines. The acts also established a Coal Industry Tribunal – outside the remit of the Conciliation and Arbitration Court and with jurisdiction over all disputes extending beyond the limits of any one State. The JCB's chairman from 1950 was the Scottish-born Sam Cochran, an accountant who had been appointed to chair the Queensland Electricity Commission in 1937.[21]

The JCB took up its duties with enthusiasm. It supervised the adoption of mechanically operated boring machines and prepared forward estimates of machinery and equipment requirements. It set up a pool from which machinery and equipment could be bought or hired and formed subsidiary companies to mine coal as an emergency operation. The Board also began prospecting for open cut mines, the first of which had been opened in New South Wales in 1944. The JCB took over three mines and bought three more. All of those purchased were either returned to their owners later or sold to private interests and the NSW State Electricity Commission. By that time, the JCB had invested about £2.5 million into a badly depleted industry. A sign of its success was that at the end of the 1950s there were 42 efficiently mechanised mines compared with just six when the JCB was established.[22] By 1958 the coal industry had steered away from nationalisation. In the words of historian M.H. Ellis, it was "tending towards control by private and semi-public coal producers acting in combination on a voluntary basis, somewhat along the lines envisioned by Mr Justice Davidson, but with statutory backing and policing of their plan by the Commonwealth and States".[23]

The JCB helped the industry to overcome a period of coal shortage in Australia in the late 1940s and early 1950s. Thereafter its strategy was to establish a dynamic, market-orientated business by completely transforming the techniques of coal mining, which had remained unchanged from the mid-nineteenth to the mid-twentieth centuries. To achieve this end, it encouraged mechanisation, rationalisation and competition. Following the establishment of the JCB, technological innovation, shaped in response to the market, became routine. After the federal government gave tax concessions for capital expenditure on mines, collieries in New South Wales increased their investment in the industry.[24] One of the JCB's greatest successes was to persuade the Miners' Federation of Australia to permit the mechanisation of mines. By 1961 about 30 per cent of daily production in New South Wales came from the mechanical extraction of pillars.[25] In 1964 the Board could proudly report that "… the Joint Coal Board has been able to guide the industry back from a state of near collapse to a state of efficiency, prosperity and growth".[26]

While the JCB had the task of devising and implementing coal industry policy, the Coal Industry Tribunal had the function of implementing those parts of state policy that related to relations between employers and employees.[27] Mine workers were represented by the Australian Coal and Shale Employees' Federation, otherwise known as the Miners' Federation, one of the most militant unions in the county in the interwar and early post-war period. Australian coalminers had early organised themselves into trade unions and had developed the idea that coal belonged to the land and the people rather than to coal mining companies. This made for much more combative industrial relations than elsewhere in the mining industry, such as metals mining, where owners and unions generally had more of a sense of partnership. The Coal Industry Tribunal was able to combine Commonwealth and State powers over conciliation and arbitration to give direct effect to the JCB's policies: to persuade workers "to attend regularly, to abandon the strike weapon and to accept managerial authority".[28] With the assistance of the JCB and the Coal Industry Tribunal, the 1950s and 1960s saw a decline in the incidence of the strikes that had so debilitated the industry in the 1940s.[29]

The JCB's remit was only for the coal industry in New South Wales. In that State black coal was mined underground in the western, southern and northern margins of the Sydney Basin. This was a large sedimentary basin extending from south of Wollongong to Newcastle and north through the Singleton area to the region around Gunnedah. The towns of Cessnock, Singleton, Muswellbrook, Lithgow and Camden were located either on or near coalfields. The industrial centres of Newcastle in the north and Wollongong–Port Kembla in the south developed close to coking coal deposits. In the early 1960s there were three main areas of coal production in New South Wales: the South Coast area near Wollongong; the Burragorang Valley, including the Camden, Oakleigh and Wollondilly collieries; and the area of Newcastle and its north.[30]

Harold Raggatt discerned four categories of coalmine owners in New South Wales in the 1960s. The first was the steel industry: the Broken Hill Proprietary Company (BHP) and its subsidiary, Australian Iron and Steel (AIS). The second category of owners was the NSW government, through its Electricity Commission and the States Mines Control Authority. The third category included two fairly large miners: Clutha Development (which would be purchased by American shipping magnate Daniel Ludwig in 1965) and Coal and Allied Industries Limited, a company formed in 1960 by the merger of J & A Brown and Abermain Seaham Collieries and operating on the Newcastle and Maitland coalfields.[31] The fourth was a category of miscellaneous privately owned mines.

Noteworthy in Raggatt's categorisation was the importance of the State government and the steel industry in the NSW coal industry. Over time, privately owned coal mines, selling part of their production overseas, would gain increasing importance. In Queensland the first coal mine had been opened on the south bank of the Brisbane River in 1843. More coal was discovered at Blair Athol in 1864, at Callide in 1890 and later in Ipswich. In contrast with New South Wales, coal mining in Queensland was regulated by a Queensland Coal Board set up under purely State legislation and including some State-owned mines.[32]

The Beginning of Coal Export Trade with Japan

Les Thiess initiated the first sales of coal to Japan after World War II. Thiess was a Queenslander who headed the large construction firm of Thiess Brothers, involved with, among other ventures, the Snowy Mountains Hydro-Electric Scheme. Les Thiess was born into a farming family in the Darling Downs in 1909. In the 1930s he and his brothers (as Thiess Brothers) moved from chaff-cutting to sinking dams and repairing roads in southern Queensland. During World War II, Thiess became involved in the construction with the US Army of major defence facilities throughout Australia. Responding to the increased wartime demand for coal, Thiess was drawn into constructing Australia's first open-cut coal mines at Blair Athol in Queensland and Muswellbrook in the Hunter Valley. Thiess was powerfully built, dogged and fired by a fierce work ethic.[33] He owned leases of open-cut coal at Callide that he transported from the port of Gladstone to Victoria in the 1940s and 1950s.[34] When an oversupply of steaming coal developed in Australia in the early 1950s, Thiess received permission from the Menzies government to export two shipments of 6500 tons of coal to Japan. But these shipments were not followed up because of the lack of continuing demand in Japan for steaming coal.[35]

Menzies' Minister for National Development, Senator William Spooner, fully supported early efforts to develop an Australian coal export trade. In the early 1950s, he asked his colleague and the Minister for External Affairs, R.G. Casey, to discuss the matter with US officials in New York. Born in Sydney in 1897, Spooner had fought in the Australian Imperial Force (AIF) at Gallipoli and on the Western Front during World War I. After the war, he founded one of the nation's most successful accountancy firms in Sydney. He also played a major part in the United Australia Party in the 1930s and in the establishment of the Liberal Party in 1944 and 1945. On 11 May 1951, he succeeded Casey in the portfolio of National Development when the latter moved to External Affairs. Spooner would hold the National Development portfolio for the next 13 years. Snowy haired, bespectacled and friendly, Spooner played an active role in development projects across

Australia, including the Snowy Mountains Hydro-Electric Scheme. He would also play a vital role in the reorganisation of the coal industry and the end of the iron ore export embargo and thus in the beginning of Australia's second minerals rush.[36]

Although Australia was a possible competitor with America as a coal supplier to Japan, the US government supported Australian coal exports to Japan two reasons. One was to allow Japan to diversify its trading partners to include Western countries like Australia, which traded in sterling. The other reason was strategic. In 1952 a US official told Spooner "the United States regarded it as important to bring Japan into the Western trade orbit. He realised that, although this would mean encouraging a competitor, it was better than Japan entering into large-scale trade with Communist China and the USSR".[37] The defeat of the Chinese Nationalist regime by the Chinese Communist Party was of crucial significance to the unfolding of Australia's second minerals rush. Had the Nationalists remained in power in China, the Japanese might have continued to purchase raw materials, like coal and iron ore, from China as they had done before and during World War II. With the advent of the People's Republic of China in 1949, Japan was forced to look for essential raw materials elsewhere – first from the United States and then from Western-orientated states in the Pacific such as Australia.[38] In other words, Communist revolution in China was of great economic benefit to Australia – one of the great ironies of post-war Australian history.

Casey's representations to the United States gained the attention of a group connected with General William Donovan, an influential consultant to the US government on security matters.[39] In 1954 the Donovan group sought to interest the Queensland and Commonwealth governments in the possibility of developing the coalfields at Nebo in Queensland to supply Japan with coking coal. It proposed to borrow $US20 million from the International Bank for Reconstruction and Development (World Bank) to build a railway and port to develop the Nebo fields, 100 kilometres south-west of Mackay.[40] But the group dropped the idea when the Japanese steel industry advised that it would rather continue purchasing expensive American coking coal than buy the inferior Nebo coal. It would not be until the Japanese-driven minerals boom of the 1960s and early 1970s that the Nebo project would be resurrected. Actual development of the Nebo field had to await the "resources boom" of the late 1970s and early 1980s. The Donovan group then contemplated the idea of developing Queensland's steaming coal deposits at Blair Athol, 340 kilometres north-west of Rockhampton. But Spooner advised "we cannot see any sound basis upon which this coal field can be developed on a substantially larger scale as an economic proposition".[41] It would not be until the 1970s that the Blair Athol mine would be developed for an export market.

US officials proposed the development of the Queensland coalfields in the 1950s with the strategic aim of providing a source of raw materials for the development of Japanese industry. While these plans were not successful, it is significant that two American companies, Utah Construction and Mining and Peabody Coal, were among the first companies to explore, and develop, mines in Queensland to supply Japan with coal. US private enterprise thus took the baton from the US government.

The interest of US officialdom in the prospects of Australia as a coal-exporting nation did not go unnoticed in Canberra. Although the Nebo and Blair Athol plans were not considered viable, Spooner actively promoted Australia as a new source of coal exports to East Asia. By that time Japan was importing three million tons of coal annually from the United States. In the 12 months to September 1954, the NSW coal industry had secured orders from South Korea for 360,000 tons of coal with an export value of over £1.25 million. South Korea needed to import the coal during its reconstruction and rehabilitation phase after the Korean War.[42]

Officials and experts advising Spooner in the 1950s were divided about the potential for long-term coal trade with Japan, the largest of the potential East Asian markets. An experienced Australian trader in Japan, Roy Duncan, advised Spooner that there would not be a market for Australian coal in Japan. But H.P. Reinbach, a coal expert working for the Victorian State Electricity Commission, thought there was a "good, substantially permanent market for coking coal in Japan".[43] However, Reinbach observed:

> The organisation for exporting Australian coal has to be improved... The Australian coal producer individually has little export experience, and requires assistance. Collaboration rather than internal competition is necessary. I believe this is possible without undue interference with the freedom of individual interests. The trading firms, taking the long view, will realise that a coordinated effort will avert losses... The only beneficiary of the principle of multiple representation is the purchaser.[44]

In the commercial sector, the Blits Trading Company Pty Ltd played an important role in re-establishing an export coal trade from New South Wales. Until 1955, most Australian exports of coal were steaming (thermal) coal, with only New Caledonia buying coking coal.[45] Jacques Blits was a Dutchman specialising in the textile trade who had migrated to Australia. After World War II, he began exporting a miscellany of commodities to Italy, Sweden and Japan. These included steel scrap, milk powder, casein and charcoal pig iron from Western Australia. While selling charcoal in Japan in the mid-1950s, Blits saw an opportunity to wean the Japanese steel mills from their then only source of imported coking coal – the American Pocahontas seam. Following Blits's representations to the Menzies government, the JCB sent a mission to explore the possibility of coal exports to the Philippines, Hong Kong, Singapore, Thailand, Japan and South Korea. In October–November 1954 the mission spent two weeks in Japan and one week in Korea.[46] Its conclusions were cautious. The mission found no market in East Asia for Blair Athol steaming coal, a temporary demand for imported coal in Korea while the country was being rebuilt and a possible demand for coking coal in Japan if Australian coal could be made available at the desired quality and price.[47]

Industrial disputes in the coal industry in the United States in the 1950s created an opportunity for the first post-war exports of Australian coking coal. The JCB took the view that fostering an export trade in coal to Japan could help improve Australia's balance of payments and make a "useful if relatively small contribution to employment in the coal industry".[48] It hoped that exporting coal could be a way to smooth out fluctuations in local demand to ensure greater regularity of production and employment in the NSW coal industry. For the JCB, the primary purpose of the NSW coal industry was to service Australian needs. An export trade was seen as a useful adjunct to coal's main role in the domestic industry – supplying the local steel industry and generating electricity.

Despite the JCB's support, there was strong opposition elsewhere in Australia to coal exports, particularly in some parts of industry and the scientific community. In 1960, for example, Australian Iron and Steel (AIS) warned the Australian government that the only mine on the South Coast of New South Wales capable of increasing production of hard coking coal was the NSW State Mine at Oakdale. AIS warned of an imminent shortage of coking coal from 1965 onwards and called for a government embargo on all long-term contracts for hard-coking coal with Japan. H.R. Brown, the chief of the coal research division of Australia's Commonwealth Scientific and Industrial Research Organisation (CSIRO), supported export restrictions, arguing that coal exports should be approved only as a temporary expedient for balancing national payments. Brown based his objections to Australian coal exports on the grounds that, of all the world's coal reserves, two-thirds were behind the Iron Curtain, one-third in America, and a mere one third of one per cent in Australia.[49] Thus, in Brown's view, the size of Australia's coal reserves did not justify exports. Reacting to this kind of talk, representatives of the Japanese steel industry became concerned that the Australian government might place an embargo on the export of coking coal. Japan's C. Itoh and Company had started arranging the importation of Muswellbrook coal from the late 1950s; it asked the Department of Trade in August 1960 whether the Menzies government intended to embargo the export of coking coal. The Japanese had been alarmed by a statement by Spooner on 18 August 1960 that "[w]e must reserve some of the hard coking coal for this industry (steel making) and that means we have less of it to export".[50]

Blits fought this current of opinion. In 1955 he arranged for the NSW Mining Company, a wholly owned subsidiary of the JCB, to sell 50,000 tons of Newdell coal blend to Fuji Iron and Steel Company and to Yawata Iron and Steel through the Japanese trading company, Mitsui.[51] Blits Trading Company, along with two other Australian companies, Heine Bros and Gollin & Company, sold more Australian coal to Mitsui and another Japanese trading company, Marubeni Ida, in the ensuing few weeks. The first of the vessels exporting the coal arrived in Newcastle harbour at the end of 1956. This was part of an early Japanese effort to assay the quality and export efficiency of Australian coal.[52]

The leading steel companies in Japan were interested enough by this first flurry of an Australia–Japan coal trade to send a survey mission to visit Australia in August 1958. The mission conducted surveys of the NSW South Coast area, including Wollongong, the Burragorang Valley and the northern coal zone, including Newcastle. Following this visit, collieries on the South Coast signed five-year contracts for the export of hard coking coal commencing in 1959.[53] Japan also began importing from the Newcastle area, taking 786,000 tons in 1957-58.[54] Even this modest increase in Australian coal imports strained infrastructure in New South Wales. In 1959–60, when NSW coal exports reached 815,000 tons valued at £3.3 million, the Japanese steel industry began to voice its concerns about the limitations of Australia's ports. In response, in mid-1959, the NSW Labor Premier, R.J. Heffron, set up a committee to advise on coal-handling facilities. He also sent the Minister for Mines, J.B. Simpson, on a visit to Japan in 1960.

Born in New Zealand in 1890, Heffron had migrated to Australia during World War I. Elected in 1930 to the NSW Parliament for the Australian Labor Party, he served as Minister for Education in the 1950s, and briefly as Secretary for Mines in 1953, before winning the premiership in 1959.[55] In that office Heffron would play a vital role in the transformation of the NSW coal industry, particularly because he had to overcome the hostility of mining unions to an export trade. One of the first and important decisions he took was sending Simpson to Japan. The Japanese steel industry surprised Simpson – who spent one full month in Tokyo – by telling him that they expected to be taking three million tons of coal from New South Wales in 1965, and perhaps more afterwards, if the coal ports of New South Wales could be enlarged enough to accommodate bulk coal carriers.[56]

Spooner followed Simpson's trip with a visit of his own in June 1960. At a meeting with representatives of the Japanese steel industry on 13 June 1960, Spooner indicated that Australian governments would have the confidence to fund better coal transport facilities if the Japanese steel industry could enter into long-term contracts.[57] The Japanese industry representatives encouraged Spooner by pointing out that Japan was short of raw materials for the production of steel, particularly coking coal and iron ore, and had lost its access to coal from North China as a result of the Korean War. The American Occupation authorities in Japan, they noted, had given the steel industry access to American coal in abundance but Australian offers would also be welcome. Chinese coal, they assured him, was not a threat to Australia. This was because the Chinese needed to use coal for making their own steel and because trade between China and Japan was subject to interruption for political reasons.[58]

Japan's interest in importing Australian coal was motivated by the resurgence of the Japanese steel industry during the 1950s. A key to efficient steel production was to minimise costs by using large-scale and capital-intensive integrated plants. The minimum efficient size of a steel mill in 1950 was decided to be between one and 2.5 million tons. In the early 1950s, the Japanese only had one within that range – Yawata Steel's 1.8 million ton plant. Other Japanese mills produced only up to half a million tons. As a result, Japanese steel was 50 per cent more expensive than that of the United States, which used much bigger operations. Japan's powerful Ministry for Trade and Industry (MITI) therefore set out to produce more efficient mills, with the result that Japan's crude steel production doubled between 1950 and 1955 and trebled from 1955 to 1961.[59] By that time Japan was producing 28 million tons of steel in modernised, large-scale mills and importing nearly five million tons of coal. Chalmers Johnson has concluded that for achievements comparable with that of MITI in post-war Japan one would look to "the wartime Manhattan Project in the United States, or to NASA's sending a manned rocket to the moon".[60] America still supplied most of Japan's imported coal at the beginning of the 1960s, 60 per cent in 1960 and 53.4 per cent in 1961. But Australia was already supplying 16.6 per cent in 1960 and 23 per cent in 1961.[61] Reflecting the increasing efficiency and competitiveness of Australian coal, overseas exports had increased from 1.4 per cent of NSW net production in 1955-56 to 10.8 per cent in 1960-61.[62]

The Problem of the NSW Coal Ports

By the beginning of the 1960s the Australian coal industry was becoming increasingly frustrated with the limitations of the nation's coal ports. Industrial leaders complained about inadequate facilities and the slow turnaround of ships at Port Kembla – the port through which the best of Australia's coking coal flowed and which supported £100 million worth of heavy industries. These industries had expanded enormously since the 1930s. By the late 1950s Port Kembla had outstripped Newcastle as a port, but no berths had been added since World War II.[63] In 1954 the Secretary of the Department of National Development, Harold Raggatt, informed Spooner that the port was not keeping up with the needs of the local steel industry and that its importance to the whole of Australia justified some interest by the federal government.[64]

Born in North Sydney in 1900, Raggatt, like Spooner, had served in the AIF during World War I. After the war he resumed his studies of geology at the University of Sydney. In the 1920s and 1930s he was active in geological surveys and mapping of the Upper Hunter and Central West of New South Wales. This period of his life gave him a strong appreciation of the NSW coal industry.[65] In 1939 the geologist W.G. Woolnough, who had been instrumental in the federal government's embargo on the export of iron ore, recruited Raggatt to succeed him as geological adviser to the Commonwealth government. Raggatt's reports to the Commonwealth government would lead to the creation in 1946 of the Bureau of Mineral Resources, Geology and Geophysics.[66] Never losing his interest in geology, Raggatt was appointed in 1951 as Permanent Secretary to the Department of National Development, a post he held during Spooner's tenure as Minister. Raggatt's decisive role in the history of Australia's second minerals rush lay in his support for the improvement of Australia's coal ports and, as we shall see in Chapter Two, in the relaxation of the iron ore export embargo in the early 1960s.

By 1959 the South Coast region of New South Wales had orders from Japan for three million tons of coking coal in the period up to 1964 as a direct result of the Japanese survey mission of 1958.[67] The manager of one of the major South Coast mining companies, Kembla Coal and Coke, told Prime Minister Robert Menzies in 1961 that his company had spent £5 million pounds in the previous five years to produce coal efficiently and economically. His company had contracts with Japanese steel mills for hard coking coal produced at its Coal Cliff colliery worth £1.25 million for a period of five years, with prospects for more contracts in the future. "Absence of reasonable harbour loading facilities", he warned Menzies, "not only makes such expansion unlikely, but actually, in the face of severe international competition, gravely imperils the chances of renewal of the present contract on its expiry in 1964".[68]

In July 1959, Japan's consul in Sydney publicly criticised Port Kembla's single coal-loading wharf as patently inadequate. Coal was being loaded at that time in Port Kembla's 50-year old outer harbour.[69] To prevent heavy surges damaging the ancient jetty or the coal loader, ships did not tie up at the jetty. They had to berth six feet away tied to heavy buoys. Vessels were loaded in that position and there was no area for storage of coal near the jetty. As a result, only one ship could be loaded at a time, resulting in long delays. The Japanese consul described the problem as serious enough to jeopardise Australia's incipient coal trade with Japan.[70] Japanese steel industry leader Saburo Tanabe inspected Port Kembla a few years later. He noted "there were two cranes with belts but they were not operating in full. The pier also looked like an old relic of the Meiji era, but this was quite understandable as no doubt they never thought Japan would ever be buying so much coal".[71] The jetty could not accommodate vessels larger than 12,000 tons loaded at a rate of 250 tons per hour. In comparison, Australia's American and Canadian competitors were then shipping somewhat better quality coal to the Japanese market in 35,000- to 45,000-ton bulk carriers that loaded at rates exceeding 2000 tons per hour.[72]

Exasperated by the limitations of coal loading at Port Kembla, mines on the NSW South Coast owned by Broken Hill Associated Smelters (BHAS) were contemplating spending £2.5 million to build private offshore loading facilities north of Wollongong.[73] The Chairman of the Southern Colliery Proprietors Association, C.W. Perdriau, supported the Japanese criticisms of Port Kembla. The collieries on the South Coast had spent many millions of pounds in the 1950s on mine modernisation. But there was no provision, Perdriau complained, in the NSW government's plans to expand the port for a modern coal loader.[74] The NSW government had begun the first stage of building a new inner harbour at Port Kembla in 1952 at a cost of £3.2 million to accommodate 17 ships each of 40,000 tons capacity. A second stage was envisaged for 1963 to make room for another 20 ships. But the new harbour was contemplated primarily for the inward shipment of iron ore to the steelworks and for the export of steel products. The original NSW port expansion program made no provision for coal in its plans. This underlined the lack of confidence in export possibilities for coal and the primacy of the local manufacturing industry in the State's considerations in the period up to 1960.[75]

The port of Newcastle also suffered severe limitations. A rock bar at the entrance of the harbour excluded ships drawing more than 25 feet at low water. The loading equipment in the basin was, like Port Kembla's outer harbour, 50 years old and storage areas were restricted. Coal was stored in trucks, which were driven to Newcastle and then unloaded. The JCB and mine owners had installed more modern coal-loading equipment at Newcastle in 1952 at a cost of £175,000 in the expectation of increasing Japanese demand. But Japan's decision not to permit the continued importation of gas or steam coals had seen the orders fall away.[76] The same had been the case in Balmain. Balmain was the export port for Lithgow and its region and the Burragorang Valley. The NSW government had invested £1 million in port improvements there for what was hoped to be one million tons of exports annually. But, by 1960, only a total of one million tons had been exported from Balmain since 1955.[77]

By the late 1950s the Menzies government was well aware of the problems at the NSW ports. But the federal government resolved to take no action because it regarded the ports as a State responsibility. The federal government doubted that the southern collieries would actually be able to increase their delivery rate and was deterred by the prior experience of falling demand at Newcastle and Balmain. The NSW government, too, was wary of taking action. In its view, "the export trade was 'chancy' and it was a serious matter to contemplate the large capital investment on port provisions when the coal export trade may not be permanent, particularly when there were so many competing demands on loan funds".[78]

The Balance of Payments Crisis

The rapid development of a coal export trade in the 1960s, like the iron ore trade, had its origins in the balance of payments crisis that developed when the Menzies government lifted the restrictions on overseas imports it had imposed throughout most of the 1950s. After lifting the restrictions, imports surged and exports were unable to keep pace. Menzies was forced to ask all his ministers to consider ways in which exports might be stimulated. In late 1960 the Commonwealth government's Permanent Heads Committee considered a range of measures to improve the balance of payments, including a submission of the JCB seeking federal funds for improvements for NSW ports. National Development Secretary Raggatt believed that such improvements warranted serious consideration for the objective of increasing Australian exports. He argued that Australia had already developed an export trade yielding £6 million a year in overseas earnings. Just to hold that export trade required improvements to the New South Wales ports. And, even if an export trade in coal did not develop, Raggatt thought that expenditure on improving the ports was justified for the benefits it would give to other industries.[79] Spooner was of similar mind, confiding to Raggatt in January 1961:

> The thing that seems to me to emerge is that the more one thinks about it the more one feels that work on Newcastle is justified from the benefit it would give the steel industry quite apart from its effect on coal exports.[80]

Spooner equivocated about putting a recommendation to Cabinet on the coal ports in the early days of 1961. This was partly because of differences of opinion within government circles about whether a coal export trade would actually develop. The continued lobbying by the JCB was a powerful influence on Spooner. On 18 January 1961, he finally instructed Raggatt to put up a submission on coal ports to Cabinet's export committee. Just as the JCB was instrumental in transforming the NSW coal industry into an efficient industry, so it was decisive in persuading governments to encourage an export dimension to the industry. The Prime Minister's Department agreed with the board's case. A minute to Menzies in early 1961 argued that:

> The important point to raise then is that in the interests of raising coal exports to, say, £10,000,000 by 1965 [it is] essential that Japanese know now that within that period the ports will be modernised because they are tying up contracts with the Americans and Canadians each of which has facilities for their big carriers.[81]

Spooner told his cabinet colleagues that it was realistic to contemplate a New South Wales coal export trade in 1965 of 2.5 million tons valued at £10 million.[82] His submission also made the case that port improvements would benefit the Australian manufacturing industry. The existing state of the port of Newcastle meant that BHP could not use its large economic carriers in the port, and improvements would reduce the costs of sending coal to Whyalla in South Australia and Kwinana in Western Australia.[83]

Cabinet agreed with Spooner's submission and Menzies accordingly sent a telegram to NSW Premier Heffron, informing him that the federal government was giving sympathetic consideration to making provision for port and loading facilities in New South Wales.[84] From discussions between State and federal officials, Spooner ascertained that the expenditure required to boost NSW coal exports would be £11 to £12 million. The work would take six to seven years to complete but would be quicker with Commonwealth assistance. After consideration, Heffron indicated to Menzies that his State government proposed to spend £10.7 million on harbour works and a further £1 million on new railway rolling stock.[85]

The NSW Premier noted that "while there is some reason to be optimistic at present about the possibility of continued coal exports, there can of course, be no guarantee that they will continue at a high rate".[86] He therefore sought a pound-for-pound contribution from the federal government for coal loading plant, equipment and associated wharfage and storage at all three NSW ports at a cost of £5.3 million, of which the Commonwealth share would be £2.65 million. Heffron assured Menzies that, if the Commonwealth made that contribution, the work on the ports would be accelerated. The State would be safeguarded to some extent against a possible future diminution of the export trade with Japan and the charges at port for handling the coal would be lower than otherwise.[87] In short, New South Wales sought Commonwealth assistance for the improvement of the ports as insurance against an export coal trade not fulfilling its early promise.

Nonetheless, the prospect of the federal government consummating a deal with New South Wales was clouded. The South Coast mining companies were still planning to build a private offshore loader north of Wollongong that would obviate the need for a government-built coal loader in Port Kembla's new inner harbour. If this happened the whole question of whether the Commonwealth should subsidise the NSW coal ports was in question. Port Kembla was the place through which the best of Australia's coking coal then flowed and Newcastle coals were the most doubtful part of an export trade. By June 1961, however, Spooner had resolved to recommend that Cabinet approve a subsidy for a coal loading installation at the inner harbour at Port Kembla. He minuted to Raggatt:

> No one can foretell the future of Port Kembla. There will always be demand for coal-loading facilities. Surely I am right when I say one morning that we will wake up and read in the papers that the present wharf has just collapsed from old age. There is a need for coal loading facilities there even though the emphasis may not be on export. But it is an issue which will have to be aired and decided.[88]

The Japanese Steel Industry Missions

Visiting Japanese missions considerably influenced the eventual decision by the Commonwealth government to subsidise improvements to the coal ports and the decision by the NSW government to proceed with the upgrades. By that year the Japanese steel industry was making rapid strides. In 1959, Japan had overtaken France in the production of crude steel. In 1960 its production exceeded 20 million tons. The expansion of the steel industry mirrored the strong development of the Japanese economy as a whole. The growth rate of Japan's gross national product (GNP) in the 1950s was far above those of Western European countries with the exception of the Federal Republic of Germany.[89] The increase in Japan's manufacturing output, particularly its steel production, was even more conspicuous. The future course of Japanese economic development had been charted by a 10-year economic plan initiated at the beginning of the 1960s to double national income by the end of the decade. The plan aimed at Japanese production of 48 million tons of steel in 1970, for which the importation of raw materials from overseas was essential.

In the 1950s the output of Japan's steel industry had risen steadily with the assistance of loans from the World Bank and other low interest loans from two state banks: the Japan Development Bank (JDB) and the Japan Export-Import (EXIM) Bank.[90] The post-war Japanese economy was characterised by an interventionist bureaucracy. There was also a high degree of cooperation across industries, between industrial firms, banks and trading houses, and between firms and the state.[91] Three pre-World War II *zaibatsu*, or industrial conglomerates, Mitsui, Mitsubishi and Sumitomo, were remodelled after the war as *kereitsu*. These *kereitsu* were specialised corporate groupings of complementary industrial firms, with a group trading house (*sogo shosha*) that would act as the banker and main source of loan finance for group companies.[92]

In 1958 Japan was dependent on overseas suppliers for basic raw materials. Over 72.8 per cent of its iron ore, 44 per cent of its coking coal and 55 per cent of iron scrap were imported. Although Japan mined some coal, its requirements for coking coal could not be met by indigenous deposits since nearly all of Japanese coking coal formed a soft coke that was unsuitable for larger blast furnaces.[93] While Japan had been plentifully supplied with high-quality American coking coal in the 1950s, the rapid development of the Japanese steel industry had made the Americans wary about increasing their supplies of high-grade coking coals to the country.

A 1956 MITI mission to the United States concluded that importing minerals from countries nearer to it in the Pacific Rim was the only option, with China excluded by political and strategic considerations.[94] Afterwards, Japan lowered its specifications for coking coal. It also became expert in blending varieties of coal so that it could use a maximum of its own coal and a minimum of imported coal. Japan compensated for the cost of freight for these raw materials by building steel plants close to the sea where large coal-carriers could unload directly using the steel mills' conveyor systems. The Japanese steel industry was structured in three tiers. At its core were three "integrated" steelmakers – firms that were involved in all stages of iron making, steelmaking and steel finishing. These three firms were Yawata Iron and Steel Company, Fuji Iron and Steel Company, and Nippon Kokan (NKK), which together accounted for two-thirds of Japanese steel production. A second tier included Sumitomo, Kobe and Kawasaki, all located in the Kansai region of Honshu. These non-integrated producers relied on the three integrated producers to provide them with intermediate iron inputs. A further tier included 44 small-scale steel mills, which focused on steel finishing operations for local markets.[95]

The reconstruction of the industry had thus far been undertaken without Australian iron ore and with relatively small quantities of Australian coal. By 1961, however, the Japanese steel industry had resolved to send two high-level missions to Australia: the first a Goodwill and Trade Mission in March; and the second a Survey Mission on Iron and Steel Raw materials in June. The Vice-President of the Tokyo Chamber of Commerce and Industry and Chairman of Fuji Iron and Steel Company, Shigeo Nagano, orchestrated the missions.[96] Before World War II Nagano had worked for Nippon Steel, a corporation that General Douglas MacArthur, as Supreme Commander of the Allied Powers in Japan, had broken up in 1950. As part of a trust-busting exercise, Nippon Steel was split into Yawata Iron and Steel and Fuji Steel.[97] In 1963 Nagano would be elected chairman of the Japanese Iron and Steel Federation, one of the most powerful components of the *Keidanren*, Japan's federation of economic organisations.[98] The Japan Iron and Steel Survey Mission that Nagano dispatched was headed by Yoshio Shimuzu, Managing Director of Nippon Kokan Steel Co., and included Saburo Tanabe, Chief of the Raw Materials Department of Fuji Iron and Steel Company, and Sadao Sugamata, head of the coal investigative team and Chief of the Raw Materials Research Office of Yawata Iron and Steel Company.[99]

The members of the coal investigative team travelled throughout New South Wales, inspecting mines on the NSW South Coast (Coalcliff, Clifton and Bellambi), the Burragorang Valley and the Newcastle area, where one of the leading miners was Coal and Allied Industries. The Japanese mission reacted positively to an Australian coal industry greatly improved by comprehensive government regulation.[100] The mission informed the JCB that the Japanese steel industry wished to reach a target annual production of 48 million tons of steel by 1970. To achieve this aim, Japan would need to import 22 million tons of coal a year, of which it hoped Australia would supply four to five million tons.[101]

In a meeting with the JCB on 29 June 1961 Sugamata announced that he was generally impressed by the efficiency of coalmines in New South Wales.[102] He described the mines in the Burragorang Valley as comparable with any of the first class mines in the United States in terms of productivity, and the hard coking coal of the South Coast as of very high quality. One of the most surprising outcomes of the mission was Japanese appreciation of the soft coking coal in the Newcastle region.[103] In Newcastle on 23 June 1961 Sugamata dramatically declared that, if the price was right and port conditions could be improved, the Japanese steel industry would be prepared to buy two or three million tons annually of Newcastle soft coking coal by 1965.[104] Few Australians or Japanese could then have believed that Newcastle would develop into Australia's busiest coal port and indeed one of the biggest coal ports in the world in the early 21st century. The first 25,000-ton coal carriers were going into Balmain in 1962 and the Japanese hoped those vessels would be able to enter Newcastle harbour by September of that year.

The Japanese steel industry's intention to import soft coking coal represented a major strategic decision. This was because Japan's government preferred its steel industry to use Japanese soft coking coal and not to import. But the steel industry estimated that Japanese mines could not meet its needs and resolved to ask the Japanese government for foreign exchange to import soft coking coal through Newcastle. A.J. Day, Senior Trade Commissioner in the Australian Embassy, Tokyo, reported to the Department of Trade on 11 July 1961:

> This, of course, is a long-range project because it will be necessary for them, in the first place, to reach the stage where local resources are insufficient but it is estimated that this could occur within three

> to five years at which time they would hope to ship from Newcastle considerable quantities of coal rising, in total, to about 3,000,000 tons per annum in five years from now.[105]

At the same time, the Japanese mission noted concerns that the steel industry would have to address before larger purchases of Australian coal commenced. Sugamata, who had also been part of the earlier Japanese survey mission in 1958, noted that there was considerable room for improvement in the productivity of the South Coast mines, most of which were working one shift rather than two. The Japanese mission was also worried that the Australian steel industry was expanding. They wondered if there was a conflict between the two goals of bigger exports to Japan and a larger Australian steel industry and whether the Australian government might place restrictions on coal exports in the future.

Another serious concern for the Japanese was the persistently poor state of the NSW coal ports. Sugamata noted that Japan had had contracts in 1960 for 700,000 tons of coal from Port Kembla but that only 640,000 tons had been exported because of problems with the coal loader and industrial troubles. South Coast mining company Bellambi Coal wanted to export 200,000 more tons out of Port Kembla, but the Japanese indicated that they would have to reserve a decision on this after considering the state of port facilities.[106] Sugamata explained that Japanese shipowners were not willing to handle increased tonnages because of the congestion of the port and that the Japanese steel industry was worried that shipowners would ask for higher freight prices. The same considerations applied to the port of Newcastle. The Japanese hoped that Newcastle could be deepened to at least 36 feet to allow ships of up to 30,000 tons capacity to enter the harbour. Finally, the Japanese were concerned about the question mark over port improvements on the NSW South Coast – whether a private offshore loader would be built and, if so, whether this would mean that the coal-loading improvements planned by the NSW government for the inner harbour at Port Kembla would be cancelled.[107]

JCB officers sought to allay the Japanese concerns. They assessed that

NSW coal reserves would be sufficient to allow for both a modest export trade with Japan and for the needs of the domestic steel industry. They also agreed to continue considering port improvements. JCB Chairman Sam Cochran emphasised that moving from one shift to two on the South Coast might be desirable, but doing so depended on continuity in demand for the coal. A way of overcoming the problem, he argued, was for the Japanese to enter into long-term contracts for the purchase of Australian coal. Cochran emphasised long-term contracts because the board was conscious that quality and price could be maintained only if plant and equipment were modern and efficient. And the financial policies that permitted this were possible only on the basis of long-term contracts.[108] Coincidentally, the Japanese had already approved a plan to promote the development of mining programs overseas through a combination of three devices: long-term contracts of 10 to 15 years duration with new suppliers; extension of loan financing from Japanese sources; and, in some cases, Japanese firms directly taking equity stakes in foreign projects.[109]

With the encouragement of the visiting Japanese missions, the Menzies government agreed to provide financial assistance to New South Wales for coal-loading plant and equipment on a pound-for-pound basis to a maximum of £1,500,000 for Newcastle, £1,070,000 for Port Kembla and £50,000 for Balmain.[110] While such assistance for public works normally took the form of repayable advances, the federal government took account of the uncertainties involved in providing for an unpredictable export trade. Consequently, it provided £1 million in the form of an advance and the rest as an interest-free loan.[111] By 30 October 1961, tenders had already been called for the removal of the rock bar at Newcastle harbour and dredging had begun at Port Kembla.

But not long after the Menzies government's decision to contribute to improvements to Australian coal ports, there were already portents of a reduced Japanese purchase of Australian coal. The JCB became so apprehensive that it dispatched Cochran to Japan in May 1962. The JCB's concern was prompted by the decision of a number of Japanese mills to postpone imports of Liddell seam coal. Part of the reason for these cutbacks was a recession in the Japanese economy. Rather like Australia's credit squeeze recession of 1960-61, this had been brought on in 1962 by the Japanese government's implementation of tight monetary policy to check the overheating of the economy and to improve an adverse balance of payments. As a consequence, Japan's government instructed the Japanese steel industry to reduce its production by 30 per cent.[112] The Japanese iron and steel industry was thereby rendered sluggish for a number of years in the mid-1960s. On top of that, the steel industry was fighting a battle with the Japanese government for foreign exchange to enable it to import cheaper supplies of Australian soft coking coal rather than having to purchase the more expensive but inferior Japanese product. Teruyoshi Tasaka, the Japanese steel industry's liaison contact with the JCB, told Cochran on 28 November 1962 that:

> Japanese iron and steel industry is at a critical turning point in the fields of use and purchase of coking coal; or perhaps it may be said that the industry finds itself at the point of complete separation from its past practices. Such drastic turn of events has presented itself only very recently, and I believe that it was almost impossible for anyone to accurately predict the situation ten months ago.[113]

In Tokyo Cochran warned the steel industry that the postponement of imports from New South Wales might cause the Commonwealth and New South Wales to rethink their expenditure on coal ports. He also informed them that Australian trade unions were opposing any such expenditure on the grounds that the export market had always been insecure because of its fluctuations. If coal exports to Japan were postponed, Cochran added, NSW collieries would find it difficult to obtain the cooperation of trade unions for exporting coal in the future.

Worse was to come. In 1963 Japan's government embarked on a program to rationalise the Japanese coal mining industry. The hard-pressed Japanese industry had come under severe pressure from competition from petroleum as an alternative source of fuel and power. One measure implemented by the government in 1963 was a request to the steel industry to buy increased quantities of soft coking coal from Japanese mines. In these circumstances, the JCB and Commonwealth and State governments held grave fears that Japan's imports of soft coking coal through Newcastle would dry up. They would drop away, moreover, just when those governments were in the middle of a project to improve coal-loading facilities at Newcastle. The decision to build these facilities had been based on projections of 750,000 to one million tons of soft coking coal being exported from Newcastle annually in the second half of the 1960s.[114] Instead, coal exports from Newcastle declined from a peak of 1.25 million tons in 1961, production slackened and miners on the northern coalfields were stood down. Tenders for the coal-loading plant had been let and a decision to award a contract needed to be made by March 1963. With a decision to proceed with the Newcastle coal loader in the balance, the JCB called for strong governmental representations to the Japanese government. The board feared that, unless the coal loader at Newcastle was installed, the chances of being able to sustain Australia's competitive advantages in the northern NSW coalfields would be low. If the NSW government pulled out of the project now, it would be very difficult to revive it later.[115]

Consequently, the Australian Ambassador to Japan, Laurence McIntyre, called on the Japanese Vice Minister for Foreign Affairs on 4 February 1963 expressing the hope that Japan would still permit imports of soft coking coal through Newcastle at the rate of 750,000 tons per year.[116] The contretemps between Australia and Japan over Japanese protection of its coal mining industry climaxed with a visit by NSW Premier Heffron to Japan in April 1963. Heffron indicated that improvements to the coal ports of New South Wales would cost his government between £10 and £12 million. He explained that – while the open cuts continued to be worked – there was little difficulty. But if it became necessary to close them down because of lack of purchases from Japan, they would be hard to re-open.

Nagano, the Chairman of the Japanese Iron and Steel Federation, mollified Heffron by promising that Japan would source all of its requirements of soft coking coal imports from Australia alone and reject offers of this type of coal from the Soviet Union. Moreover, Japan's steel industry had asked MITI for approval to import the same quantity of soft coking coal in 1964 as it had imported in 1963.[117] Japan, he predicted, would recover from its temporary ills and, by the time the deepening of Newcastle harbour was completed, it would be purchasing more than sufficient imports of soft coking coal to justify the cost of the work carried out on the port. Rather than being criticised, Nagano predicted, Heffron would be praised for improving Newcastle harbour.[118]

In the long-term, as will be seen, Nagano's prediction turned out to be right; his assurances helped smooth over the difficulties in the developing Australian–Japanese coal trade. Newcastle exports recovered from the setback experienced between 1963 and 1965. Though imports of soft coking coal from Newcastle were limited to 1.5 million tons under long-term contracts in the second half of the 1960s, spot purchases were allowed over and above the 1.5 million ceiling so that 2.5 million tons were often imported annually.[119] In the south, the dredging of Port Kembla and the construction of a wharf were completed by June 1963 and the new coal-loading installation was employed for the first time in December 1963 to load 29,488 tons of coal into the Japanese ship *Aino.*[120] By mid-1963, a new coal-loading facility at Balmain was also completed and available for use. And in 1964 the NSW Maritime Services Board awarded a contract to A. Goninan & Co. Ltd for a new coal-loading facility at Carrington Basin in Newcastle harbour, with a completion time of 100 weeks.[121]

The Moura–Kianga Field: Thiess, Mitsui Peabody

In the late 1950s, the Australian coal industry was beset by overproduction. Thiess saw this first hand when the Muswellbrook Coal Company cancelled a contract with Thiess Brothers to work Muswellbrook's open cut mine.[122] At Thiess's own Callide mine there was a small but constant demand for steaming coal by electricity producers and railways in Victoria, but the bulk coal loader at Gladstone Harbour was in operation only intermittently. The trade in Callide coal to Victoria ceased in 1958. Thiess had earlier received some orders from overseas markets in East Asia for Callide coal but had had no follow-up. He became convinced that he needed to blend the Callide product with other types of coal and therefore decided, personally, to finance a search for quality coal in Central Queensland. He hired F.W. Whitehouse, an Ipswich-born geologist, to help him.[123]

Whitehouse started off near the town of Blackwater in the Bowen Basin. This was an area that had been identified as possibly coal-bearing by a British study commissioned by the Queensland government in 1949.[124] Whitehouse began his investigations in 1957, at about the time the Queensland Labor government had been replaced, after 25 years, by a Country–Liberal Party government with Frank Nicklin as the Premier, Ken Morris as his Deputy, Tom Hiley as the Treasurer, Gordon Chalk as Minister for Transport, and the cane farmer, Ernie Evans, as the Minister for Mines. At that time the Queensland economy had not yet emerged from its colonial structure. As historian Roger Stuart described it:

> The state emerged from that period with a narrowly based, dependent and highly decentralised economy. It was based overwhelmingly on primary (essentially pastoral) and primary produce processing industries and it was dominated by southern-based finance, pastoral and mining capital. Thus the surplus generated by the state's primary production flowed largely to Sydney, Melbourne and London, thereby reinforcing their dominance, rather than being reinvested in Queensland to establish a more independent and diversified economy.[125]

It was against this backdrop that the Nicklin government set about replacing the old Labor order in Queensland, an order that had spanned most of the period from World War I to the mid-1950s, by exploiting the emerging potential of minerals such as coal and bauxite. It took time, however, for the Nicklin government to be persuaded of the coal industry's potential. Thiess played a crucial role in this conversion.

When drilling around Blackwater proved disappointing, Whitehouse prospected further south at Kianga Creek in the Dawson Valley. It was here in 1945 that a government geologist named John Reid had reported the presence of coal. After drilling for several months, Thiess resolved to request a mining agreement from the Nicklin government so that he could test the coal deposits and plan their development. But Ernie Evans, the Minister for Mines, was preoccupied with supporting the development of Queensland's newly discovered and immense deposits of bauxite.[126] More importantly, he was not yet convinced of the viability of coal exports from Queensland. Thiess therefore had to bide his time with his discoveries of coking coal. When George Pearce, Member for Rockhampton and the federal government whip, invited Thiess to accompany him on a trade mission to Japan, Thiess engaged Roy Duncan, the veteran Australian trader in Tokyo, to act as his agent. After Duncan introduced Thiess to representatives of the Japanese steel industry, the Queenslander was elated when they told him that they were interested in examining his finds.[127] On returning to Australia, Thiess grouped together 14 separate Thiess companies into Thiess Holdings, which he floated as a public company on the Sydney and Brisbane stock exchanges with 450,000 shares made available to the public. Foreign direct ownership of Thiess Holdings was quite small.[128]

In the second half of the 1950s, Thiess recovered 10,000 tons of coal from Kianga to send to the Japanese steel mills for testing and Whitehouse in the meantime proved three seams containing what he estimated to be 13 million tonnes of coal.[129] The Japanese trading company Mitsui, now in alliance with Thiess, sent a geologist, Hiroshito (Harry) Okano, to undertake a survey. Okano accompanied Whitehouse on a trip by Jeep to Kianga. During this survey, the Australian and Japanese geologists concluded that the coal in Kianga was not suitable for export to Japan but that the 60-kilometre area between Kianga (soft coking coal) and Baralaba (semi-anthracite) might well contain medium volatile hard coking coal. The Queensland government geologists concurred with this conclusion.

Moreover, the Australian Trade Commissioner in Tokyo, Neville Stuart, pointed out that the Japanese were actively looking for alternatives to America's excellent but expensive coal. They were interested, he thought, in the prospects of Kianga coal or coal that could be found to the north of it. As it happened, Okano's gut feelings about the coal north of Kianga were right. While prospecting at Moura in 1957, 20 kilometres to the north, the Japanese and Australian geologists discovered hard coking coal, some of which was extracted to send to the Japanese mills for testing.[130] Mitsui then dispatched its chief geologist to supervise further drilling, which found indications that the Moura-Kianga field contained somewhere between 300 and 400 million tons of coking coal. In December 1959 Thiess flew to Tokyo where Mitsui and the Japanese steel mills confirmed that they were definitely interested in the prospects of developing the Kianga–Moura coalfield.[131] But even with this Japanese interest, Thiess struggled to convince the Queensland government. At length, he persuaded Ernie Evans and his Under-Secretary, George Clarke, to travel to Japan to meet Mitsui representatives and visit the two biggest steelworks, Fuji and Yawata. That visit convinced both Evans and the Queensland Treasurer, Tom Hiley, to back Thiess by granting him exclusive prospecting rights for coal in his exploration areas.[132]

In 1961, when the Japanese steel mission visited Australia and the iron ore experts had flown to Western Australia (see Chapter Two), Shimuzu flew north with the coal experts, Sugamata, Kats Tanaka, chief of Yawata's coal division, and K. Nisshio, coal import manager of Fuji Iron and Steel.[133] A limitation of the Moura-Kianga area was that coal had to be hauled from Kianga to the port of Gladstone via Rockhampton, a distance of 320 kilometres. A direct line between Kianga and Gladstone would be half that length.[134] The Japanese agreed to take 150,000 tons of coal on short order and a further 250,000 tons by the long rail route pending the construction of a more direct rail line. By that time Thiess Brothers and Mitsui had estimated that the total capital costs of a Moura-Kianga project, including further testing, constructing loading facilities at the port and a new railway line from Kianga to Gladstone, would be £15 million. They were confident of raising the necessary finance between them, but the rate at which they could do so would not permit them to reach their desired target of exporting two to three million tons annually until at least 1964.[135] This could be compared with the 2.5 million to three million tons of exports that the NSW government hoped to be exporting by 1965.

Mitsui recommended that Thiess Holdings form an alliance with a major foreign miner and nominated the Peabody Mining Company of America. Peabody had expressed an interest in forming a partnership with Thiess Holdings given Mitsui's interest in the project. In the meantime, Mitsui indicated that it would invest in a direct rail link between Kianga and Gladstone when its access to credit improved. After talks in Tokyo, Thiess Holdings reached an agreement with the Japanese steel mills for the sale of 2.4 million tons of coking coal over a period of five years. With news of the agreement, the Menzies government agreed to provide assistance to the port of Gladstone to improve its coal-handling facilities. Menzies accepted Nicklin's submission that federal assistance would help achieve greater export earnings on a nationwide basis and contribute to the regional development of Gladstone, which was then a small town dependent on seasonal demand at its meatworks.[136] Central Queensland, Nicklin advised Menzies, was a dull spot in national development that could be stimulated by mining investment.[137] Australia's second minerals rush would transform Gladstone over the next half-century into Australia's fifth largest multi-commodity port, the world's fourth largest coal exporting terminal and now a major liquefied natural gas (LNG) hub.

Thiess Holdings obtained a coal-mining lease from the Queensland government subject to a royalty of 2.5 cents per ton above one million tons per year. After Thiess formed a joint company with Peabody, the sales contract with Mitsui was increased to 3.4 million tons of coal over seven years. When Mitsui & Co. formally entered the venture, which then became Thiess Peabody Mitsui (TPM), the agreement with the State government was amended to 5 cents per ton royalty above one million tons per year.[138] When discussions on the Moura-Kianga railway began, the Queensland railway system was in crisis and, at first, the Queensland government contemplated that the consortium would build a private railway line.[139] But later the Queensland government saw government-owned railways as a way of gaining extra revenue. Freight charges for coal haulage would offset losses on other lines and cross-subsidise the State railway network. In an agreement made in 1965 the State agreed to build the railway itself at a cost of $27.5 million.

The railway would be run by the State, which would provide $A12.7 million capital. The remainder would be financed by the company through security deposits that would be repaid from freight charges. The company was obliged to lodge a $15 million security deposit with the Treasurer of Queensland. Queensland repaid TPM the deposit from rail freights. In this ingenious scheme devised by Queensland Transport Minister Gordon Chalk, the Japanese coal buyers ultimately paid for the railway, rolling stock and even depreciation. Indeed, buyers of coal even paid for the interest that TPM borrowed from the United States as a security deposit. Thus the Queensland export coal industry began on a very different basis to the industry in New South Wales, which was reformed and developed through comprehensive federal and State intervention.[140] In Queensland, the State government encouraged coal industry development largely through foreign capital. It also enabled coal to be hauled on the State railway system subsidised by the new mining companies. The Queensland government would gain much of its revenue in the second minerals rush through rail charges on coal.[141] The arrangement proved a boon to the State government and underpinned the unbroken political hegemony of the non-Labor parties in Queensland between 1957 and 1989. But it imposed an increasing burden on coal companies that complained of the high cost of coal transport in Queensland. As will be seen in Chapter Three, this contrasted with the situation in Western Australia where mining companies built their own railways in much more remote locations and achieved some of the lowest transport costs in the world.[142]

Utah Development Company

In February 1962, as the TPM contract was being negotiated, Mitsubishi, another major Japanese trading company, revealed the details of the contract to American company Utah Construction and Mining Company (Utah).[143] The American company was then also prospecting for coal in Central Queensland. From building railroads, Utah had diversified into construction and, after World War II, mining. It acquired an iron ore mine in Utah and coal deposits in Arkansas. It participated with Cyprus Mines Corporation in developing the vast Marcona iron ore deposits of Peru, and on its own developed a uranium mining and milling complex in Wyoming and steaming-coal deposits in New Mexico. By 1960 it had an estimated net worth of $US62 million.[144] In 1959 Utah recruited an American geologist, Richard Ellett, to explore in Australia for minerals that could be mined on a large scale, transported on bulk carriers and marketed by agencies dealing in bulk raw materials. Ellett's two priorities were iron ore and coal, while Utah's philosophy was to embark on long-term mining prospects. It was prepared to forgo opportunities for high short-term profits that might not recur in return for the stability provided by long-term contracts protected by cost-escalation clauses.[145] Such clauses allowed one party to pass on increases in costs to another party. As we will see, however, there were no cost escalation provisions in the original contracts for iron ore.

Australian companies also expressed an interest in Central Queensland coal in the first half of the 1960s. In 1964 Sir Edward Warren, Chairman of Coal and Allied Industries, contemplated the possibility of a joint venture with the Mount Morgan Company, which operated a small anthracite coalmine in southern Queensland. A Mount Morgan geologist, Brian Vitnell, found the anthracite in southern Queensland unpromising and quickly examined the prospects of coking coal in Central Queensland. But Warren declined to pursue the coking coal prospects in Central Queensland, thus leaving the field to Utah. The historian of Coal and Allied Industries, Christopher Jay, concluded:

> The long-term cost to Coal & Allied would be enormous. It missed

> out on the opportunity to become a major producer in both New South Wales and Queensland, to capitalise on its export experience and to grow to a size and balance sheet making it effectively invulnerable to takeover. This was the single most important strategic decision to be made for the company in the 1960s. Had it gone the other way, the whole history of the coal industry in the next three decades would have been fundamentally altered.[146]

As a supplier of iron ore to Japan – through Marcona – Utah was much more conscious of the growing potential of the Japanese market than most other Australian companies in the early 1960s. As soon as he arrived in Australia in 1960, Utah's mining geologist Ellett made a comprehensive examination of the Mount Goldsworthy iron ore deposit in the Pilbara in Western Australia and encouraged Utah to become a partner in a consortium to mine it. Learning of Thiess Brothers' exploration of the Moura–Kianga area in Central Queensland, Ellett obtained Queensland government approval to explore in the Bowen Basin in November 1960.[147] Though the presence of coal in the Bowen Basin was generally known from the early years of the twentieth century, the Great Depression had inhibited the investment of capital in the Queensland coal mining industry until Thiess's renewed efforts in the late 1950s.

Surmising that economically strippable coal was most likely in an area of gentle dips, Utah's geologists, led by the South Australian Don King, also obtained prospecting rights to 1750 square kilometres around Blackwater. This was near where Thiess had first begun prospecting several years before. King confirmed the presence at Blackwater of coal deposits. By October 1962 he had found that these extended continuously as far north as Goonyella, an area that he reported "gave promise of an important coking coal discovery".[148] By 1964, the year in which Coal and Allied passed up an opportunity to investigate coal deposits in the Bowen Basin, Utah was well advanced in negotiating arrangements with the Japanese trading company, Mitsubishi Shoji Kaisha Limited, to develop the deposits in the northern part of the Bowen Basin. Two years later, in 1966, Utah and Mitsubishi established Central Queensland Coal Associates (CQCA) with Utah holding an 85 per cent stake. In 1968 the consortium would be granted mining rights to over 6330 square kilometres of the Bowen Basin to a distance about 240 kilometres north of Blackwater through German Creek to Norwich Park and Goonyella. As mining historian Alan Trengove described it, "[t]he event may not have appeared spectacular at the time, but it was the greatest single coup in the history of Queensland mining".[149]

In the meantime, Utah worked as the sole developer of a mine near Blackwater that could be linked by a short spur-line to Queensland's Central Railway and thence to the port of Gladstone over a distance of 320 kilometres. Homes for miners employed at the Blackwater enterprise could be built at the existing township of Blackwater, although its infrastructure would have to be expanded. The total cost of establishing the Blackwater mine and supporting facilities was less than $30 million. But even with Blackwater's advantages as an initial mine site, Utah studies showed that its coal would have to be exported for less than $US10 per tonne to compete with higher quality American coal. As Japanese mines on the island of Kyushu were closed after 1963, Utah exploited the opportunity to replace the Japanese low-rank blending coal with Blackwater coal. In May 1965 Utah signed a letter of intent with the Japanese steel mills to sell more than 13.5 million tonnes of coking coal over the 10 years from 1968 to 1978. This $US130 million contract was the largest coal purchase made by the Japanese steel industry to that time and ranked with the biggest iron ore purchases negotiated in Western Australia's Pilbara (see Chapters Two and Three).[150] Under the terms of an agreement with the Queensland government, Utah was entitled to export 100 million long tons of coking coal from its leases in the area.

Larger agreements were to follow in 1969 when CQCA signed long-term contracts for the delivery of 85 million tonnes of coal from Goonyella and Peak Downs, the latter becoming the largest single coking coal mine in the Western world in the 1970s. The two mines were the first CQCA operations to be opened in the early 1970s and would be linked to a new port, Hay Point, constructed 18 kilometres south of Mackay, the centre of the Queensland sugar industry. Hay Point had a number of advantages: its distance of 200 kilometres from Goonyella, proximity to the major city of Mackay, its capacity to accommodate large bulk carriers in deep water and the availability of land. Completed in October 1971 at a cost of $30 million, with installations capable of loading nine million tonnes of coal annually on to bulk coal ships, Hay Point had the largest capacity of any coal port in Australia.[151]

Linking the mines with Hay Point was a 228-kilometre railway built and operated by the State but financed entirely by Utah by way of a $36 million security deposit. This sum was again amortised by freight charged to the company by Queensland. The CQCA deals were sharply criticised by the Labor Opposition because of the low price of the coal negotiated. But Utah argued that the price was necessary to permit capital investment in the coalfields. Utah pointed out that its substantial investments in Australia – 42 per cent in mining investment and 58 per cent on infrastructure such as new townships like Moranbah, railways and port installations – relieved Queensland of a major part of its financial obligations.[152] As the historian of Utah Brian Galligan has commented, the Queensland government compensated for the low royalties charged to Utah with two advantages. One was the requirement under its lease agreements for Utah to stockpile and make available to the State steaming coal produced as a by-product of mining coking coal. The other was that freight rates for hauling coal were set on a sliding scale adjusting to tonnages and subject to escalation for movement in costs. At the time the CQCA agreements were made, it was not possible to predict the timing, costs and capacity of new mines that might be built and hence the costs of new railways which might have to be constructed. Fixing freight rates for subsequent mines was consequently left to the government's discretion. With the booming of coal prices in the 1970s "this discretion to determine rail freights for additional mines would become a licence [for Queensland] to take money".[153]

The Australian Coal Rush

One of the signal developments of the Australian coal industry in the 1960s was the very quick growth of the export market. By providing a steadily growing market, Japan's rapid economic growth provided a major stimulus for Australian black coal mines. Japanese demand was driven by the phenomenal growth of its steel industry, in which output increased 271 per cent from 1960 to 1971. In 1955-56 New South Wales was the only State exporting black coal – a tiny 204,000 tons valued at about $A1.5 million. By 1965-66, Australia as a whole was exporting 7.9 million tons of coal valued at nearly $64 million – 6.2 million tons from New South Wales and 1.7 million tons from Queensland (TPM). By the mid-1960s, New South Wales alone was exporting at double the rate of the most optimistic forecasts of the 1961 Japanese steel mission. In the space of 10 years between the mid-1950s and mid-1960s, the Australian coal industry had brought about a major change in the pattern of world trade in coal. Japan was taking 83 per cent of its coal imports from the United States in the mid-1950s, but by 1965 the United States was supplying 40 per cent of the Japanese market and Australia almost as much, at 39 per cent.[154]

The increase in Australia's coal exports between 1961 and 1971 was even more dramatic – a sevenfold increase from 2.8 million tons in 1961 to 20.5 million tons in 1971. These 1971 exports were made up of 12.7 million tons from New South Wales and 7.8 million tons from Queensland. By 1970 the Australian coal industry had become one of the great Australian industries in its own right, producing more than 50 million tons of coal each year with a pit-top value of over $A250 million.[155] Accentuating Australia's competitive advantage vis-à-vis the United States after 1970 was that American coal export prices increased from about $US14 per ton free on board (f.o.b.) in 1970 to about $US20 per ton f.o.b. in 1971.[156] Free on board prices excluded transportation costs. This price increase was because of the introduction in 1970 of an American Coal Mining Health and Safety Act, railway rolling stock shortages and labour problems. On a cost insurance and freight basis (c.i.f.), this meant that Australian coal could be landed in Japan at half the cost of American coal.[157]

For a period from 1960 up to 1966-67, coal was Australia's most important mineral export. This changed in 1967-68 when the value of Australian iron ore exports began to rise rapidly. Coal was still Australia's second most valuable mineral export in the late 1960s and early 1970s. In 1970-71 coal constituted 4.6 per cent by value of Australia's exports and 14.2 per cent of all Australian exports to Japan, by which time the value of iron ore exports, at $A329 million, was nearly double the value of the coal exported to Japan in that year. In the 10 years to 1972, more than $247 million was invested in New South Wales on coalmine development, plant and equipment and in the year 1970-71 alone, about $50 million was spent.[158]

By the beginning of the 1970s Australia had entered the ranks of the largest coal exporting countries in the world. In 1970 Australia exported 18 million tons or 42 per cent of the coal sold by Australian industry. The United States remained the largest world exporter of coal in 1970, sending abroad 63.3 million tons or more than three times the amount exported by Australia. Other large world exporters were Poland, which exported 28.4 million tons, West Germany and the Soviet Union, although their large export trade was offset by imports. Coal exports from New South Wales represented 40 per cent of black coal produced in the State, with 27 per cent consumed for electricity generation and 23 per cent for the Australian iron and steel industry. The proportion of Queensland coal production exported overseas was higher at 69 per cent, with 21 per cent consumed for electricity generation.[159]

In New South Wales the mining companies' increasing focus on export business is exemplified through the particular case of Coal and Allied Industries. When J & A Brown and Abermain Seaham merged in 1960 to form Coal and Allied Industries, the company was not particularly interested in exporting. According to Christopher Jay, Edward Warren saw the best prospects of Coal and Allied in selling steaming coal to the NSW Electricity Commission. Born in 1895 in Broken Hill, Warren was one of the great figures of the NSW coal industry. Starting as an office boy in Abermain Seaham, he was decorated for his military service in World War I and later became Chairman of the NSW Combined Colliery Proprietors' Association, an office he held for 23 years until 1972. Appointed by the NSW government as a member of the Coal Conservation Committee in 1951, he worked on the employers' side to improve industrial relations in the coal mining industry and also greatly encouraged the mechanisation of the industry.[160] In 1972 he was awarded the First Order of Merit of the Order of the Sacred Treasure of Japan, the first Australian to be so honoured.[161]

The burgeoning Japanese export trade eventually enticed Coal and Allied to diversify from simply providing for the needs of the Australian steel industry and NSW power stations. When the Japanese mastered the art of blending hard United States coking coal with cheaper Australian soft coking coals, a long-term contract with Japan transformed Coal and Allied's Liddell mine into the top export coalmine in New South Wales. Investment of £1 million a year for the three years before 1964 saw output per man lift to 8.43 tons as compared with a State average of 7.43 tons. Coal and Allied almost doubled total sales of coal between 1963-64 and 1968-69, based on two activities: sales of steaming coal to power stations on the northern coalfields and increased sales of soft coking coals to Japan. The company benefited from the Japanese steel industry's expansion of its coal imports, and from the fact that American coal cost on average $US4 per ton more, landed in Japan, than Australian coal. The completion of the coal-loading facilities at Newcastle in 1967 had also helped to make Newcastle coal competitive with other coal.[162]

Problems of Internationalisation – the Queensland and NSW Coal Industries, 1970-72

In September 1969 the now American-owned Clutha Development Company approached the Liberal Premier of New South Wales, Robert Askin, and his Minister for Mines, Wal Fife, for approval to build two railways to new and privately constructed coal installations in the northern and southern parts of the Sydney coal basin.[163] One railway would traverse 130 kilometres from Clutha's Newdell (Foybrook) mine to Port Stephens; and the other would follow a route of 61 kilometres from the Burragorang Valley Mine area to South Clifton, midway between Thirroul and Helensburgh.[164]

At that time New South Wales was supplying about 30 per cent of Japan's requirements for coking coal and Clutha was exporting nearly one-third of the NSW coal trade.[165] In a meeting in December 1969, the owner of Clutha, American multimillionaire Daniel K. Ludwig, expressed his concern to State ministers about the limitations of New South Wales coal ports, notwithstanding improvements between 1963 and 1967. Ludwig wanted to fulfil Clutha's coal contracts to Japan, enter the European market for coal and even start exporting to the United States in ships ranging between 100,000 and 200,000 dead weight tons (DWT).[166] The American magnate had well in mind what Queensland had been able to achieve through agreements with coalmining companies to build new railways and ports. With its proposed new coal-loading facilities, he hoped to double Clutha's exports to Japan and take a sizeable share of the European market. New South Wales as a whole, Clutha estimated, would more than double its annual coal-loading capacity from 16 million to 40 million tons per annum. Clutha undertook to provide, on the Queensland model, all the capital, estimated at $128 million to construct the facilities and to take on itself all of the risk.[167]

The State bureaucracy in Sydney was sceptical. A committee of senior NSW officials opposed Clutha's proposals, largely on the ground that it would deny the State Treasury a share of coal freights on State railways, which, at $25 million annually, was a far greater contribution to the State budget than coal royalties.[168] The cabinet overrode this advice and passed special legislation to facilitate Clutha's plans. But the grand plan ran afoul of community campaigns against Clutha that reflected growing concerns about the impact of mining projects on communities and the environment.

The coal industry now faced other obstacles. While the expansion of the coal industry had been dramatic from 1958 to 1970, there was a decline in Japanese demand for New South Wales coal from the second half of 1970 into 1971. This was caused by a slump in Japan's steel industry. Coal production also fell at this time because of industrial disputes in New South Wales. These saw the quantity of coal exported fall below contracted rates. Then in January 1971 the national wage case saw a six per cent increase to the mining industry's wage bills. Other hurdles included increases in charges levied by State authorities for loading and transporting coal and a variety of increased taxes and royalties. This series of events reduced the NSW coal industry to a state of chronic anxiety and prompted the NSW coal owners to band together to send a joint message to the Japanese steel mills for a meeting. Between 1966-67 and 1969-70, shipments of coal to Japan from New South Wales almost doubled. But between 1969-70 and 1970-71, New South Wales coal exports to Japan dropped 1.7 million tons to 9 million tons, while Queensland exports to Japan jumped 1.2 million tons to 6.8 million tons.[169] Between 1968 and 1972, NSW coal owners secured no new contracts with the Japanese steel mills, while Queensland made three major long-term contracts with large tonnages: CQCA's Goonyella and Peak Downs in 1969; Thiess Brothers' South Blackwater coal in 1969; and CQCA's Saraji mine in mid-1972.[170]

In these circumstances, the Japanese steel mills agreed to a visit by a NSW coal mission. The mission represented all eleven NSW coal exporters – Austen and Butta, Bellambi Coal Co., Bloomfield Collieries, Clutha Development Pty Limited, Coal and Allied Industries, Coalex Pty Ltd, Gollin and Co. Ltd, Huntley Collieries Pty Ltd, J & K Johnstone Holdings

Pty Ltd, Kembla Coal and Coke Pty Ltd and RW Miller Holdings. The first coal trade mission ever sent to Japan by a supplier country, it set itself the objective of obtaining from the Japanese steel industry the best estimates of steel production in the years to 1980 and the quantities of coal that the Japanese were likely to import until that year. It also sought to demonstrate the importance of New South Wales as a supplier of coal to Japan, the unreasonableness of coal prices in relation to costs and the inadequacies of the escalation provisions in NSW coal contracts.

The Japanese steel industry gave the mission a lukewarm welcome. It pointed out that the August 1970 statement by US President Richard Nixon indicating the need for wage and price controls in America had induced the worst economic crisis in Japan since the end of World War II. Exports of steel had plummeted and Japan's previously optimistic estimates of production had been revised downwards. The NSW mission nonetheless persisted in seeking price increases varying between $US1.50 and $US2 per ton. But the Japanese steel industry felt threatened by the attempted cartelisation of a key input for one of its most important export industries and asked the mission to go back to Australia and then to return to Japan to negotiate as single companies rather than as a bloc. In these separate negotiations the Japanese steel industry conceded increases ranging only between US45 cents and US65 cents, a disappointing result for the mission. By 1972, opinion in the NSW coal industry began to favour Commonwealth intervention. For example, Bellambi Coal believed that Commonwealth intervention was necessary to prevent the selling of Queensland coal at less than its market value and at less than the cost of NSW production. Clutha also pushed for federal intervention to enable NSW coal to compete with Queensland open-cut coalmines, specifically requesting a federal directive for the Queenslanders to mine a mixture of underground and open-cut coal.

The disquiet in the NSW industry about the competition from low-cost Queensland coal coincided with governmental uncertainty about the adequacy of Australia's coal reserves. In March 1971 the Bureau of

Mineral Resources argued that if the exploitation of coking coal proceeded unchecked, the Australian steel industry might have only limited reserves of coking coal by 2000. It urged that Commonwealth export controls on black coal were now more necessary than for iron ore. Cochran's successor as Chairman of the JCB, B.W. Hartnell, agreed, arguing:

> Whilst I recognise the importance of the coal export trade and the mutual advantages it has brought to Australia and Japan, it is necessary that the current and prospective volume of exports of the low volatile coking coals in particular be kept under review to ensure that the future industrial development of this country is not inhibited by suitable coal at acceptable prices.[171]

In 1971 Hartnell had reached the conclusion that one of the major problems with the coal industry was that internationalisation had proceeded too far, with 40 per cent of the market for Australian coal coming from overseas. As he advised the Secretary of the Department of National Development: "It could well be that this proportion is already dangerously high in terms of maintaining stability in the industry".[172]

Conclusion

The transformation of the NSW coal industry and the rapid development of its Queensland counterpart formed a key early component of the second minerals rush in Australia. The transformation of the black coal industry began in the late 1950s and continued into the 1960s when Japanese steel companies began to buy Australian coal in increasing quantities. The expansion of the Australian black coal industry relied on explosive growth in Japan, which affected both coal and iron ore. In the case of New South Wales, it took Commonwealth and State intervention to convert an undercapitalised, inwardly focused industry into an internationally competitive dynamo. This NSW example challenged the assumption that commodity-exporting industries uniformly required government deregulation in order to compete in international markets.

In the NSW coal industry, comprehensive government regulation was necessary to modernise a poorly performing industry to enable it to compete internationally. The development of the Queensland coal industry followed discoveries of coal that could be easily mined by open-cut methods. As was the case with iron ore in the Pilbara, the presence of coal in Central Queensland had been generally known. But Australian, American and Japanese companies had to take the risk of exploring the Bowen Basin in the late 1950s and early 1960s and then had to convince Japanese steel mills and Australian governments to support the development of the deposits. A NSW export trade with Japan preceded the Queensland trade. Both industries then developed in tandem. New South Wales and Queensland both had established rail networks, so laid to incorporate other lines. Queensland's more decentralised network was much more able to accommodate new lines operated by the government. In Western Australia's Pilbara, by contrast, rail lines had to be private as there was no existing rail network.

The development of the coal industry helped to transform the Australian economy, which had been beset in the 1950s by a systemic inability to balance its export and import accounts without imposing import restrictions. By the early 1970s coal had become a significant export earner and, with iron ore, had considerably improved Australia's balance of payments. Coal mining was also transforming Australia in other ways – by beginning Queensland's transformation from a Labor-dominated State overwhelmingly dependent on primary and pastoral production to a Country-Liberal Party-dominated State that would become substantially dependent on the new resource industries of coal and bauxite.

2

Unlocking the Pilbara: The End of the Iron Ore Export Embargo 1960-1964

From the late 1930s to 1960, Australia's iron ore was reserved exclusively for the Australian steel industry. In the 1940s and 1950s Broken Hill Proprietary Company (BHP) held leases over virtually all of the suitable and accessible deposits of iron ore in Australia. The company's leases over Iron Knob and Iron Monarch in South Australia's Middleback Ranges gave it access to the highest-grade iron ore in Australia.[1] With its takeover of Australian Iron and Steel (AIS) in the mid-1930s, BHP also gained leases over iron ore deposits on Cockatoo Island in Yampi Sound off the north-west coast of Western Australia (WA).[2] Deposits of iron ore were also identified at Savage River in Tasmania but these were considered economically inaccessible until the 1960s.

BHP had been originally formed in the 1880s to mine silver, lead and later zinc at Broken Hill, which by 1910 was Australia's tenth-largest city, selling ores worth £80 million annually. But its glory days were behind it. Like the Western Australian gold-mining town of Kalgoorlie, Broken Hill was able to survive because of the tenacity of its residents and because its ore body was so rich that it continued to yield profits during the twentieth century. While BHP diversified from its mining roots at Broken Hill in the first half of the twentieth century, other mining companies, such as the Zinc Corporation, continued to mine Broken Hill's rich lode.[3] In the course of establishing lead smelters at Port Pirie in the early twentieth century, BHP began acquiring leases over iron ore deposits in South Australia to use the ore as a flux for smelting lead. To make use of the rest of its iron ore, BHP then engaged in "vertical" integration by constructing an iron and steel plant at Newcastle.[4] When it took over AIS, BHP acquired another steel plant at Port Kembla in New South Wales. With collieries near Newcastle and Wollongong, and access to manganese, an essential alloying element in steel making, BHP gained a virtual monopoly over an Australian iron and steel industry. This industry produced iron from iron ore, converted the iron into steel and rolled the steel into various finished shapes. From all of its steel plants, BHP had expanded its production from one million to nine million tonnes of steel annually between 1946 and 1976.[5] Thus from its roots in mining in the late nineteenth century, BHP became a behemoth of Australia's manufacturing industry in the first half of the twentieth. In the long boom from 1945 to 1973, BHP was the largest company listed on the Australian stock exchange in terms of invested capital and market value.[6]

In the mid-1930s, at about the same time that AIS obtained iron ore leases on Cockatoo Island, a British company, H.A. Brassert, obtained a lease from the WA government on another Yampi Sound deposit on Koolan Island. By 1937 the Japanese had been negotiating with the Soviet Union and India for iron ore supplies while also looking at Australia as a potential supplier. At that time steel plants were closing in the United Kingdom owing to a shortage in raw materials; Norway had placed an embargo on all exports of iron, steel and scrap; and Japan was actively negotiating to take the entire iron ore output of Malaya.[7] Iron ore deposits had first been discovered on Koolan Island by pearlers operating in Yampi Sound in the 1870s. But the deposits were left untouched until 1907, when the Australian Prospecting Association of Charters Towers, Queensland, reinvestigated the island's minerals potential.[8] After Brassert obtained the lease on Koolan Island, the company aimed to sell one million tons of iron ore per year to the Nippon Mining Company of Japan commencing from 1938.[9]

In 1936 the Australian government had stressed that it would not involve itself in the Brassert iron ore export negotiation. But as the world rearmed, iron ore and scrap became widely sought after and the deteriorating strategic situation in East Asia began to alarm the Australian government.[10] In particular, the prospect of Japan gaining the rights over some of Australia's iron ore reserves aroused considerable anxiety in Canberra. A United Australia Party – Country Party coalition government led by the Tasmanian Joseph Lyons had just sparked a trade war with Japan by discriminating in favour of British textiles. Japan retaliated by reducing its purchases of Australian wool and wheat. In response to these developments, Lyons' cabinet discussed Brassert's iron ore plans in a meeting in Canberra in March 1937. As a result, cabinet authorised the Prime Minister and the Minister in Charge of Development and Scientific and Industrial Research, Senator Alexander McLachlan, to arrange a survey of Australia's iron ore reserves. Before the technical survey was concluded, McLachlan recommended to cabinet on 28 July 1937 "that the export of iron ore from Yampi Sound be limited to 20 million tons and that the export of all other iron and iron ore be totally prohibited".[11] By that time, the Australian government had become seriously concerned about Japan's aggressive intentions. Japan, having first occupied Manchuria in 1931, had then invaded China in July 1937.[12] At the end of that month, the Japanese held the entire area surrounding Beijing and were commencing an invasion of Shanghai. Later in 1937, the Australian Trade Commissioner in Japan, Longfield Lloyd, warned his government:

> It is a considered opinion that the fulfilment of the Brassert–Japan Mining Agreement could have a practical effect of affording a foothold to Japan upon Australian territory for a very long time and the position is so exceedingly serious as to justify the compulsory cancellation of the Brassert scheme by any means whatsoever.[13]

Lloyd listed three possible ways to prevent the mining agreement going ahead – by using the Immigration Act (the restrictive White Australia policy); by the exclusion of exports on declaration of insufficiency of iron for Australian or British Empire needs; or by using the Commonwealth Defence Act.[14]

In December 1937 McLachlan's report on the iron ore survey commissioned by cabinet pointed to a belief in many quarters in Australia that "existing known bodies of iron ore are insufficient for future Australian requirements, and that exclusive rights should not be granted to any foreign company over a commodity which is so vital to national interests, especially those of defence".[15] The report also raised the possibility that any foothold that the Japanese might gain in Yampi Sound could give rise to future political difficulties with Japan and possible interference by her on the grounds of "protection of national interests", the grounds that had been the basis of Japanese military intervention in Korea, Manchuria and China in the 1930s.[16] The report was submitted just as the Australian Waterside Workers Federation was beginning a nationwide campaign against the selling of pig iron to Japan.[17]

McLachlan's report concluded that there would be grounds for the prohibition of the export of iron ore to Japan if it could be established that Australian resources were in danger of being depleted.[18] The Commonwealth's geological adviser, W.G. Woolnough, obliged his government by presenting a technical case to deny the Japanese a foothold in Yampi Sound. Walter George Woolnough was born on 15 January 1876 in Grafton, New South Wales. He was educated at Sydney Boys' High School and then at the University of Sydney where he came under the influence of the Australian geologist, Sir Edgeworth David. After lecturing in geology and mineralogy at Adelaide University and then Sydney University, Woolnough was appointed foundation professor of geology at the University of Western Australia, a post he held from 1913 to 1919. In 1927 Woolnough was appointed geological adviser to the Commonwealth government and held this post until his retirement in 1941.

Despite his experience in Western Australia, Woolnough was sceptical about the potential of the State's iron ore reserves.[19] In his view, workable deposits of iron ore had to be situated close to water transportation and adequate supplies of good quality coal. While many other sources of iron ore were known in Australia, the only deposits that satisfied Woolnough's criteria were those at Iron Knob in South Australia and at Yampi Sound in Western Australia. He explained:

> The world trend at the present time is towards rapid increase in the development of iron ore industries in those countries possessing the two absolute essentials of adequate ores, and above all, suitable fuel supplies. Australia possesses the latter. As pointed out above, however, the ore supplies appear to be definitely limited. Nevertheless, the fuel factor alone makes it certain that our juvenile iron industry must expand rapidly. At present we are using over two million tons of iron ore a year. In view of the expansion of the iron and steel industries which had taken place in the last few years, and the practical certainty of further large expansion within the next few years, it is certain that if the known supplies of high grade ore are not conserved, Australia will in little more than a generation become an importer rather than a producer of ore.[20]

Overriding protests, Prime Minister Lyons informed the Japanese Consul-General in Australia on 18 May 1938 that the best expert advice available to the government, namely from Woolnough, was that "the accessible iron ore deposits of Australia which are capable of economic development are so limited as to compel their conservation for Australian industrial requirements".[22] The Lyons government also announced that an embargo on Australian iron ore exports would take effect from 1 July 1938.[245] Two years later, Woolnough conducted another survey of iron ore resources with the States that concluded that Australia had only about 350 million tons of accessible, high-grade iron ore.[23]

For many years after Japan's defeat in 1945, federal and State governments remained staunchly committed to the embargo. The WA government, in parallel with the federal government, refused to issue licences to prospect for iron ore. Instead, it ordained that any deposits would be awarded not to the discoverer, but to a company that could guarantee to bring manufacturing to the State. In 1957, when announcing that ore was being imported from New Caledonia because of the scarcity of Australian supplies, Menzies' Minister for National Development, William Spooner, described it as "most alarming" that Australia had only enough iron ore to last another 50 years.[24]

The decision of Commonwealth and State governments in the early 1960s to relax the 22-year-old embargo of iron ore exports was one of the most strategic economic policy decisions of the period after World War II. The decision was not reached easily and was made in the face of much opposition. But, together with the decision to negotiate a trade pact with the Japanese in 1957 and the contemporaneous decision to improve coal ports, the relaxation of the embargo helped open up Australian trade and changed its trajectory away from Britain and Europe towards Japan and East Asia. Thus the decision to expand the NSW coal ports and to dismantle the iron ore embargo were two of the key stimuli for Australia's second minerals rush.

Relaxing the Embargo

No major discoveries of iron ore were publicised in Australia before 1960. But in the 1950s BHP found low-grade iron ore in the Northern Territory and a joint federal–state team collaborating with BHP prospectors discovered higher-grade iron ore in the Constance Range in Queensland.[25] Small deposits of high-grade ore were also known to exist at Mount Goldsworthy on the northern fringe of the Pilbara and at Tallering Creek and Koolyanobbing further south.

In the 1950s Western Australia was a lightly populated State heavily reliant on the agricultural sector. It had enjoyed little of the employment growth that had accompanied the development of manufacturing industry in the eastern States.[26] It was for this reason that WA governments were interested in the possibility of mining as a stimulus to the development of manufacturing. With this in mind, the WA Labor government arranged to test the known iron ore deposits at Mount Goldsworthy in the Pilbara, and at Tallering Peak north-east, and Koolyanobbing, east of Perth.[27] By the late 1950s several Australian mining companies were agitating for the relaxation of the Commonwealth's iron ore export embargo. Heeding these calls, the WA Labor government made several unsuccessful requests to the Menzies government to allow it to export small quantities of iron ore in the middle and late 1950s. In 1959, in anticipation of the end of the embargo, retired federal treasurer, Arthur Fadden, secured for himself the position of purchasing agent for WA iron ore with the Japanese steel industry. In the same year, a Liberal–Country Party government was elected in Western Australia on a policy platform aimed at industrial development and hence at helping Western Australia catch up with the other Australian States.[28]

The new WA Premier, David Brand, was a former storekeeper who had been responsible, as WA Minister for Works, for establishing a petroleum refinery in the Kwinana industrial area of Perth in 1952.[29] As premier from 1959, Brand wanted to encourage further developmental projects in his State and was strongly supported in this by Charles Court, the energetic Minister for Industrial Development, Railways and the North West. Born in England in 1911, Court migrated to Perth with his family when he was six months old. Schooled at Perth Boys School, he studied accountancy at night, practised as an accountant in the Depression years and in 1938 established the firm Hendry, Rae and Court. Returning to Western Australia after military service in World War II, Court won the State seat of Nedlands for the Liberal Party. On the formation of the Brand government in 1959, he was a small-town accountant with no experience of international finance and trade. But he had an acute awareness of Western Australia's history, its industrial backwardness and the ghost towns that remained as warnings of the transience of mining booms. The most intelligent and hard-working of Brand's cabinet, Court would acquire a comprehensive understanding of the mining industry and of international finance and trade and gain the respect and trust of the Japanese steel industry. As Minister for Industrial Development in the 1960s he was determined that mining should transform Western Australia from "dependence to pride".[30] Despite his *laissez faire* rhetoric, however,

Court was an interventionist in mineral development, arguing that the State should decide which companies would be allowed to develop iron ore and on what terms. These terms usually required a commitment to build a steel mill or at least to downstream processing. He also argued that the State should decide the areas in which companies could operate. Above all, Court would seek to devise an orderly "plan" for the development of the iron ore deposits of Western Australia.[31]

Not long after he was elected premier in 1959, Brand went to see Spooner, to push the federal government to lift the embargo on the export of iron ore. Spooner, who was himself beginning to examine the prospect of exporting coal to Japan, let Brand know that the Bureau of Mineral Resources was conducting a survey of Australia's iron ore resources following one conducted only two years earlier that had recommended the retention of the embargo. Spooner held out the hope that if Western Australia applied to the Commonwealth for authority to export a small amount of ore – specifically ten to 15 million tons spread over a period of five or ten years – it might be given an exemption from the ban.[32] But Brand and Court were not interested in half measures. To put pressure on the Commonwealth government, which they doubted would actually lift the embargo, Brand and Court boldly called tenders for the export of ore from Mount Goldsworthy and Koolyanobbing.[33] At this stage the WA government was not contemplating a large-scale export industry but rather one that would be a useful adjunct and stimulus to the development of local manufacturing industry. Its early plan for iron ore resembled the Joint Coal Board's initial stance on the development of coal exports in New South Wales, namely that the export trade should be an adjunct and aid to a domestically focused industry. The announcement of the tenders explained that the WA government was interested in selling only a limited amount of iron ore to gain urgently needed funds for long-term works such as improving country water supplies and developing ports "without prejudice to the ultimate establishment of an integrated steel industry in Western Australia".[34] Brand explained to Menzies that "… it is our view that we should press ahead immediately to obtain a firm indication of the demand for our iron ore and the conditions under which sales could be negotiated if an import license is available".[35]

The WA initiative was supported by the retired chief Commonwealth mineral economist, John Dunne. Dunne penned an article in the *Australian Financial Review* on 15 October 1959 entitled "Isn't it time we took the prohibition off iron ore?"[36] Brand's notification to Prime Minister Menzies of the tenders was followed by a WA government press announcement and a flurry of excitement from the Japanese steel industry.[37] The retired federal treasurer Arthur Fadden also lent his support to Brand, arguing that calling for tenders would present the Commonwealth with an irresistible case for re-examining the embargo.[38]

The Australian government, however, was reluctant to have its hand forced by the impatient Western Australians. Menzies angrily informed Brand in October 1959 that the Western Australian initiative did not have Commonwealth government approval. He followed this with an emphatic statement in the House of Representatives that no export licences would be granted for Western Australian iron ore. He also instructed his Minister for National Development, William Spooner, to tell a visiting Japanese trade delegation that there was no prospect that the iron ore embargo would be lifted.[39] But despite this declaration to the Japanese, Spooner was himself moving cautiously in the direction of change on iron ore policy on the advice of his departmental head, Harold Raggatt. Raggatt, as we have seen, was actively advocating Commonwealth support for upgrading the NSW coal ports. Paradoxically, Raggatt, a geologist handpicked by Woolnough to succeed him as geological adviser to the Commonwealth, would play a vital role in reversing the embargo that his patron had helped put in place. For by the end of the 1950s, Raggatt was encouraging Spooner to authorise another survey of Australia's iron ore resources following the inconclusive survey of 1958. When the survey was completed in March 1960, the Bureau of Mineral Resources found Australia's reserves of accessible iron ore to be little more in 1960 than they had been in

1938.[40] But this time the bureau urged a change in federal iron ore policy to encourage exploration. With this support, Spooner recommended to cabinet that the embargo be retained on Australia's three known accessible deposits of high-grade ore – in the Middleback Range of South Australia; the Cockatoo and Irvine Islands at Yampi Sound in Western Australia; and in the Koolyanobbing area of Western Australia. But outside these areas, Spooner proposed that mining companies be allowed to export one million tons annually up to a total of half of the deposits. And he recommended that iron ore deposits up to two million tons be authorised for export without any limit on the annual rate.[41]

Senior officials advising Menzies were sceptical about Spooner's proposal. One of these officials was Peter Lawler, senior economic adviser in the Prime Minister's Department. Lawler noted that Australia produced some of the cheapest steel in the world but it did not export. He suggested the possibility of overseas steel companies mining ore bodies not already committed to BHP as long as they were used for production of steel and not for the export of ore. In March 1960, influenced by Menzies' scepticism, cabinet deferred a decision and asked Spooner to consult with the local steel industry, i.e. BHP.[42]

The chairman of BHP, Colin Syme, discouraged a change in iron ore policy. Born in Perth on 22 April 1903, Colin Syme was educated at Scotch College Perth, the University of Western Australia and the University of Melbourne from which he graduated with a law degree. After making his mark as a lawyer in the financial world in the 1930s he became a BHP director. By the late 1930s, Syme was at the centre of the company's management as it consolidated its position in heavy industry, steel fabricating companies and the joint venture, Commonwealth Aircraft Corporation. Syme chaired BHP from 1952 to 1971 during a transitional period when the company was diversifying from steel-making into mining activities. But he was no prophet of an iron ore export industry in Australia.[43] He argued to Spooner on 2 April 1960 that the Bureau of Mineral Resources assessment of iron ore reserves was that they were not "of major magnitude in relation to future requirement".[44] The partial easing of the embargo would not, he suggested, establish a new export trade in iron ore. He reinforced his point by referring to the fact that BHP was then importing some additional iron ore from New Caledonia.[45] Despite this lukewarm response from the BHP chairman, Spooner tried to get his recommendations confirmed by cabinet in May 1960.[46] But by this time, doubts were growing among senior ministers and officials. Officials advised John McEwen, Minister for Trade and acting Prime Minister, that BHP was warning that even a strictly limited amount of exports might be unwise in the absence of further iron ore discoveries.[47] McEwen accepted these cautious views and instructed that the matter should go back to cabinet for further consideration.

John McEwen, commonly known as "Black Jack", was born on 29 March 1900 in Chiltern Victoria, the son of a pharmacist from Ireland and his second wife. After his parents' early deaths, McEwen was raised by a maternal grandmother and left school at 13 to help the household.[48] He enlisted in the AIF too late to see active service in France but qualified to farm a small block in rural Victoria under the Soldier Settlement Scheme. McEwen's farming drew him into politics and the Country Party. Winning the seat of Echuca in 1934, he served as Minister for the Interior before World War II and as Minister for External Affairs in the early stages of the war. When the federal coalition returned to power after almost a decade of Labor rule in the 1940s, McEwen served as Minister for Commerce and Agriculture from 1949 to 1956 and then as Minister for Trade from 1956 until 1971. McEwen used his powerful position in Menzies' cabinet to negotiate the Agreement on Commerce between Australia and Japan in 1957. The aim of this agreement was to help Australia's rural industries, particularly wheat and wool. It would not be until the 1960s and the second minerals rush that a substantial minerals trade would develop between Australia and Japan, particularly in coal, iron ore and bauxite. McEwen had oversight of the overseas trade dimension of the minerals boom, but he shared responsibility for the minerals industries with the Minister for National Development, a portfolio always held by the Liberal Party. It was perhaps this demarcation of federal ministerial responsibilities that prevented the minerals industries developing their own marketing authorities, such as the Australian Wheat Board that had served the wheat industry so well.

The event that most helped Spooner's case for relaxing the embargo on iron ore exports was the deterioration of the Australian economy in 1960 and 1961. After the disastrous balance of payments crisis that followed the Korean War wool boom in the early 1950s, the Menzies government had balanced imports and exports through a regime that licensed imports. The import licensing system lasted throughout the 1950s until, on 22 February 1960, McEwen announced its abolition. Imports worth up to £800 million a year, McEwen declared, would no longer be licensed.[49] In support of the decision, McEwen commented that "[o]ur external trade and payments position is one of considerable strength. Our overseas reserves are high and our export earning prospects good".[50] In the ensuing months, however, imports rushed in, exports could not keep pace and Australia's external reserves plummeted. Prime Minister Menzies, who was forced to introduce a corrective mini-budget in November 1960, emphasised that a major element of remedial action had to be "the opening and developing of overseas markets, not only for wool, wheat and meat, but also for processed and manufactured goods".[51]

Adding to Menzies' worries in 1961 were the European Economic Community's (EEC) bid for rural self-sufficiency through its Common Agricultural Policy and the British government's request to be admitted to the Common Market.[52] These developments threatened Australia's traditional markets for rural exports and the relationship with its biggest trading partner, Britain. Menzies clung to his vision that economic redemption lay in the expansion of agricultural exports and the development of trade in manufactured products. He did not highlight possibilities in the mining industry, which, after its heyday in the nineteenth and early twentieth centuries, had fallen into stagnation.[53] In the early 1950s, the best days of Australian mining appeared to have passed. The main mining fields – the Sydney coal basin in New South Wales, Broken Hill's silver, lead and zinc, and Kalgoorlie's gold fields – were all, as Geoffrey Blainey has observed, discoveries of the nineteenth century. To be sure, rich lead, zinc and silver deposits had been discovered at Mount Isa in Queensland in the 1920s, but Mount Isa's development had been severely hampered during the lean times of the Depression, although it was resuscitated when huge quantities of copper were developed after 1956.[54]

As the Menzies government wrestled with its economic problems, the WA Premier David Brand wrote to Menzies on 20 October 1960.[55] He informed him that the WA government's objective was to persuade BHP to establish a steel industry in Western Australia by railing iron ore from Koolyanobbing to the coast. To that end Brand sought Commonwealth assistance to standardise the railway between Koolyanobbing and Perth. If this could be achieved and if BHP agreed to build a blast furnace at Kwinana, Brand proposed to lease the Koolyanobbing iron ore deposits to BHP. BHP in turn would agree to use the Koolyanobbing ore to establish the beginnings of a steel industry in Western Australia with a capacity of 500,000 tons per year. Federal aid for the Koolyanobbing to Perth railway was Brand's main objective in writing to the prime minister. But he also took the opportunity to repeat his request for the lifting of the iron ore export embargo. Thinking of iron ore deposits like Mount Goldsworthy and Tallering Peak, Brand sought to persuade Menzies to allow exports from places such as these that would never be of commercial value unless sold outside of Australia. He therefore asked Menzies to grant his State an export licence to export 10 million tons of iron ore from Mount Goldsworthy. The £45 million proceeds of the exports, he advised, would be used to develop a deep-water port at Port Hedland.[56]

Menzies' advisers, however, remained staunchly opposed to allowing iron ore exports. Sir John Bunting, Permanent Secretary of the Prime Minister's Department, cautioned against lifting the embargo.[57] Bunting noted that Australia's known reserves of iron ore were not much larger in 1960 than they were in 1958 when the Department of National Development had then recommended against lifting the embargo. What had changed was that, in 1960, the Department of National Development was pinning its faith on discoveries and beneficiation augmenting Australia's limited iron ore reserves. Bunting said Spooner had invoked BHP's chairman as supporting his department's position. But Spooner, in Bunting's view, was twisting Syme's words. "At best", Bunting argued, "it might be said that they [BHP] have not directly opposed carefully limited export of iron ore from small deposits, but even in this they have displayed a marked lack of enthusiasm".[58]

In Bunting's opinion, Spooner's argument that exporting iron ore would improve Australia's balance of payments was also exaggerated. It would take time to develop the primitive Port Hedland. Then, even if exports reached two million tons per year, this would only amount to £10 million pounds in gross receipts and less in net receipts for Australia. Bunting therefore recommended against lifting the ban. He argued, moreover, that permitting exports from small deposits would be a troublesome policy to administer for little impact on the balance of payments.[59] Bunting's advice not to lift the embargo reflected scepticism that there were large undiscovered iron ore deposits in Australia, and a disposition towards the encouragement of rural exports and manufacturing as a way to improve the balance of payments. This was the conventional view at the time – that improvements to the balance of payments would come largely from manufacturing and rural development rather than from mining.

In the end, when the matter went to cabinet on 24 November 1960, after deferral of a decision in March, the result was a compromise. Cabinet agreed that the embargo would remain on the three known high-grade iron ore deposits (the Middleback Range in South Australia and Yampi Sound and Koolyanobbing in Western Australia), but that exports would be permitted from smaller, or as yet undiscovered deposits at the rate of one million tons annually and of no more than half of the deposits overall. Cabinet thereby hoped to encourage exploration and to add, if only modestly, to Australia's foreign exchange reserves.[60] Having partially relaxed the embargo, however, the government wanted to manage iron ore reserves to protect what governments and industry still saw as Australia's industrial future. This meant leaving room to modify, and if necessary reverse, the "policy of export if such action was necessary in the national interest".[61] For this reason the cabinet insisted on Commonwealth control over the quantities of iron ore exported. In a separate decision, the Menzies government agreed to support the standardisation of the railway to Koolyanobbing. At a final cost to BHP of $90 million, the big steelmaker would in 1968 commission a blast furnace with a capacity of 800,000 tons per year, a sinter plant and additional wharf facilities to utilise the Koolyanobbing ore. However, BHP's undertaking to construct an integrated steel plant by 1978 would later be jettisoned.[62]

Mount Goldsworthy

While the Commonwealth embargo on iron ore exports was in place, Western Australia had imposed a counterpart ban under which all iron ore deposits were reserved to the Crown and were unavailable for pegging by individuals or companies. In March 1961, following the partial relaxation of the federal export embargo in late 1960, Brand announced the end of his State's counterpart ban. In doing so, he divided Western Australia's iron ore into three categories: the known deposits of high grade ore, such as Mount Goldsworthy in the Pilbara and Tallering Peak further south, that were not covered by mineral leases; the numerous known medium and low-grade deposits; and a third category of hitherto unknown deposits that could be taken up as in what were known as 130 square kilometre "temporary reserves". While reserving the Crown's rights in the first two categories, Brand announced that the State invited individuals or companies to apply for temporary reserves in the third. A holder of temporary reserves had the exclusive right to explore in his area but not to mine it. If economic deposits of iron ore were discovered, the holder of temporary reserves could then apply to the State for mineral leases and then mine them under conditions agreed by the government.

In March 1961, the Western Australian government also called for new tenders to mine and export up to 15 million tons of iron ore from Mount Goldsworthy in the first category of deposits and now a ministerial reserve.[63] As a result of the tender it awarded a new mineral lease to a consortium of one local and two American mining companies. Consolidated Goldfields Australia Ltd, a subsidiary of Consolidated Goldfields Ltd of the United Kingdom, had suggested to Cyprus Mines Corporation of Los Angeles that they submit a joint tender. In turn, Cyprus Mines had recommended that the Utah Construction and Mining Company of San Francisco be invited to participate due to their joint involvement in Marcona Corporation's deposits in Peru. American capital and know-how would become an important element in all the foundation Pilbara iron ore projects. But this American contribution would go hand in hand with Australian management of mining companies, substantial Australian capital and, later, Japanese equity participation in mines.[64] To win the Goldsworthy tender the three associates, known as Mount Goldsworthy Mining Associates (MGMA), had agreed to establish a port (ultimately at Port Hedland), to build a three-mile causeway and also to build a 193-kilometre standard gauge railway and two townships, one at Goldsworthy and the other at the port.

The agreement was ratified by the Western Australian parliament in the *Iron Ore (Mount Goldsworthy) Agreement Act* 1962. Starting with the Goldsworthy agreement, the Brand government made all major new iron ore projects subject to individual agreements approved by act of parliament. These agreements covered all recognised aspects. They included timetables to submit proposals and implement them, royalty rates and contributions to local infrastructure. Environmental issues were not covered. In this respect, the WA government procedures echoed those of the Queensland government for bauxite and coal mining. Royalties were levied as a percentage of net f.o.b. revenue: 7.5 per cent for lump ore and 3.75 per cent for iron ore fines.

MGMA submitted an application to the federal government for an export licence on 22 September 1962. They had by that time spent a considerable amount of time testing the deposit.[65] This emboldened them to seek to export the whole 64 million tons of the deposit at the rate of four million tons per annum over 21 years. While exporting at this rate would exhaust the Goldsworthy deposit, they counted on finding additional iron ore later. The consortium's request was completely outside the terms of the federal government's iron ore policy. But rather than refuse the request, Spooner asked his cabinet colleagues to relax further the iron ore export embargo. By this time, huge new deposits of iron ore had been publicised in Australia. Even without the application from MGMA, Spooner argued, there would be strong pressure for complete relaxation of the ban on the grounds that "the deposits recently discovered in W.A. are so large that the original reason for the export policy, i.e, conservation of our limited resources, has now vanished".[66]

But Spooner was not in favour of completely removing all Commonwealth controls. This would leave the federal government without any say in the WA iron ore deposits. One danger he foresaw was that of companies being formed to mine the iron ore becoming, like H.A. Brassert in the 1930s, virtual subsidiaries of the Japanese steel companies to which the ore was to be sold. In this scenario the deposits would be developed to meet Japanese objectives rather than Australian ones. Another danger was that, in their effort to obtain an export market, new iron ore companies might offer to sell their ore at an uneconomic price.[67] A further worry was about BHP, with its well-established position in South Australia and Yampi Sound, and at other WA properties at Koolyanobbing and Deepdale near Robe River in the Pilbara. With these assets BHP might be able to undercut the competition of other mining companies. For these reasons, Spooner felt that some Commonwealth control over iron ore needed to be retained.[68] His solution was to remove restrictions on either the total quantity or yearly tonnages of iron ore exported from the smaller deposits, now defined as five million tons. Proposals from all other deposits would be considered on their merits on a case-by-case basis.

Cabinet approved his recommendations but insisted that the Minister for National Development should retain the right to disapprove applications to export if the Commonwealth was dissatisfied with the price negotiated for Australian iron ore.[69] This was the first time that the Commonwealth had introduced a "reasonableness of price" condition on the export of iron ore. Cabinet also left in place the total prohibition of exports from Yampi Sound, Koolyanobbing and the Middleback Range.

MGMA, as the first of the major Pilbara mining companies to begin export operations, were forced to confront the fact that no one in Australia had any experience in marketing iron ore overseas. In the absence of any Australian expertise, the group turned to Marcona Corporation, which was jointly owned by the two American partners in MGMA, Utah and Cyprus. Marcona Corporation, the operator of an iron ore mine in Peru, thus became the marketing agent in Japan for Mount Goldsworthy iron ore exports.[670] In this capacity, Marcona facilitated a long-term contract with the Japanese steel mills for the export of 16 million tons of iron ore over seven years beginning in April 1966. But the Japanese were so concerned about the deficiencies of Goldsworthy's nominated port, Port Hedland, that they would buy ore only if MGMA accepted responsibility for shipping. Hence the original sales contract with the Japanese was negotiated on a cost insurance and freight (c.i.f.) basis rather than free on board (f.o.b.). Goldsworthy's first shipment of ore to Japan was 20,000 tons in 1966. Following the Goldsworthy precedent, prices for all the original Pilbara iron ore contracts were denominated in US dollars and there was no provision for cost escalation.

Prospectors and Geologists in the Pilbara

The most famous of the Australian prospectors in the Pilbara region of Western Australia was Langley George (Lang) Hancock. Born in Perth in 1909, Hancock was reared on two pastoral properties in the Pilbara at Mulga Downs and Hamersley Station.[71] After serving in the army in

World War II, he turned his attention to prospecting for minerals and made a number of finds. One of these led to the mining of blue asbestos at Wittenoom Gorge. Hancock sold the asbestos mine to Colonial Sugar Refining Limited (CSR) in 1943, while retaining a minority interest. In 1938 he forged a partnership with another Western Australian, E.A. (Peter) Wright, whose experience in business and finance and reputation in Perth gave Hancock a platform for his development ambitions in Western Australia.[72] In 1942 Hancock purchased his first aircraft, a two seater Klemm Swallow, which he used as an aid to prospecting and to commute between his property and local community centres. Hancock shot to national prominence in the 1960s with his sensational account of how he had found huge amounts of iron ore in the Pilbara in 1952.

As Hancock told it, he was flying his plane across the Hamersley Range in November of that year when he was caught in a heavy thunderstorm. Forced to dive low beneath the clouds, he flew through a gorge at the Turner River where he noticed rain streaming down cliffs like walls of iron.[73] Neil Phillipson, one of Hancock's biographers, regards this account as a fiction. He pointed out that the Turner River, over which Hancock claimed to have flown in 1952, flows north to the coast and not south and that if Hancock had been flying south, he could not have been flying through the Turner River and noticed any iron ore enrichments.[74] Phillipson's confusion can be explained by the fact that there was a second "Turner River" flowing south-west into the Hardy River, which in turn flowed into the Ashburton River. The WA government changed the name of this second "Turner River" to "Beasley River" to avoid confusion.[75] It is more likely that the iron enrichments that Hancock saw on this 1952 flight were on the walls of the gorge of the Beasley River. Phillipson also disputes that a thunderstorm occurred in the Pilbara on the date claimed by Hancock. But the Pilbara is so vast and was so undeveloped in the 1950s that many thunderstorms went unrecorded.[76] Thus Hancock's account of his discovery in the Hamersley Range may well be true, or substantially so. In any case, what is clear is that Hancock came to know of the presence of vast amounts of limonite ore in the Hamersley Range at some time in the 1950s. But he had little incentive to do much about it until the export embargo was eased.[77] Limonite iron ore is one of the main varieties of iron ore but it has a lower iron content than hematite ore, which can be as high as 70 per cent.[78] Throughout the 1960s Hancock advocated mining as the future of Australia and enthused about new technologies, such as peaceful nuclear explosions, that could assist large mining operations. In the course of pursuing his dreams, he took up the fight against those who stood in the way of his developmental ambitions, whether they were State and federal governments and their bureaucracies or multinational companies. Completely lacking faith in Australia's democratic system, Hancock sought a return to *laissez faire*, with ministers of the Crown, bureaucrats and businessmen in their boardrooms in Sydney and Melbourne reduced to minor roles. Robert Duffield wrote that Hancock "wants *action*, action freed from the confines of consensus, action decided by men who have not only vision but the power to implement it".[79]

The area of the Hamersley Range, where Hancock found the ore, was not virgin territory. Aboriginal peoples had lived in the area for at least 30,000 years and it had been traversed by prospectors searching for gold in the 1880s and 1890s. By 1890 more than a thousand men were working the Pilbara and Ashburton goldfields, which had resulted in a railway from Port Hedland to Marble Bar, closed in 1951.[80] But by 1961 this work force had largely dissipated and the Indigenous people were working in the pastoral industry in conditions akin to slavery.[81] The region's entire population (excluding Aboriginals who were not counted in the census at that time) was only 3423 in the small townships of Port Hedland, Roebourne, Marble Bar, Nullagine and Wittenoom, all of which were linked by unsealed roads. This region of 170,000 square miles extended from an inhospitable coastline to scree slopes, rocky escarpments and arid spinifex plateaux and then through the endless miles of the Great Sandy Desert to the Northern Territory border. Bred in an area that he was confident was rich in minerals, Hancock joined the push for the end of the export embargo. When Brand's government first called tenders for export from the Mount Goldsworthy deposit in 1959, Hancock and Wright proposed that, in exchange for permission to sell 10 million tons of iron ore, they would spend £2.5 million in developing Port Hedland into a port capable of handling 10,000-ton ships, compared to the then maximum of 5,000 tons.[82]

One of the contributions of the two men to the later development of the iron ore export industry was their persistence in persuading the London-based Rio Tinto Company to take an interest in iron ore. The history of the Rio Tinto company dates back to 1873 when a London banker, Hugh Matheson, purchased the Spanish Rio Tinto copper mines, 50 kilometres from Cadiz.[83] Taking its name from the red river of Andalusia, the mine had supplied Phoenicians, Greeks, Carthaginians and Romans with copper and iron for centuries.[84] After injecting capital into the venture and building an 80-kilometre railway, the Rio Tinto company became one of the greatest mining companies in the world in the 1880s on the back of its Spanish property. But Rio Tinto was in eclipse in 1948 when it recruited to its ranks the lawyer Val (later Sir Val) Duncan.

Val Duncan had served as a British staff officer advising Generals Bernard Montgomery and Dwight D. Eisenhower on the allied landing in France in 1944. He had also helped at the policy level to restore civil order in the Netherlands and Germany after the war.[85] Duncan would later, as head of Rio Tinto, become one of the two or three most influential British businessmen of the second half of the twentieth century. Having been appointed managing director of Rio Tinto in 1951, Duncan sold out of Rio Tinto's Spanish operation and invested the proceeds in a worldwide expansion that included the Rhodesian copper belt and Canadian uranium.[86] A visit to Australia in 1954 encouraged him to take an interest in uranium in the Northern Territory and to set up a subsidiary in Australia – the Rio Tinto Mining Company of Australia Limited. The subsidiary earned a substantial income from its ownership of the Mary Kathleen uranium mine in Queensland and used the profits to spend some £1.7

million exploring for other minerals in Australia throughout the 1950s.[87] Hancock and Wright, who were often known as "Hanwright", came to Rio Tinto's Melbourne offices on 22 June 1959 and struck an agreement under which the two men would jointly earn 2.5 per cent royalties from gross sales of all minerals, except manganese, from their mining titles in the Pilbara. This was before the lifting of the iron ore embargo when manganese and not iron ore was the main attraction for Rio Tinto.[88]

In 1961, after the export embargo was eased, Hancock persuaded Rio Tinto to send a company geologist, the Swiss-born Bruno Campana, to accompany him to examine his iron ore finds in the Pilbara. By following a chain of flat-topped mounds over 100 kilometres, Campana confirmed that Hancock had indeed found at least 1000 million tonnes of limonite ore, ranging in grade from 50 to 57 per cent iron content.[89] While inferior in iron content to hematite, limonite ore was being mined in other parts of the world as feedstock for blast-furnaces. Campana's report encouraged Rio Tinto to join with Hancock and Wright in staking a claim with the WA government to 10 temporary reserves, each limited to 130 square kilometres.[90] Rio Tinto had been anxious to secure a larger exploration area, but the WA government had refused on the grounds that it was reluctant to grant a monopoly in the Hamersley Range to one company. But the WA government granted Rio Tinto and Hancock and Wright exclusive rights to prospect in the areas covered by the temporary reserves and to a mining lease on terms negotiated with the State government if payable ore was discovered.

In 1961 Hancock was worried that other companies would pre-empt his finds and he pressed Rio Tinto's Duncan to come to Perth to put firm mining proposals to the Brand government. Duncan obliged and on 4 November 1961, together with Hancock and Wright, he met with the WA Minister for Mines, Arthur Griffith.[91] The same evening Duncan drafted a "heads of agreement" for the exploration and development of the temporary reserves for submission to Brand. Duncan asked the WA government to consider granting Rio Tinto a mineral lease by the end of

1962 in three phases. The first phase would involve exploration and drilling to determine the extent and grade of the deposits; a preliminary survey for a railway, probably to Depuch Island; and investigations overseas on the export potential of the ore. In this first phase Rio Tinto undertook to raise the finance for the development of Depuch Island as a port, but asked the WA government to be responsible for the construction of a mining township and ancillary services and also to be responsible for the dredging of the port.[92] In the second phase, Rio Tinto would proceed with the financing of the operation and full-scale construction operations. The third phase in Duncan's heads of agreement concerned a steel industry. Acknowledging the Brand government's strong interest in the industrial development of Western Australia, Duncan declared:

> Rio Tinto has expressed the view that as they are not steel people it would not be possible for them to give any categorical undertakings at this stage that they would establish such an industry. Rio Tinto, however, appreciates the attitude of the Government and would intend in due course to endeavour to persuade one of its many steel associates around the world to set up a steel works in Western Australia provided that conditions are propitious.[93]

Duncan suggested that if Rio Tinto did not proceed to establish steel works based on the Hamersley ore, then it would be willing to surrender part of its leases to the government.[94] But he clearly intended that "in due course" would be none too soon.

Duncan's gambit ended in failure. By this time, the WA government was at loggerheads with Hancock and had rejected his tender to mine and export up to 15 million tons from Mount Goldsworthy. In addition, the WA government was not confident that Rio Tinto, as a mining company, was the right vehicle to develop a Western Australian steel industry from the Hamersley Range's iron ore riches.[95] According to Court, because Hancock had helped with the Brand government's election policy, he thought he could "take over the Mines Department":

> It was as blatant as that, and I had to remind Lang that governments had to observe certain laws and procedures. I had to tactfully remind

> him that he wasn't the only person interested in iron ore … he wasn't the only person that had a great love of the State and Australia and wanted to see all those things happen.[96]

Rio Tinto's involvement in the Pilbara might have ended there but for the merger in 1962 of the British companies, Consolidated Zinc Corporation Limited and the Rio Tinto Company Limited, to form the Rio Tinto-Zinc Corporation Limited (RTZ) Company. Their Australian subsidiaries, Consolidated Zinc Pty and the Rio Tinto Mining Company of Australia, formed Conzinc Riotinto of Australia Limited (CRA). Consolidated Zinc's history in Australia went back to the early twentieth century, when it had pioneered recovering zinc from the tailings of the silver–lead ore mined in Broken Hill in New South Wales. W.S. Robinson, W.L. Baillieu and Englishman Francis Govett, had launched the Zinc Corporation in 1905 to treat the tailings, operate lucrative mines in Broken Hill, and take up smelting activities in Britain and Australia.[97] CRA thrived after the merger and in the 1970s would become Australia's largest mining house and a world-ranking mining company in its own right until the mid-1990s when it merged with its parent company. Consolidated Zinc's former Australian head and the new head of CRA, Maurice (later Sir Maurice) Mawby, persuaded Val Duncan to join forces with the American Kaiser Steel Corporation in prosecuting Rio Tinto's Pilbara plans.

Mawby was born in Broken Hill in 1904, the son of a grocer's assistant who had migrated there from England. He attended the local school and later studied at Broken Hill's technical college where he gained a diploma in metallurgy in 1927 and a diploma in geology in 1934. In 1928 he joined the Zinc Corporation where he was appointed mill foreman and was greatly influenced by industrialist and financier W.S. Robinson and Robinson's son, L.B. Robinson.[98] During World War II, as technical secretary of the Commonwealth Copper and Bauxite Committee, Mawby participated in a government mission to North America and Mexico to study metallurgical practices.[99] Mawby combined a deep understanding of geology and mining practices with a shrewd understanding of Australian politics. He was friends with State and federal parliamentarians from both sides of politics and would form a particularly close bond with Charles Court in Western Australia and Frank Nicklin in Queensland. In 1947 Mawby was general manager of Consolidated Zinc's subsidiary Enterprise Exploration Pty Ltd. He described his approach to searching for new mineral fields in remote parts of Australia thus:

> The disadvantages of operating in the Kimberleys [sic] and Pilbara district of Western Australia, Central Australia, and the Gulf Country of Queensland, are fully realised, but such operations as Mt Isa, which is now one of the major mineral fields in the world, shows what may be possible and what we hope to repeat.[100]

After becoming CRA chairman in 1962, Mawby assigned exploration in the Pilbara to Haddon King, the Canadian-educated director of exploration for the former Consolidated Zinc Pty. King instructed his geologists to examine every sizable hematite body in an area of about 27,000 square kilometres, a much larger area than the temporary reserves granted by the Western Australian government to Hancock, Wright and Rio Tinto.

When the 71-year-old Kaiser Steel geologist, Tom Price, visited the Pilbara in March 1962, he enthused on the sheer abundance of banded iron-ore formations containing 30 per cent ore of a kind in America that was providing an increasing source of pellets. Price told the Australian press:

> There are untold millions of iron ore in the Pilbara deposits. I think this is one of the most massive ore bodies in the world. There are mountains of ore there … it is just staggering. It is like trying to calculate how much air there is.[101]

Price's visit to Australia was instrumental in persuading the WA government to grant additional temporary reserves to Rio Tinto and to agree to make a fresh start on negotiations to develop iron ore in the Hamersley Range.[102]

In September 1962, after exploring several thousand kilometres in the Hamersley Range, two of King's geologists, Bill Burns and Ian Whitcher, came across a mountain of rich, blue-black ore between three or four miles long and up to 4000 feet wide that turned out to have iron content of 66 per cent. In doing so they had made one of the greatest mineral finds in world history.[103] The location of the find was forty kilometres outside the Hancock/Wright/Rio Tinto temporary reserves and only a kilometre away from an adjacent BHP temporary reserve. This massive hematite ore body, subsequently named Mount Tom Price, contained 600 million tonnes of high-grade ore and on its own doubled what had been the official estimates of Australia's *total* reserves of high-grade iron ore in 1960.[104] In the following month, October 1962, Mawby wrote to a colleague that the "whole future of Western Australian iron ore is so encouraging and the deposits so large that all of us are merely passing phases in what will ultimately be a very important and well established industry".[105]

In 1957, not long after Lang Hancock claimed to have noticed limonite ore in the walls of the Beasley River gorge but about five years before the discovery by CRA geologists of the Mount Tom Price hematite deposit, another Australian prospector, Stan Hilditch, came upon a similar hematite deposit in the Pilbara. This deposit was located east of Mount Tom Price and some 400 kilometres from the coast. Hilditch was born in 1904 near Newcastle but grew up in Kalgoorlie. His formal training included three years study of geology at the Kalgoorlie School of Mines. Hilditch survived financially during the Great Depression of the 1930s by trucking cattle to markets in Western Australia. After World War II, the Sydney entrepreneur Charles Warman agreed to back Hilditch in his mineral exploration in Western Australia. With his wife Ella, Hilditch roamed the Pilbara in an old Thames truck on a shoestring budget. In 1957, after travelling for about six years, the prospector stumbled on a massive iron ore deposit at what later became known as Mount Whaleback, the principal ore body of the future Mount Newman iron ore operation.[106] When informing Warman of the find, Hilditch received a lukewarm response – the refrain was that iron ore was a cheap commodity and that the find was too far from the coast.

Despite Warman's lack of enthusiasm, he and Hilditch secured 358,000 hectares of temporary reserves covering the Mount Newman discoveries.[107] Later, Warman and Hilditch offered Rio Tinto an option on the Newman deposits, but the company had more than enough to develop with its deposits in the Hamersley Range.[108] Similarly, when the pair approached Western Mining Corporation (WMC), it was deterred by Mount Newman's distance from the coast.[109] Finally, the two men were introduced to executives of American Metal Climax (AMAX) Company. AMAX was best known for its Climax mine in Colorado that produced most of the world's molybdenum, a metal used in stainless steel.[110] AMAX sent a mining engineer and geologist from the United States to examine the deposits with Hilditch and subsequently agreed in 1963 to purchase the temporary reserves taken out by the two men. AMAX bought the temporary reserves on a royalty formula designed to earn Hilditch and Warman $10 million over the life of the project. While a considerable amount of money, this was nothing like the bonanza obtained by Hancock and Wright from Rio Tinto.[111]

The Australian entrepreneur who was most responsible for stimulating international interest in iron ore deposits at Robe River in the Pilbara was Garrick (later Sir Garrick) Agnew. Born in Perth in 1930, Agnew enrolled in engineering at the University of Western Australia in 1949 and while there trained as a competition swimmer, representing Australia at the London and Helsinki Olympic Games in 1948 and 1952. Having secured a sporting scholarship, he moved to the United States to study psychology at the Ohio State University in Columbus. A "big healthy man with a brown face, big forehead and curly hair", he was a restless man with an original turn of mind.[112] When he came back to Western Australia, he turned his attention to prospecting for minerals, forming Garrick Agnew Pty Ltd and going into business as an iron ore broker.[113] In the early 1960s Agnew developed an interest in the red-faced hills and mesas at Robe River, an area that had been surveyed half a century earlier by A. Gibb Maitland. An officer of Western Australia's colonial government, Maitland had noted the distinctive red-brown capping of the Robe River mesas in 1898 and had reported to his government presciently that they might contain iron ore.[114]

Despite the fact that the deposits in Robe River were smaller and had lower iron content than other major hematite deposits in the Pilbara, Agnew persuaded the Japanese trading company Mitsui to look at them. In 1961, Mitsui's executives Teruji Nomura and Takashi Imai became the first people outside Garrick Agnew's circle to fly into a primitive airstrip to visit the Robe River site. The Japanese businessmen noted that the Robe River ore was of a lower grade than the deposits in the Hamersley and Ophthalmia Ranges but also that it could be more easily mined and was nearer to the coast than the latter deposits.[115]

Agnew's company formed a partnership with a Delaware-based company, Howe Sound, to register a new firm in Perth called Basic Materials, which acquired prospecting rights to the deposits examined by Agnew at Robe River. Iron ore can be exported in three main forms – lump ore, iron ore fines and pellets. Lump ore is defined as anything from six millimetres to 30 millimetres in size and is preferred because, in a blast furnace, its particle size enables oxygen to circulate around the raw materials to melt them more efficiently. Anything less than six millimetres is defined as iron ore fines (or powdered ore). Before iron ore fines can be introduced into a steel-making blast furnace, they must be processed into either sinter or pellets; otherwise the fine particles block the upward movement of gases through the blast furnace. Sinter production is achieved by mixing the fine iron ore with a binder and then heating it to about 1200 degrees centigrade. Pellets are produced by binding the iron ore fines through an industrial process. The executives from Mitsui saw the potential of the Robe River area to export iron ore pellets. At the suggestion of Imai, Agnew approached the major American iron ore miner and pelletising expert Cleveland Cliffs, a company headquartered in Ohio where Agnew had studied psychology.[116] In 1962 Cleveland Cliffs' Vice-President, Bill Dohnal, came to Western Australia to view Agnew's deposits first hand and at once saw potential in them. He contemplated the possibility of exporting three million tons of iron ore pellets a year. After years of struggle, an iron ore project at Robe River would eventually be established in the early 1970s.

The Formation of Hamersley Iron

Consolidated Zinc Pty and the Rio Tinto Mining Company of Australia came together to form Conzinc Riotinto of Australia Ltd (CRA) in 1962. The British-controlled, but Australian-managed, company announced that it would sell 2,500,000 five-shilling shares to the public at 16 and sixpence each. This would bring the Australian public's stake in CRA to 10 per cent. The next 15 years would see CRA grow from a company with $A139 million capital in 1963 into Australia's largest mining house in 1975, with $A1700 million in capital. Its subsidiaries were involved in a miscellany of mining activities, including lead and zinc at Broken Hill, copper in Bougainville, iron ore in the Hamersley Range and bauxite in North Queensland. Over the years, CRA Chairman Mawby and his team of Australian managers had gained much experience in the Australian mining industry and an understanding of the workings of Australian politics that the British Rio Tinto chief Val Duncan lacked. In March 1962 Mawby confided to a colleague, R.C. Atherton, that he doubted whether the Brand government would have given Rio Tinto all the temporary reserves it had but for the knowledge that Consolidated Zinc would soon be joining the company.[117] Moreover, in the course of developing bauxite deposits at Weipa in North Queensland, Consolidated Zinc, then under Mawby's stewardship, had developed close links with the Kaiser Corporation in the United States. Kaiser's participation, Mawby hoped, might help to convince the Brand government to grant mining leases on the potential of a CRA–Kaiser Steel partnership ultimately to manufacture steel.

In July 1962 Val Duncan and other members of RTZ and CRA went to Oakland, California, to negotiate the formation of Hamersley Iron, a new company in which CRA held 60 per cent of the shares and Kaiser Steel 40 per cent. Hamersley Iron, later incorporated under the Victorian *Companies Act*, would in the next two decades become one of the largest iron ore companies in the world. Though American involvement was important in its success, Mawby and CRA were clear from the start that Australians would oversee its management. The Australian mining engineer Struan Anderson, who had managed the Rum Jungle uranium project in the Northern Territory for Consolidated Zinc, subsequently became the managing director of the new company.[118]

Mawby was aware that close association between Rio Tinto and Hancock and Wright had been one of the reasons behind the Brand government's rejection of Val Duncan's 1961 bid to obtain a mining agreement from the WA government. Mawby and Struan Anderson therefore decided to buy out the two Australian entrepreneurs before approaching the WA government again. To buy them out, the company had to resolve the question of royalties. Hancock and Wright felt that, as they had introduced Rio Tinto to the former goldfields in the Pilbara, West Pilbara and Ashburton areas, they were entitled to royalties from any ore mined in the area and not just from the areas covered by temporary reserves.[119] As the price of buying out Hancock and Wright, Hamersley Iron agreed to give the two men royalties on any ore mined from the vicinity of the three old goldfields, including the Mount Tom Price area. As Struan Anderson explained to Mawby on 28 August 1962:

> The areas we finally obtain from the Government almost certainly will extend to cover one found just outside present reservations. We could hardly deny Hancock and Wright royalty on production from these immediate extensions which we feel certain the Government will give to us, and I am sure that whatever amended areas we finish up with in our coming deal with the government must be subject to royalty.[120]

Given the immensity of its projected mining operation, Hamersley Iron contemplated building alliances with other organisations. Australia's big steelmaker BHP had to be ruled out because the WA government opposed an alliance with Hamersley Iron. The federal government, having already faced enough criticisms over BHP's steel-making monopoly, was also opposed.[121] Hamersley Iron also contemplated alliances with other companies, including MGMA and Cleveland Cliffs. But in the end, Anderson advised Mawby that:

> We know that a joint port (and perhaps some joint railway) arrangement could save perhaps a little in operating costs but I think it is essential to realise that if, by avoiding such a sharing arrangement, a competitor might be kept out indefinitely or his start delayed, we would then gain more from the other direction, i.e. we sell more tonnage.[122]

On 7 January 1963, Struan Anderson set out Hamersley's proposals in a letter to Arthur Griffith, the Western Australian Minister for Mines. Hamersley Iron was aware that the Brand government had made BHP's lease to Koolyanobbing ore conditional on building a blast-furnace at Kwinana.[123] But Anderson argued that Hamersley Iron should be relieved of the obligation to build a steel industry for a considerable time. He gave three reasons. The first was that the company would need time to export enough ore to recoup some of the capital required to mount an export operation in such an isolated area. The second reason was BHP's dominant position in steelmaking in Australian and overseas markets. The third reason was Hamersley Iron's "belief that revolutionary new metallurgical developments are likely to take place in the next decade or so, developments in which we are interesting ourselves and which could well put a newcomer into a favourable position, at the right time, to start up against well established competition".[124] Hamersley Iron would involve itself after 1968 in the possibility of processing ore into a metallised product in a process subsequently given the name HImet. But further moves into manufacturing and exporting steel products were inhibited by competition from the Japanese steel industry, already becoming the most efficient in the world, and by the less propitious economic circumstances in the 1970s.[125]

By this time, Hamersley Iron had come to accept, through a study of the Goldsworthy agreement, that the WA government had virtually no capacity to contribute to the costs of infrastructure for an iron ore industry. As Mawby advised Val Duncan on 30 August 1962, the WA government would adhere closely to the pattern of the Goldsworthy agreement and he therefore advised that "the suggested basis for an approach to the Government may have to differ considerably in form to the 'Duncan plan', though not in the underlying objective".[126] Consequently, Hamersley Iron itself would have to take responsibility for the construction and maintenance of a railway, port, towns, water and generation of power. Struan Anderson elaborated:

> Consideration of competitive overseas operations, current and projected, reveals that nothing short of very high and specialized standards of equipment, particularly in respect of railway and port facilities, would suffice. With the high degree of mechanisation necessary to ensure satisfactory costs, the towns would not be large, but high-class staff and low staff turnover would be essential; for these reasons and to meet severe climatic conditions, we contemplate building specialized and unconventional housing with a high proportion of married accommodation.[127]

The Brand government accepted Hamersley Iron's reasoning, largely because Charles Court as Minister for Industrial Development had now taken over responsibility for new iron ore projects in the Pilbara. Court was much more favourably disposed to Hamersley Iron than the cautious Griffith had been. On 30 July 1963 the Western Australian government signed an agreement with Hamersley Iron, which would have to spend £500,000 on preliminary investigations of the iron ore deposits and on the construction of railway, port and town sites. It would also have to spend not less than £30 million on mining, railway and wharf facilities by 24 March 1968. Within 10 years of beginning the export of iron ore, Hamersley Iron was required to spend at least £8 million on a secondary pellet processing plant with a production capacity of two million tons per year. And after 20 years of iron ore exports, Hamersley Iron would have to submit complete plans for an integrated iron and steel industry requiring the investment of at least £40 million. The Iron Ore Hamersley Range Agreement Act, which in retrospect embodied a great deal of wishful thinking on the government's part, was enacted on 13 November 1963 and gave Hamersley Iron temporary reserves totalling 7008 square kilometres (including the Mount Tom Price area) from which the company could ultimately secure a mineral lease on an area not exceeding 777 square kilometres.

On 17 December 1963 Hamersley Iron requested permission from the Menzies government to export 200 million tons of iron ore over a period of 21 years. In making this request, Mawby had a lengthy conversation with Menzies in which the two men agreed:

> ... because of our geographical positions, we must become closer to Japan, which is now our largest customer. Our economies are more or less complementary and, divided as we are by the uncertainties of Indonesia and the appalling poverty, ignorance and backwardness of the other South Asian countries (perhaps Malaya excepted) and India, Japan and Australia must keep together.[128]

The scale of the Hamersley project, which dwarfed that of MGMA, sparked some debate in Canberra. Spooner's preference was for a number of iron ore companies to operate in the west. But he wondered whether competition between them would lead to cost-cutting and therefore a loss of export income to Australia. But he ruled out any effort, governmental or otherwise, to rationalise iron ore mining activities and was wary even of using export controls to secure "reasonable prices".

By 1964 Hamersley Iron and MGMA were pursuing their plans with energy and AMAX was investigating Mount Newman. Meanwhile, BHP was taken aback by the development of its competitors. As it was, BHP was excluded from exporting ore from the key areas in which it had leases – in Yampi Sound, Koolyanobbing and the Middleback Range in South Australia. But in a belated and audacious bid to break into the emerging export industry in late 1964, BHP negotiated with the Brand government to develop low-grade iron ore deposits at Deepdale on the mouth of the Robe River in the Pilbara.[129] In its agreement with the WA government, BHP agreed to build a town, mine, port and processing plant to develop the ore. It hoped, thereby, to earn the right to export not only processed (pelletised) ore from Deepdale but also unprocessed iron ore from those areas which federal policy still specifically prohibited. The Brand government, with a parliamentary majority of only one, was anxious to demonstrate as quickly as possible the value in royalty payments of the new iron ore companies, which as yet had made no export sales. BHP was in a stronger position than its competitors to export quickly and the government could see no good reason not to accede to its request. As Court explained to Struan Anderson, he thought it only fair to lift the BHP export ban if and when BHP fulfilled its initial Deepdale commitment: that is, when the company had developed an approved port site, railway from port to quarry and the necessary quarry and associated development.[130]

At that time, towards the end of 1964, Hamersley Iron was preparing to negotiate with the Japanese steel industry. In its view, had it been left to BHP, Australia probably would never have developed an export market for Australian iron ore. Hamersley Iron had helped to open up the Pilbara on the assumption that BHP's ore deposits and liquid funds earned from steel-making would not be deployed against it. So when Hamersley Iron heard of the BHP proposal it was furious. Struan Anderson wrote to Charles Court on 12 October 1964 to prosecute his case. He remonstrated that BHP had been behind the iron ore export ban for years to protect its monopoly position. The iron ore export embargo should not be lifted from BHP's properties, he insisted, until after the investment in major new non-BHP iron ore ventures in the Pilbara had been amortised.

Mawby, the chairman of CRA, fired his biggest shot with Menzies. Securing an audience with the prime minister on 22 October 1964 to discuss "something more important to Australia than any issue that has ever arisen in the country's mining history", and speaking on behalf of a new company poised to secure an export deal with Japan, he accused BHP of aiming to prevent any other large mining/industrial company from being established in Australia.[131] It had, he said, conducted little exploration in Australia in the 1950s, while continuing to publicise that Australia had no surplus iron ore for export.[132] Its support for the continuation of the embargo and delay in finding new ore, he charged, had unwittingly aided the development of competitor iron ore projects in South America and South and Southeast Asia. BHP's chairman Colin Syme, he alleged, had tried to undercut BHP's rivals by advising the Japanese not to concern themselves with newcomers in the Australian iron ore business since they would soon satisfy all their needs from BHP. Finally, Mawby mentioned the pressure that BHP was exerting on the Brand government by threatening to persuade the Queensland government to give it rights to iron ore in the Constance Range if Western Australia made its terms on the development of iron ore too stringent.[133]

Menzies accepted Mawby's submission and cabinet decided on 11 December 1964 that the remaining export restrictions should be "firmly maintained". To Brand and BHP Menzies justified the decision to keep the iron ore embargo in place for Yampi Sound, Koolyanobbing and the Middleback Range on three grounds other than conservation.[134] The first was that cabinet's decisions in 1960 and 1963 to maintain the embargo in these areas were taken because the deposits stood in a special relationship to the Australian steel industry. The second was that when the federal government provided assistance for the standardisation of the railway between Perth and Koolyanobbing, it had been understood that the iron ore deposits would be used by the steel industry to be established in Western Australia. Finally, the companies that had taken advantage of the 1960 relaxation to explore and develop new deposits could, with justification, claim that their chances of profitable operations would be adversely affected by such a fundamental change in the Commonwealth's export policy as the proposal involved. To BHP's chagrin, it would not be until 1966 that the remnants of the iron embargo would be abolished. By that time, BHP had taken another path into the Pilbara iron ore industry, as a partner in the development of iron ore deposits at Mount Newman.

CRA and Hamersley Iron saw off another major challenge to their development plans in 1964. The American entrepreneur Daniel Ludwig, owner of Clutha Development in New South Wales, was a self-made millionaire who had built his fortune via a family company, National Bulk Carriers, the parent of a group of companies involved in global sea transportation and shipping construction.[135] Having signed his royalty agreement with Hamersley Iron, Lang Hancock invited Ludwig to investigate iron ore discoveries he had made at Nimingarra and Yarrie near Mount Goldsworthy. Impressed, Ludwig took up temporary reserves there and at Wittenoom. Ludwig's boldest initiative was to write to the Brand government in 1964 to propose the "Ludwig Plan". Its essentials were that the Ludwig organisation would explore and develop the iron deposits around Koodaideri over which Hancock had temporary reserves; establish a beneficiation plant to upgrade the ore; finance and build a super-port at Cape Keraudren which could accommodate 166,000-ton ships and could service other iron ore projects, like Hamersley Iron; and lead the financing and construction of roads and railways to bring together in one operation all the infrastructure for the iron ore mines in the Pilbara.[136] Ludwig estimated fewer ports in the Pilbara and potential capital savings of $150 million.[137] Although the scheme was one of breathtaking boldness, it was strongly opposed by Hamersley Iron, which could see its interests being subsumed under the Ludwig behemoth. Court agreed, seeing the whole proposal as a "Ludwig benefit", and persuaded the WA government to reject it.

Hamersley Iron's $100 million loan

Hamersley Iron then faced a formidable challenge: to secure long-term contracts for the sale of ore that would justify the lending by financial institutions of the capital required to construct the project. After Hamersley Iron's engineers and accountants had made their assessments, they arrived at a total cost of $US88 million for the first stage of the project, a figure that would later climb to $US105 million. This made the Hamersley project "the largest investment in an initial mine project in Australian history" ($1.3 billion in today's dollars).[138] Hamersley Iron needed to borrow $US100 million and outlay about $US200 million in total. In 1965 no Australian company had ever tried to borrow such a large sum and no Australian financial institution could accept a risk of such magnitude.[139] Two rival Japanese trading houses, Mitsubishi Shoji and Marubeni-Ida, each sought to negotiate sales on behalf of Hamersley Iron.[140] CRA preferred the former and Kaiser Steel the latter. Anderson solved the problem by inviting the two Japanese companies to act jointly on Hamersley's behalf as "introducers" to the Japanese steel mills. The arrangement earned the nickname "Marubishi". At that time the Japanese steel mills were taking relatively small quantities of lower quality ore from India, Malaysia and Goa. Hamersley Iron's bold plan was to convince the mills to switch to long-term contracts with Australia for immense quantities of higher-grade ore to be transported in giant carriers over the relatively short sea route between Australia and Japan.[141]

In September 1964 negotiations reached a critical point when the steel mills invited price proposals from all the new Australian companies, including Mount Newman. In 1964 Japan was importing 28 million tons of iron ore from all sources but there was no guarantee that it would be able to take ore from all the new Australian projects that were trying to get started. The imperative to reach a deal quickly prompted Struan Anderson to offer to trim the production schedule for the Hamersley project from an original 36 months down to 19 months, hoping that this astonishingly short schedule would persuade the Japanese. His ploy succeeded. In December 1964 the Japanese steel mills signed a letter of intent with Hamersley Iron to import 65.5 million tons of direct shipping ore, over 16 years beginning in August 1966. Worth £270 million, it was the largest sales contract ever made by a company operating in Australia.[142] The decision to build the Hamersley project in 19 months has been described as "one of the most courageous acts in Australian industrial history".[143] Reflecting back on the decision in 1988, Russel Madigan remarked that "if we had been doing this today, we wouldn't even have completed the environmental impact study in 19 months".[144]

The sales agreements provided Hamersley Iron with the means to borrow $US120 million from banks in the United States. This sum represented two-thirds of the capital needed to construct the direct shipping ore project and a processing plant to produce pellets from finer ore. CRA and Kaiser Steel would contribute the remaining capital. Kaiser Steel approached its bank, the Bank of America, which in turn assembled a consortium of 11 other banks to examine the company's business case. The negotiations were complex and two of the banks withdrew because of concern about Japan's long-term capacity to take Hamersley ore and the real possibility of a glut in the market as Mount Newman and later the Robe River projects came on stream. Another concern was the introduction by the US Federal Reserve Bank of a policy to discourage American banks from making overseas loans. Court and the Federal Treasurer, Harold Holt, were enlisted to gain from the US government the necessary assurances that American institutions would be able to invest in the Australian iron ore projects. These official representations helped persuade the American financial institutions to lend Hamersley Iron the funds that it needed.[145]

Conclusion

In 1960 the 22-year-old federal embargo on iron ore exports from Australia was still in place. By 1966 it had been completely lifted and two companies, MGMA and Hamersley Iron, had obtained sales contracts with Japanese steel mills and raised the necessary finance to build large mining operations in Western Australia. The lifting of the embargo was not pre-ordained. In 1960 Australia's exports were overwhelmingly from the rural sector and any expansion was expected to come from manufacturing rather than mining. Even the WA government, which had pushed so hard to end the export embargo, initially saw iron ore only as the basis for a future steel industry in the West or as an adjunct to other industrial development. For decades political and mining leaders believed that Australian reserves were barely sufficient for Australia's own steel industry. Thus when the extent of the Pilbara iron ore deposits were revealed in 1961-62, it came as a huge surprise. Lang Hancock and his partner Peter Wright were instrumental in unlocking the Pilbara's potential and Hancock's account of his aerial discovery of iron ore enrichments in the Pilbara in 1952 is probably close to the mark. Hancock's main contribution was enticing Rio Tinto, through Bruno Campana, and later CRA to investigate his claims and further explore the Hamersley Range. CRA geologists later discovered the massive hematite body at Mount Tom Price. And prospectors Stan Hilditch and Garrick Agnew were attracting overseas companies to investigate Pilbara iron ore deposits that later became the basis of successful mining operations. Lifting the iron ore export embargo was one of the most dramatic economic policy decisions made by an Australian government and arguably ranks with the far-reaching economic reforms of the Hawke government in the 1980s in forging the modern Australian economy. The immediate consequence of the raising of the embargo was the fast-tracked development of one of Australia's great export industries, as will be seen in Chapter Three.

Hamersley Iron train, Mount Tom Price, 1960s. NAA: B942, IRON ORE [10]

Port Hedland from the air, 1974. NAA: A6135, K1/7/74/63

3

The Pilbara Iron Ore Industry, 1964-1972

The original gold rush of the nineteenth century involved thousands of individual prospectors migrating to goldfields in Victoria and New South Wales. By contrast, the iron ore boom of the 1960s and 1970s was triggered by large companies, all of them joint ventures, which mobilised large amounts of finance and utilised a relatively small workforce in highly capital-intensive operations. The magnitude of the iron ore rush can be seen in the immensity of the capital investment, the feats of engineering necessary in building mines, railways and ports, and the swelling tonnages exported from 1966 onwards. By the late 1960s iron ore had already overtaken coal as Australia's main mineral export and the Pilbara had become a world-class supplier, furnishing about half of Japan's needs. By this time, Japan itself had developed one of the world's most dynamic and efficient steel industries.[1] As the boom reached dimensions undreamed of at the beginning of the 1960s, the federal and WA government sought to regulate it in different ways, the Commonwealth by influencing prices and Western Australia by deciding which companies would develop specific areas of the Pilbara. The WA government and its Minister for Industrial Development, Charles Court, viewed both federal intervention and Lang Hancock's various schemes as a threat to its plan for the orderly development of the Pilbara reserves. It resolved both of these issues by persuading the Commonwealth government to dismantle its iron ore price guidelines in 1967 and by imposing, against Hancock's opposition, a plan for the Pilbara's development in 1971.

The CSR-AMAX Partnership: The Mount Newman Iron Ore Company, 1964-65

In June 1964 the American firm AMAX approached the Melbourne financier, Ian Potter, about funding to develop the iron ore deposits at Mount Newman. As we have seen, these deposits were discovered in 1957 by Stan Hilditch and his backer, Charles Warman, who sold the temporary reserves to AMAX (American Metal Climax) in 1963. Potter was born on 25 August 1902, the third of four children of a Bradford wool merchant with business interests in Britain and Australia. His family settled in Australia after World War I and Potter went to Sydney University where he obtained a Bachelor of Economics degree. In 1929 he joined the staff of Melbourne stockbroker, Edward Dyason, and in 1936 founded his own stockbroking firm, Ian Potter & Co.[2] In the following decades Potter would become one of Australia's leading financiers. Following AMAX's approach to Potter, he persuaded one of Australia's largest companies, the Colonial Sugar Refining Company Ltd (CSR), to join AMAX in its Mount Newman venture with a 35 per cent stake and an option to go up to 45 per cent.[3]

The general manager of CSR was Sir James Vernon. An industrial chemist who had risen through the ranks to become general manager in 1958, Vernon was close to the Menzies government, which in 1963 appointed him as chairman of a committee of economic inquiry into the Australian economy. Vernon's company, CSR, whose virtual monopoly of sugar production was well established by the 1930s, had diversified in the 1940s and 1950s. With Britain's move in the early 1960s to join the European Economic Community, Vernon saw that CSR had to find new markets.[4] One of his great successes was to sign long-term sugar contracts with Japan in the 1960s, a dress rehearsal for his later negotiations with Japan over iron ore and bauxite. Vernon's lieutenant in this exercise was the Queensland-born Gordon Jackson, who would succeed Vernon as General Manager of CSR in the 1970s and early 1980s. In the period from 1964 to 1966 Vernon would play a crucial role in establishing the

Mount Newman Iron Ore Company (Mt Newman) as a viable project with majority Australian equity.

CSR had been closely linked with the Pilbara through an association with Hancock in asbestos mining at Wittenoom, 160 kilometres from Mount Newman. The company was also about to embark on a venture to develop the Gove bauxite resource in the Northern Territory.[5] Under Vernon's leadership, CSR took up the option to assume 45 per cent equity in the Mount Newman Iron Ore Company in late 1964. On 26 August 1964 Mt Newman, beginning as an AMAX–CSR partnership, signed an agreement with the WA government with similar conditions to those in Hamersley Iron's agreement, including the obligation to build a steel industry. But the planning of the two companies had to be modified in 1965 when the Johnson Administration in the United States decided to restrict American loans financing overseas projects. AMAX consequently agreed that it would lower its stake in Mount Newman to an equal partnership with CSR.[6] CSR's 50 per cent equity in the Mount Newman enterprise gave the project enormous prestige in Australia.

On 11 October 1964 Mt Newman applied to the Menzies government to export 210 million tons of ore over a period of 21 years, commencing at the end of the decade. Menzies' cabinet approved the proposal on 11 November 1964 subject to the Minister for National Development's approval that the prices eventually negotiated were reasonable.[7] Not long afterwards, on 30 January 1965, the Japanese steel mills accepted the Mount Newman Iron Ore Company's offer to sell 100 million tons of iron ore over 21 years and proceeded to negotiate a price with the newly formed company.[8]

Like Australia's coal companies, Mt Newman had to negotiate with the united front of the Japanese steel industry.[9] The company negotiated a price of $US14.31 cents per iron unit f.o.b. up to five million tons per annum and $US13.785 cents per unit f.o.b. for quantities in excess of five million tons. It also negotiated discounts on the total payable by the Japanese depending on the quantity in excess of five million tons in any particular year.[10] The prices were 4.6 per cent lower than those obtained by Hamersley Iron on quantities up to five million tons per year and nearly 15 per cent lower for quantities over five million tons per year.[11]

The Japanese steel industry regarded the Mount Newman deal with its large quantities of high quality iron ore at a low price as a coup. Indeed, by March 1965, other smaller mining companies in Australia were becoming concerned at the low prices negotiated by Hamersley Iron and Mt Newman in particular. Sir Gordon Lindesay Clark, the chairman of Western Mining Corporation – one of the smallest WA iron ore operations but wholly Australian-owned – complained bitterly to Australian Treasurer Harold Holt that the prices negotiated by the two largest iron ore companies would represent a loss to Australia of $US10 million per annum.[12] He impressed on Holt the need for a federal iron ore policy that prevented the smaller iron ore suppliers from being frozen out of the development of other deposits. The Commonwealth Treasury agreed with Clark, warning Holt of the possibility that "Australia's iron ore resources are being sold to foreigners for a song".[13]

The changing climate of opinion in government and business circles about the iron ore industry came at an unfortunate time for Mt Newman. It had yet to have its iron ore prices approved by the federal government. By that time the Minister for Trade and Industry, John McEwen, and David Fairbairn, William Spooner's successor as Minister for National Development, had become concerned about the erosion of Australian iron ore prices that meant less foreign exchange for Australia.[14] The English-born grazier, David Fairbairn, was educated at Geelong Grammar and then Cambridge University. After war service in World War II, he took over his family's pastoral property near Albury in New South Wales. After he was elected to the House of Representatives for the Liberal Party, he was appointed Minister for Air in 1962 and then as Minister for National Development from June 1964 until March 1971. Among the least known of Menzies' ministers, he nonetheless played an important role during the mining boom, particularly in the federal regulation of the iron ore industry. On federal regulation, he tended to side with his more interventionist Country Party colleague, McEwen.

Fairbairn agreed that prices for Australian iron ore had been effectively cut back as a result of commercial bargaining with the Japanese steel mills. He felt that MGMA and Hamersley Iron, whose prices he had already approved, should be allowed to finalise their contracts.[15] But he would not vouchsafe the even lower Mount Newman prices that he had not approved and planned to submit to cabinet. McEwen wanted far more drastic action. McEwen recommended to his cabinet colleagues on 8 April 1965 that Fairbairn should instruct Hamersley Iron and MGMA not to conclude their contracts with the Japanese, pending a complete overhaul of Australian iron ore policy. If this were not done, he warned, "it would scarcely be possible to influence the prices negotiated by other companies in the future".[16] Cabinet, however, baulked at taking this step. On 8 April 1965 it left the decision to Menzies and Fairbairn, who on the following day decided that the Hamersley and Goldsworthy contracts should be allowed to proceed.[17]

The moment of decision for Mt Newman came when Fairbairn placed the pricing issue before cabinet on 18 May 1965. He believed the Menzies government should take action on the Mount Newman contract, especially since its operations were not scheduled to commence until 1969, while Hamersley Iron and MGMA were due to start exporting in 1966.[18] Fairbairn put to cabinet the option of asking Mt Newman to renegotiate on the basis of Hamersley Iron's prices. In the bigger picture, he speculated on whether the Commonwealth could ask the States to insist on higher royalties, encourage the sellers to form a consortium or to take an even larger step and promote the formation of an international iron exporters' cartel.[19]

When cabinet considered Fairbairn's submission on 18 May 1965, opinion was divided.[20] Accordingly, it authorised Menzies to consult with Vernon, who sought to justify the negotiated price for Mount Newman's first 100 million tons of ore, one of the largest such contracts in the world at that time. The price in his estimation was only one or two per cent lower than Hamersley Iron's, "all things taken in".[21] Vernon pointed out to Menzies that the Australian stake in Mt Newman through CSR's 50 per cent equity meant that £A310 million out of £A418 million in export earnings would remain in Australia over 22 years.[22] He conceded that it would have been advantageous to set up an Australian exporters' consortium when the embargo on iron ore exports was lifted, but he doubted whether such an organisation could be established now. Most importantly, he insisted that if the Australian government decided that the Mt Newman had to renegotiate its deal with Japan, the whole Mt Newman enterprise would fail.[23]

Vernon's stature as a businessman and adviser to government, and the efficacy of his representations to Menzies, were crucial to the ultimate success of the Mount Newman project. Menzies conceded Vernon's arguments. In effect, as McEwen later objected, Menzies made the decision outside the cabinet room and his cabinet colleagues endorsed the Mount Newman deal on 26 May 1965. "A highly controlled and unified demand", they lamented, "had in this case operated against a scattered and disorganised supply and so forced down the price".[24] But they confirmed Menzies' decision that the government's approval of the sale of 100 million tons of Mount Newman ore was "beyond recall" and that federal endorsement of the price should not be withheld.[25] At the same time, they decided that there should be entirely new arrangements for the sale of iron ore and an arrangement by which the operating company, the State and the Commonwealth were in close consultation from the beginning. The government was now concerned that the uncoordinated development of iron ore industry was leading to the duplication, overcapitalisation and underpricing of Australia's mineral resources. Consequently, it resolved to introduce guideline prices for iron ore to take effect in the following year.[26] But the Australian government's decision to use its powers over trade to regulate the iron ore rush in the interests of stabilising iron ore prices brought it into conflict with Japan and Western Australia.[27]

The Beginnings of Cleveland Cliffs' Robe River Project, 1964-1966

In 1963 the specialist American pellet producer, Cleveland Cliffs, demonstrated its commitment to developing the limonite iron ore at Robe River by buying out all the capital stock and shares of the Garrick Agnew vehicle, Basic Materials, together with Agnew's prospecting rights at Robe River.[28] Cleveland Cliffs hoped for an agreement from the Japanese steel mills to purchase about 35 million tons of Robe River pellets. To do so it asked for the assistance of the WA government to construct a railway, town and port to enable the ore to be transported.[29] But Court informed the company that Western Australia did not have the financial capacity to assist and, in any case, could not discriminate in favour of Cleveland Cliffs when it had not helped with the infrastructure of the other Pilbara companies.[30]

Court, however, had good reason to be enthusiastic about the Robe River venture from the beginning. Robe River iron ore deposits were near the Deepdale deposits held by BHP, and Court had earlier conceived a grand plan to compensate for the State's inability to finance infrastructure. Cleveland Cliffs and BHP, he thought, should come together to build a railway, town and port for their joint projects. He also thought that BHP should have a period of grace before actually developing its Deepdale deposits, provided that the Australian company joined in the building of infrastructure which would get the American project up and running. Court also saw economic benefits for Western Australia in secondary processing – that is, of pelletising – limonite ores. Constructing an iron pelletising plant with an annual 3.6 million ton capacity would cost $US125 million, employ 325 people directly and 390 people indirectly, and benefit Australia to the extent of $US22 million annually in employment, supplies consumed and income taxation.[31] Court also saw how Robe River could advance his goal of developing the lower grade limonite deposits as well as the higher grade hematite ores.[32] In this sense, Court's thinking was not unlike that of the Joint Coal Board, which was advocating the balanced development of Australia's underground and open cut black coal resources.[33]

Throughout 1964 Cleveland Cliffs explored the possibility of a joint venture with BHP without success.[34] As the year went on the American company came under increasing pressure to submit a proposal to the Brand government on how it planned to develop its temporary reserves. In developing this proposal Cleveland Cliffs argued that Robe River had to be treated as a special case. This was because the other companies were planning to ship large quantities of ore early on in the development of their projects while their processing commitments to the State – building pellet processing plants or possibly steel plants – came much later. The Vice-President of Cleveland Cliffs, William Dohnal, argued that the reverse was true for Cleveland Cliffs, which would need to make large expenditures early on in its project without being able to compensate by selling large amounts of iron ore until much later.[35] Because of special factors – the lower iron content of its limonite ores and the necessity to build a capital-intensive pellet plant to process the ore – Cleveland Cliffs asked for a concession on the royalties and port charges which it would be obliged to pay to the State government.[36] The Brand government accepted Dohnal's argument and gave Cleveland Cliffs until 31 December 1965, with possible extensions until the end of 1969, to notify the State government that it had secured a satisfactory sales agreement for the ore and finance for the operation.[37] The government ratified the agreement later in November in the *Iron Ore (Cleveland Cliffs) Agreement Act* 1964.

In the following year Cleveland Cliffs decided to vest all of Basic Materials' rights in Western Australia into a wholly owned subsidiary, Cliffs International, which in turn assigned them to a Western Australian subsidiary, Cliffs–Western Australia Mining Company. In March 1965 Cleveland Cliffs formed an arrangement with Mitsui & Co., one of the largest general trading companies in Japan and part of the Mitsui group. Mitsui, as we have seen, was also involved in developing coal deposits in Central Queensland with Thiess Brothers and Peabody and had been one of the first companies to inspect Garrick Agnew's Robe River discoveries in the early 1960s.[38] Largely because of Mitsui's early participation, Robe

River would become the Pilbara iron ore project with the largest Japanese equity. This resulted from Cleveland Cliffs' long established policy of involving consumers' interests in pelletising operations to give them more stability. While an alliance of this nature was hard to achieve with the steel mills themselves, Mitsui and Co. was closely associated with the Japanese steel industry in a purchasing and selling capacity and, moreover, was a substantial shareholder in two of Japan's biggest steel companies, Yawata and Fuji. Yawata and Fuji in turn had interlocking shareholdings with Mitsui.[39] The alliance with Mitsui proved a great advantage for Cleveland Cliffs in eventually securing large mining contracts with the Japanese steel mills.

In the latter stages of 1964 and the first half of 1965, the Australian iron ore companies negotiated their own sales contracts with the Japanese steel mills and Cleveland Cliffs joined them. But the American pelletising specialist found the negotiations heavy going and on March 1965 sent a cable to Court pressing the WA government to express concern to the Japanese steel industry at the low prices offered for pellets and iron ore fines.[40] While the other iron ore companies were exporting lump ore, Robe Cleveland Cliffs at Robe River was dependent on selling lower-grade fines and pelletised ore from the Robe River limonite, the prices of which it thought were being particularly penalised. Brand and Court firmly resisted this invitation for State government intervention in the commercial negotiations. Court informed Cleveland Cliffs that the Brand government was a private enterprise government. Iron ore prices, he insisted, were something that companies should negotiate individually.[41] An extraordinary and dangerous situation would arise, Court warned, if the State government negotiated iron ore contracts and eventually found itself in possession of an agreement for the export of iron ore and pellets. The inevitable question for the State would then be to determine which company got the business. Cleveland Cliffs accepted this position.[42]

Despite the difficulties in facing down the united front of the Japanese steel mills, Cleveland Cliffs vice-chief Dohnal was able to secure a deal on 23 August 1965 with the assistance of Mitsui. The arrangement was for the sale of 71.4 million tons of Robe River pellets over a 21-year period commencing at the end of the decade at a price of US 19.1 cents per unit of iron for an aggregate value of $US900 million.[43] Despite this very significant sales contract, the problems for Cleveland Cliffs became insurmountable in the first quarter of 1966 when BHP decided not to participate in the joint development with Cliffs of their respective Robe River deposits. At that point, Cleveland Cliffs decided that the Robe River project – with estimated capital costs of $US200 million (about $2.4 billion in 2015 dollars) – could not proceed. Such were the looming problems in developing the Robe River deposits that Cleveland Cliffs persuaded BHP to look at the possibility of both companies joining another iron ore project in trouble – the Mount Newman project.

An Industry in Flux, 1966–1967

Though the federal government had reluctantly approved the Mount Newman contract with Japan in 1965, it determined simultaneously to introduce tougher regulation of the iron ore industry by setting guidelines for iron ore prices. Commonwealth intervention and problems with raising finance and securing contracts left the infant iron ore industry in a state of uncertainty in the period from 1966 to 1967. In the mid-1960s the Brand government hoped that four projects would form the sinews of an iron ore province in the Pilbara – Hamersley Iron owned by CRA and Kaiser Steel in the Hamersley Range; the Mount Newman Iron Ore Company owned by CSR and AMAX in the Ophthalmia Range; MGMA owned by a consortium of one local and two American mining companies in the northern Pilbara; and BHP and Cleveland Cliffs at Robe River.

The Goldsworthy and Hamersley projects made their first export sales in 1966, but the Mount Goldsworthy operation, whose iron ore reserves were much more limited than other projects, would enter serious discussions on merging with another iron ore company in 1966. The projects of Cleveland

Cliffs and the Mount Newman Iron Ore Company were also hampered by problems in raising finance. Hamersley Iron too suffered major setbacks in 1965 and 1966. In 1965 the Menzies government intervened in its plans to build a pellet plant and in 1966 imposition of federal iron ore guide prices hampered Hamersley's future sales agreements.

On 24 August 1965 Hamersley Iron had sought permission from the Menzies government to export 16 million tons of pellets at the rate of one million tons per year from 1968 onwards.[44] The prices for Hamersley pellets were even lower than those negotiated by Cleveland Cliffs for Robe River pellets in August 1965. As CRA and Hamersley Iron chief Maurice Mawby explained to the Minister for National Development, David Fairbairn, this was because the Japanese mills were making an allowance to Cleveland Cliffs, which had to embark on building *ab initio* its railway, mine and port. Hamersley Iron, by contrast, was well on the way to constructing its facilities.[45] The Hamersley pellet request placed the Menzies government in a dilemma. On the one hand, Hamersley's pellet prices were clearly not equal to the best that could be secured from the Japanese. On the other hand, refusing the deal might jeopardise an Australian objective to become the dominant iron ore supplier to Japan. Cabinet made a decision to approve the deal.[46]

Hamersley Iron was not so fortunate when it returned to the federal government in October 1965 with an additional request.[47] By then Hamersley Iron had discovered that it was not economical to establish a pellet plant with an annual throughput of less than two million tons. It had therefore negotiated with the Japanese to sell 24.4 million tons of pellets by offering a small reduction in price from 18.5 cents to 18.32 cents per unit.[48] By this time the complaints about low Australian iron ore prices were reaching a crescendo in Canberra.[49] In the belief that Hamersley Iron would continue to build its pellet plant, Fairbairn recommended that cabinet refuse Hamersley's second pellet contract.[50] On 30 November 1965 the Menzies government, with McEwen in the vanguard, endorsed Fairbairn's recommendation. It was the first time that an Australian government had intervened to stop a commercial deal of such magnitude except for defence reasons.[51]

The decision of the Menzies government stunned CRA. Mawby protested to Menzies that it meant the loss of not only eight million dollars of additional pellet sales, but also the annual income of $US18 million from the whole Hamersley pellet project.[52] He added:

> It seems entirely unreal to me that this great industrial enterprise [the Hamersley pellet plant], bringing in 18 million dollars a year and doing so much to develop the North West, can founder on a token price reduction (in accordance with established business practice) which, in fact, vastly improved the economics of the project.[53]

Mawby expressed himself at a loss to understand the "reasoning behind this extraordinary, and what appears to be anti-Australian decision".[54] Mawby's emphasis on the anti-Australian nature of the decision reflected an increasing defensiveness on the part of CRA about its British ownership at a time when Australia's links with Britain were declining. Eventually, Mawby and other senior CRA managers solved Hamersley Iron's problem. They renegotiated with the Japanese to sell the company's original 16 million tons of pellets over 10 rather than 16 years, thus giving it the throughput to make construction of a pellet plant economical.[55] The federal government, however, would still not allow Hamersley Iron to sell the additional 8.64 million tons of pellets at a lower price than the federal guideline price. Hamersley Iron thereafter nursed grievances over the Menzies government's refusal of its second pellet contract. In May 1966 the Australian government, now led by Harold Holt, introduced guideline prices for the sale of Australian iron ore above the prices negotiated by Mt Newman in 1965. This meant that Hamersley Iron would have to sell future tonnages of iron ore at a competitive disadvantage. The Japanese steel mills detested the Commonwealth guideline prices and Charles Court actively opposed them on the grounds that they would mean that the mills would give further iron ore business to countries other than Australia. Australian companies like BHP and CSR approved of the guideline prices but Hamersley Iron persistently argued that the Australian government was disadvantaging it by having approved the 1965 Mt Newman deal and then introducing floor prices in 1966 above those same Mount Newman prices.

By the beginning of 1966, when Hamersley Iron was in the midst of these difficulties, the Mount Newman project was also in dire trouble.[56] The financing of Mount Newman was at that time based on AMAX and CSR contributing in equal measure $US50 million of equity, on their securing loans of $US45 million in Australia and $US55 million in the United States and on their having $US10 million in working capital – a total of $US160 million with repayment of debt spread over 12 years.[57] The company hoped that this sum would finance the building of a railway from Mount Newman to Port Hedland to accommodate 60,000-ton iron ore carriers. By the end of 1965 it had become apparent that the original estimates for the project had increased from $US150 million to about $US200 million. CSR could not meet the increased capital costs of the Mount Newman project, and its debt financing agreement with Australian insurance companies was dependent on CSR's maintaining 50 per cent equity in the project.[58] Accordingly, the Japanese steel companies suggested that the Mount Newman Iron Ore Company should approach the Mount Goldsworthy Associates and Hamersley Iron for an agreement to share rail and port facilities, although warning against an actual merger with either company.[59]

While AMAX was predisposed to an actual merger between the Mt Newman and the Mount Goldsworthy operation, both CSR and the Holt government were opposed. AMAX then approached Kaiser Steel, the US partner in Hamersley Iron, requesting a deal. AMAX suggested that the Mount Newman project might not build its own railway all the way to Port Hedland but instead share Hamersley Iron's port facilities at King Bay (later renamed Port Dampier).[60] This was the first occasion, after Daniel Ludwig's 1964 plan, that Pilbara mining companies had considered a plan for sharing railway and port facilities. The basis of the suggested cooperation would be that Mt Newman would build a railway to join the

Hamersley Iron railway and also use Hamersley Iron's facilities at King Bay. For its part Hamersley Iron outlined its conditions for agreeing to do so. It would carry iron ore mined in Mount Newman in return for per ton freight to King Bay and a lump sum of $US50 million against those charges. Hamersley Iron arrived at the $US50 million lump sum on the basis of its estimate of Mount Newman's capital savings by using King Bay rather than Port Hedland. In Australia James Vernon negotiated for the Mount Newman project and Maurice Mawby for Hamersley Iron. But Vernon baulked at this high price for cooperation and offered Mawby an alternative. He suggested that Hamersley Iron should sell its port at King Bay for $60 million to a Harbour Board established under Western Australian law with joint representation on the board by Hamersley Iron and Pilbara Iron (the company operated by CSR in the Mount Newman Joint Venture).[61] But Mawby was determined to resist this suggestion. On 21 February 1966 negotiations between Hamersley Iron and Mt Newman broke off. In the following month Brand flew to Canberra to inform Holt that there were real difficulties in maintaining the 50 per cent Australian equity in the Mount Newman project.

Prime Minster Harold Holt immediately convened a crisis meeting with John McEwen and David Fairbairn. The three ministers decided to intervene in Western Australia. Holt authorised McEwen to try to persuade Frank Coolbaugh, head of AMAX, and Val Duncan, head of RTZ, to agree to Hamersley Iron sharing rail and port facilities with the Mount Newman Iron Ore Company.[62] Duncan was, however, reluctant to aid his most formidable likely competitor in the emerging Australian iron ore trade, and Vernon did not want to consent to terms from Duncan that would have left Mount Newman as a satellite of Hamersley Iron. But in declining McEwen's good offices, Duncan made the intriguing suggestion that CRA might join CSR and AMAX as a partner in the Mount Newman Iron Ore Company. If this had eventuated, CRA and Rio Tinto might have succeeded in dominating the two largest Australian iron ore projects as early as the mid-1960s.[63] But McEwen dismissed Duncan's suggestion.

The decision by Pilbara mining companies to build their own railway and port facilities rather than share them drew some strong criticism at the time on the grounds of inefficiency. In the longer term, however, the decision resulted in the Pilbara having three main ports rather than one. Moreover, the large iron ore companies argued that it was more efficient for them to have exclusive control of their own networks than to allow for multiple users. This greater efficiency of single-user rail networks would be demonstrated half a century later during the China-inspired iron ore boom in the early 2000s.

Since a deal with Hamersley Iron was now off the table, Vernon was again instrumental in constructing a viable solution for Mt Newman. His idea was to form a consortium that included BHP and Cleveland Cliffs in the Mount Newman joint venture. The idea was that each of the four companies – AMAX and Cleveland Cliffs and BHP and CSR – would take a 25 per cent stake in the Mount Newman joint venture. Vernon hoped that the four companies together, by producing more iron ore than was contemplated in 1965, would together be able to afford the higher capital cost of building a mine, railway and port at Port Hedland. If the groups raised a total of $US100 million in the United States and $US100 million in Australia, the equity contributions of the two Australian companies would be $US25 million each, with the remainder borrowed from banks and institutions in Australia.

This plan was aborted in August 1966 when Cleveland Cliffs withdrew.[64] By that time the four companies involved had reached the conclusion that the capital costs of the Mount Newman enterprise were such that there was no prospect of including a pelletising plant, one of the main attractions for Cleveland Cliffs joining the Mount Newman venture. On top of that, the argument for parity in Australian and American ownership of Mount Newman meant that Cleveland Cliffs would have to scale back its role to that of investor rather than its accustomed position of managing pellet projects. Cleveland Cliffs was also mindful of Court's insistence that pellets would have to be produced at Mount Newman if Cliffs were to retain its long-term rights to develop its Robe River deposits. Since Cleveland Cliffs had always considered Robe River an important part of its future, the American pelletising company decided to decline Vernon's invitation to participate in the Mount Newman project.[65]

Cleveland Cliffs turned anew to the Robe River project. In the meantime, by September 1966, the Mount Newman project enterprise came under increasing strain when AMAX hinted that it might have to withdraw unless an agreement were made within weeks. AMAX's main concern was that further lost time would see the cost of the project increase to a level unacceptable to AMAX's shareholders.[66] The Mount Newman Iron Ore Company's problems at this juncture stood in contrast to those of Hamersley Iron. The latter had by then completed its railway and commenced exporting ore but it struggled to persuade the Japanese to purchase additional ore at Commonwealth guideline prices. The next few months of 1966 would be decisive for both the Mount Newman joint venture and Hamersley Iron.

The Emergence of Hamersley Iron and the Mount Newman Mining Company as Iron Ore Giants, 1967–1972

The first step in restoring the viability of the Mount Newman Mining Company, as its name became in 1967, was renegotiating the sales contract that CSR and AMAX had signed with the Japanese in 1965.[67] Although Cleveland Cliffs had left the Mount Newman project, the joint venture had been immeasurably strengthened when BHP decided not only to participate but also to manage it.[68] As Australia's monopoly steel-maker, BHP had decades of experience mining iron ore, although for domestic purposes and not export. But even with BHP's leadership of the consortium, large problems still had to be overcome. The main difficulty was that the original terms of sale to the Japanese steel mills, involving the sale of between five and eight million tons per annum, were unprofitable. The rate of production volume was too low to persuade financial institutions to provide loan funds. Consequently, the reconstructed joint venture set about renegotiating the contract with the Japanese steel mills in October 1966. The final shape of the joint venture was American Metal Climax 25 per cent; BHP (via Dampier Mining Co.) 30 per cent; Pilbara Iron (CSR and five Australian insurance companies) 30 per cent. There was also 10 per cent participation from two Japanese trading companies, Mitsui and C. Itoh Iron Pty Limited, and five per cent by the British Selection Trust Ltd.[69] The reconstructed Mount Newman Mining Company sought to persuade the mills to take a larger quantity of ore and a higher ratio of "fines" to lump ore than was originally proposed. The aim was to have a 15-year amortisation period as opposed to the 20 years of the original contract. In the early stages of the contract the company would thereby be assisted by a higher price and the larger overall cash flow that would result from higher export tonnages.[70]

At this point Rio Tinto Zinc and Kaiser Steel, the overseas owners of Hamersley Iron, made a dramatic intervention. While the Mount Newman consortium's negotiations with the Japanese steel mills were underway, Val Duncan, head of RTZ, and Eugene Trefethen, head of Kaiser Steel, instructed CRA to make an offer to the Japanese Steel mills. The suggestion was that Hamersley Iron and Goldsworthy would together supply to Japan from their now operating mines the tonnages of iron ore that Mount Newman had contracted in 1965 to supply from their incomplete operation.[71] This would have put the Mount Newman operation out of business before it started. If the Japanese mills had accepted, and the Holt government approved, the Goldsworthy-Hamersley offer, there would have been only one major iron ore operation in Western Australia – Hamersley Iron. But in an effort to maintain competition between competing suppliers operating from two major ports (Port Hedland and Dampier), the Japanese mills declined the Hamersley-Goldsworthy offer and accepted the revised Mount Newman proposal.

At that point, the Holt government faced the difficult decision of either disallowing the revised Mount Newman contract, which was still below federal iron ore guide prices, or approving it and alienating Hamersley

Iron. A.J. Campbell, Acting Secretary of the Department of Trade and Industry, worried that time was running out to settle the Mount Newman problem:

> The longer the business is left "unfixed" the greater the risk that the Japanese (despite their obvious desire to keep Mount Newman viable and therefore to benefit from the low prices) will turn elsewhere: or that AMAX will become frustrated with the whole deal; or that BHP will become frustrated and weaned away.[72]

On 25 October 1966 the Holt government agreed, taking the view that the revised contract could not result in a less favourable outcome than the sale of 100 million tons under the 1965 contract. Holt's cabinet approved the revised contract even though it was still below Commonwealth guideline prices.[73]

The revised contract enabled CSR to overcome the considerable obstacles in the way of providing the Australian share of the financing of Mount Newman. Vernon and Potter persuaded Australian life insurance offices to invest in the project and, as a sweetener, to take up equity in Pilbara Iron on the basis of a revolution in Australian financing. As the *Australian Financial Review* described it, "these institutions have broken with Australian tradition by agreeing to make a substantial investment in a mineral project on the basis of anticipated cash flow rather than asset backing".[74] Ian Potter and Co. helped raise $44 million of medium- and long-term debt for CSR's share of the joint venture.[75]

Predictably, Hamersley Iron was galled by the apparently favourable treatment that the Mount Newman project was receiving. It continued to be hampered by Commonwealth guide prices and the Japanese acceptance of a large Mount Newman contract with which it struggled to compete. The failure of the Australian government to approve deals proposed by Hamersley Iron led to Mawby instigating high-level complaints, notably from Edgar Kaiser, President of Kaiser Industries and spokesman for the American minority interest in Hamersley Iron. Kaiser advised Holt that the Australian government's inflexibility in implementing iron ore guidelines had meant that Hamersley Iron and Australia had lost business to Peru.[79] He argued that a

> basic inequity exists in the guideline price due to the fact that Mt Newman base and incremental prices are substantially below guideline prices, Hamersley prices and Mt Goldsworthy prices. As a result, the Japanese resist placing orders at guideline prices, which are in excess of Mt Newman prices.[77]

Kaiser appealed to Holt to endorse a package that would improve Australia's foreign exchange by $500 million, provide royalties to Western Australia, result in $100 million corporate taxation and provide a second pellet plant at King Bay. [8]

Court added his strong voice to Hamersley Iron's criticism of Commonwealth guideline prices for iron ore.[79] By August 1967 William McMahon, the federal Treasurer, and his department and Fairbairn and the Department of National Development had become sympathetic to their criticisms. Only McEwen thought that it was a mistake to abandon the guidelines – it would in his view make the government look foolish and jeopardise future negotiations.[80] The effectiveness of the campaign by Hamersley Iron and Court against the federal guidelines, however, outweighed the arguments of those who fought against them. The most decisive was that the Japanese were not giving Australians any more iron ore business while the guidelines were in place. This consideration allowed Fairbairn to persuade cabinet to let him abolish the guidelines on 25 August 1967.[81] Two months later, with the guideline process no longer an inhibition, Fairbairn was able to announce the approval of Hamersley Iron's application to sell 40 million more tons of iron ore.[82] But the federal government's protracted negotiations with Australian iron ore companies over prices had one significant domestic consequence. The government persuaded iron ore companies and other Australian mining companies to form a peak body, the Australian Mining Industry Council (AMIC), that would represent the mining industry to government with one voice. AMIC would later become known as the Minerals Council of Australia (MCA).

The November 1967 decision to approve Hamersley's exports was a significant boost for what was now a public company with nearly 20 per cent Australian equity. This had been achieved by the largest flotation in Australia to that time. An even bigger breakthrough came two years later in October 1969 when the Australian government approved letters of intent under which Hamersley Iron would sell a further 112 million tons of lump ore and fines to the Japanese steel mills in the 15 years from 1972 with an option to sell a further 63 million tons. The deal bought Hamersley Iron's total booked sales in 1969 to 272.5 million tons worth $A2133 million. The breakthrough, which would take Hamersley Iron's annual production of iron ore to 28 million tons in 1972, brought the company into the ranks of the world's top iron ore mining companies, ranking with Brazil's Comphania Vale do Rio Doce (Vale).[83] By the early 1970s Hamersley Iron was the third largest iron ore producer in the world.

Not long afterwards, the Japanese steel mills announced on 17 November 1969 their agreement to purchase an additional 60 million tons of iron ore from Mount Newman. Including a $350 million agreement to sell BHP 70 million tons of iron ore over 20 years, the Mount Newman Mining Company had a comparable 276 million tons on its order books worth an estimated $1890 million. Hamersley Iron and the Mount Newman Company thus both became world-ranking iron ore companies, with Hamersley Iron shipping 34.7 million tonnes of iron ore in 1976 and Mount Newman 29.1 million tonnes in the 12 months up to 1978. Together the two companies made up the bulk of Australian iron ore sales in the mid-1970s (81.4 million tonnes in 1975).[84] By the late 1970s Australia was the second largest producer after Russia and had become the leading world exporter of iron ore.[485]

The Resurrection of Robe River, 1967-1972

Having pulled out of the Mount Newman venture in August 1966, Cleveland Cliffs turned its attention to reviving the mothballed Robe River project. Cleveland Cliffs informed Court that it was now thinking about developing its Robe River deposits.[86] But this would be without BHP and using a port based at Cape Lambert, 145 kilometres from its base of operations at Mount Enid, as opposed to its first nominated site at Cape Preston, which was 48 kilometres closer. Cleveland Cliffs' changing ideas about Robe River left Court bewildered. He had been disappointed at Cleveland Cliffs' failure to consummate an agreement with BHP in 1966. Now he was surprised that the American company thought it was capable of developing its Robe River deposits without any assistance from the Australian company towards infrastructure costs.[87] Court was also aware that the Japanese mills regarded their legal and moral commitment to Cliffs under the 1965 sales agreement as having run out. The response of Cleveland Cliffs to this situation and to Court's pressure was to write to all members of the Japanese steel industry on 11 November 1966. The company undertook, within six months of their acceptance of the letter, to send them a "firm and final decision" as to the development of the Robe River deposits, with deliveries anticipated for 1971, a pilot pellet plant completed by January 1967 and capital cost reductions on pre-1966 projections.[88]

To get the Robe River project restarted, Cleveland Cliffs had to surmount two basic problems: reducing the capital costs of the project by $US40 or $US50 million; and finding some means of replacing the planned participation of BHP. Cape Lambert was attractive as a port site because it could be developed at lower cost to accommodate much larger ships than Cape Preston.[89] But the change of port sites disturbed Court. He had some hopes of being able to employ peaceful nuclear explosives, obtained under the US government's "plowshare" programme, to develop a deepwater port at the more isolated Cape Preston.[90] But despite Court's doubts, Cleveland Cliffs believed that it could start Robe River on the basis of exporting four million tons of pellets annually as well as 3.5 million tons of prepared sinter fines. Selling "prepared" sinter fines was an important development. Fines would be "prepared" just before the stage when they would normally be pelletised.[91] Earlier in the Robe River project, Cleveland Cliffs had thought it unlikely that it would be able to sell any of its limonite ore in unprocessed form as fines since they contained between nine and 10 per cent water. But research in 1966 and 1967 had demonstrated to Cleveland Cliffs that it would be possible to export some of their limonite fines provided that they were in prepared form. Meanwhile in Japan, Saburo Tanabe, now Managing Director of Fuji Steel, was pressing for the termination of the 1965 Robe River pellet contract. Indeed, Tanabe advised Cleveland Cliffs to abandon pelletisation altogether and just to export lump ore and fines.

In January 1967 the Japanese steel mills gave Cleveland Cliffs an official extension to submit a revised proposal to them by 28 June 1967. But this was heavily qualified. Fuji Steel had taken an adamant stance in favour of sintering fines as opposed to pellets, while Yawata Steel preferred pellets.[92] By that time the Brand government had become fearful that Cleveland Cliffs had not made sufficient progress on engineering planning and in securing finance.[93] In response to Brand's warnings, the President of Cleveland Cliffs, Stuart Harrison, noted that one of the biggest threats to the project was the mills' apparent preference for sinter fines instead of pellets. Harrison therefore pressed the WA government to highlight to the mills its strong support for pelletisation (and industrialisation). Court agreed and made a confidential approach to the Japanese steel mills, stressing the importance that the Western Australian government placed on pellets as oppose to fines.[94]

Another milestone passed on 28 June 1967. Cleveland Cliffs had still not submitted a detailed plan on the development of Robe River to the Japanese mills. Senior representatives of the Japanese steel industry therefore notified Court that as far as they were concerned their 1965 sales contract with Cleveland Cliffs no longer existed – it had legally ended.

But the Japanese mills did not want to publicise this fact. They wanted to allow Cliffs to hold on to their financial arrangements with their American bankers. They also did not want to let other mining companies know how easy it was to get out of a major contract with the mills.[95] At the same time, the steel mills indicated that they would be willing to enter into fresh negotiations. Cleveland Cliffs in the meantime was searching around for a replacement for BHP in meeting some of the enormous financial costs of the project. Contemplating alliances with such companies as Gulf Oil, International Utilities and Mobil, Cleveland Cliffs eventually decided on a partnership with Daniel Ludwig, the American multimillionaire owner of National Bulk Carriers. Through a subsidiary, Sentinel, that held reserves of iron ore at Nimingarra and Yarrie, Ludwig agreed to take up to a 50 per cent ownership of the Robe River project.[96]

By August 1967 the Brand government was again beginning to worry about the slow progress on Robe River. If the Japanese steel industry did not receive a concrete proposal from Cleveland Cliffs, Court warned, the steel mills would wind up the negotiations on Robe River, either abandoning it or letting it take its place in the queue of other iron ore projects.[97] Spurred on by Court's unrelenting pressure, Cleveland Cliffs submitted a proposal to the mills on 31 August 1967. The company proposed to sell to the Japanese 4.2 million tons of pellets and 3.5 million tons of fines annually over a period of 21 years with Cape Lambert as the nominated port site. It stipulated a period of four months to complete arrangements for finance, partners and other details.[98]

By this time the Australian government had abandoned iron ore price guidelines and the Japanese mills felt obligated to Court for his crucial role in reversing federal iron ore policy. But it was not easy for the Japanese steel industry to reward Court by accommodating the large tonnages that Robe River was seeking. Another factor that gave the Japanese steel industry pause was continuing uncertainty about Cleveland Cliffs' partners and what its sources of finance were. Devaluation of the British pound in 1967 had raised the cost of financing and the US government was applying increasing pressure against the lending of American capital abroad. The steel mills, moreover, were still smarting from the failure of the 1965 Robe River contract from which Cliffs had expected to walk away without paying any compensation or damages. Thus in the latter stages of 1967 a stand-off developed where the mills pressed Cleveland Cliffs to line up their finance before signing a contract, while Cliffs desired a contract to finalise that finance.[99]

To break the deadlock Court accepted an invitation from Saburo Tanabe to meet him in Sydney when the Japanese industrialist visited in November 1967.[100] Court explained to Tanabe that Robe River was the fourth and last of the original Pilbara iron ore projects (after Goldsworthy, Hamersley and Mount Newman), which the Western Australian government had supported to achieve the fuller development of the Pilbara. Court informed Tanabe that he supported the negotiations of Cleveland Cliffs. He expressed the opinion that development of the Robe River project would benefit the Japanese by giving them a third major iron ore port after Dampier and Port Hedland and greater security for their iron ore supplies.[101] By this time, with the Hamersley and Goldsworthy projects launched and Mount Newman underway, Court's views had great weight with the Japanese steel industry, greater than any federal minister's. Such was his influence that Tanabe agreed to try to coordinate with all the Japanese steel mills sufficient annual tonnages to enable the Robe River project to commence. Tanabe undertook to persuade the mills to reach an understanding with Cliffs by the end of January 1968 subject to the mills' testing of the viability of Robe River limonite fines.[102]

Tanabe was as good as his word and by February 1968 had assembled a counter-offer. The cooperation between Court and Tanabe highlighted the sense of partnership between Japan and the Western Australian government in the development the iron ore industry. The State government and Court in particular were of critical importance in establishing the Pilbara iron ore industry, more so than the federal government, which was too remote to appreciate the entire situation. In assembling the counter-offer, the mills had to assess the cost of handling ore with 55 to 57 per cent iron content as opposed to the 64 per cent iron content obtained from ore in places like Goldsworthy, Hamersley and Newman. Another factor they had to consider was the loss of productivity from their investment in sinter plants from using a lower grade iron ore with 10 per cent moisture. In a letter of 5 February 1968 seven major Japanese steel mills proposed to accept annual tonnages from Robe River of 3.6 million tons of pellets and 2.6 million tons of fines subject to full-scale sintering tests of Robe River fines in Tokyo.[103] Though Hamersley Iron's Rus Madigan was "staggered" by the immensity of the offer, the tonnages and prices were not yet sufficient to make Robe River financeable.[104]

For the rest of 1968, Cleveland Cliffs sought to increase the tonnages and prices. Negotiations bogged down as the Japanese steel mills sought assurances on the quality of Robe River pellets. During this time Cleveland Cliffs became nervous about the ability to hold its then major partner, Daniel Ludwig. At the same time, Tanabe's confidence in Robe River's prospects waned. In late 1968 Tanabe recommended that Robe River be delayed for a few more years. For Cleveland Cliffs and Charles Court that would have meant the end of the project and they persisted in pressing the Japanese mills for an agreement. Finally, on 28 April 1969, the Japanese steel mills signed what at the time were the world's biggest iron ore sales – contracts worth $1.37 billion in 1969 dollars (worth over $15 billion in today's dollars). The mills agreed to purchase 87.7 million tonnes of pellets over 21 years and 73.4 million tonnes of sinter fines over 15 years.[105]

The sales contract permitted work to begin on a construction project costing $274 million (about $3 billion in today's dollars) and completed in 1972. After the Ludwig organisation withdrew from the project in 1969, Australian participation, through an Australian public company, Robe River Limited, increased. The final composition of the Cliffs Robe River Iron Associates was Cleveland Cliffs through its Australian subsidiary (30 per cent); Mitsui Iron Ore Development Company (30 per cent); the Australian company Robe River Limited (35 per cent); Mount Enid Iron

Ore Company (5 per cent). The combined Australian equity through Robe River Limited and Mount Enid Iron Ore Company (a Garrick Agnew vehicle) was 40 per cent and Japanese equity was the highest of any Pilbara iron ore project. Ludwig separately entered into an agreement with the WA government to develop smaller resources at Nimingarra, east of Mount Goldsworthy. In 1971 these deposits were acquired by MGMA, as part of an increased production to eight million tonnes per year.

A State in Miniature

Building the infrastructure of an iron ore industry in the Pilbara – ports, mines, townships for workers and railway systems to transport the ore – was a colossal ambition. No funding, State or federal, was available for infrastructure in a region that lacked facilities taken for granted elsewhere in Australia. What was required was "something like creating a State in miniature" in Western Australia's northwest.[106]

When Rio Tinto first contemplated the feasibility of operating a large mining operation in the Pilbara hinterland in the early 1960s, inland freight costs were a major disincentive. It was not clear at that time whether owning a private railway would even be possible and the known costs of rail transport elsewhere in Australia were discouraging – sixpence a ton-mile from Broken Hill to Port Pirie and about the same from Mount Isa to Townsville. However, a 1962 study by Rio Tinto engineers indicated that a privately built railway in the Pilbara could operate at twopence a ton-mile. These cost estimates made iron ore mining operations located hundreds of kilometres inland a viable proposition.[107]

Armed with its 1965 sales contract, Hamersley Iron set about the herculean task of building a 1.435-metre standard-gauge track from the mining site at Mount Tom Price to the port over a distance of nearly 300 kilometres. Then the heaviest track ever to be built in Australia, it was designed to support two diesel–electric locomotives each hauling about

150 wagons carrying 15,000 tons of ore. Building the railway required installing 650 metal culverts at 375 locations on the main line, constructing 16 bridges, excavating 1.6 million cubic metres of rock, and hammering more than five million dog-spikes into the track.[108] Historian Alan Trengove has described completing the main Hamersley railway in about one year as "one of the major engineering achievements in Australian history".[109] With its later extension from Mount Tom Price to Paraburdoo, the Hamersley railway covered a distance of 386 kilometres, longer than the distance from Melbourne to Albury. Despite the necessity for a large outlay on operating and maintenance, the Hamersley railway was a vital ingredient in the rise of Hamersley Iron as one of the world's great iron ore mining companies. Transporting ore in the 1970s at about one cent per tonne mile compared favourably with rail freight costs in North America of between 2.5 cents and six cents per tonne mile and about six cents for bulk cargo on Australian state railways.[110]

Building the Mount Newman railway broke even greater records because of its longer distance of 450 kilometres, though over more even terrain. Workers were gathered from all around the world, from Labrador to Thailand, with Thursday Islanders who were experts in building railways in northern Queensland in the front-line.[111] The workers laid 62,000 tons of track constructed from BHP steel, built 25 bridges and nine miles of culverts, and made deep cuttings through hard rock. One reckoning was that enough timber was employed on the Mount Newman railway to build 10,000 average homes. The thirsty crews were thought to have consumed the equivalent of 1.5 million middies of beer.[112] The construction of the Hamersley and Mount Newman railways, as well as the earlier Goldsworthy railway and somewhat later Robe River line, marked a spectacular period of rail building in Australian history. WA Premier David Brand remarked that, while railway transport appeared to have been in eclipse in Australia at the middle of the twentieth century, about 2000 kilometres of track was laid in a short period of about six years in Western Australia.[113] The railway systems built by the pioneer Pilbara companies in the 1960s and early

1970s form the basis today of the Rio Tinto and BHP railway systems, two of the biggest privately operated networks in the world.

The mining companies looked at facilities able to handle ships of 40,000 to 60,000 ton capacity to transport Pilbara iron ore and there were none of that size. Hamersley Iron began its studies of possible ports in 1962 and examined sites including Exmouth, Depuch Island, Maud Landing, Cape Keraudren and Port Hedland. From the start, Struan Anderson aimed at avoiding sharing a port and railway system with competitors for the sake of delaying the start of those competitors or keeping them out altogether in the hope of gaining from selling a higher tonnage of iron ore.[114] In mid-1962 he assigned the Consolidated Zinc subsidiary, Central Engineering Services (CES), to oversee port planning for Hamersley Iron. CES had to deal with two problems: first, that the Pilbara region, with its high tides, was swept by cyclones in the summer months; and, secondly, that the region did not then have any ports capable of handling ships larger than a few thousand dead weight tons (dwt).

CES chose the Royal Netherlands Harbourworks Company, which was building a port for bauxite in Weipa for Comalco, to undertake a new study. The Dutch company narrowed the choice to King Bay (now known as Parker Point at the Port of Dampier). But CES recommended Cape Lambert rather than King Bay, whose rocky hinterland and three kilometres of salt flats combined with the disadvantage that a railway would have to take an uphill gradient on the last stage of its approach. Despite these disadvantages, Anderson preferred King Bay on the grounds that it was sheltered by land mass and islands and had more potential for expansion with a greater amount of land space for harbour works. In the end, an independent arbitration panel headed by James Bissett, the general manager of the British Phosphate Commissioners, gave a decision in favour of King Bay in December 1963.[115]

Some experts in the early 1960s doubted that it would be possible to produce a viable iron ore port on a coast as subject to cyclones as the Pilbara was. But the CES engineers felt they could succeed if they took some essential precautions. One was to batten down all moveable parts on the wharf. Another was to design for waves to pass underneath the wharf rather than break over it. Because waves at Dampier were known to reach 11 metres, the lowest part of the platform was therefore fixed at 12 metres. The consequence of this was that, when King Bay was dredged to a depth of 15 metres, the platform would be 27 metres high. This in turn meant that the Japanese-engineered, 500-tonne ship loader, which poured iron ore into the holds of berthed ships, almost reached a height of tall buildings at 50 metres.[116]

Construction of the port at Parker Point, Dampier, commenced in January 1965 and was completed in 1966. CES engineers recognised early that East Intercourse Island off King Bay was an excellent site for a second berth. Kaiser Engineers and Constructors Incorporated carried out the construction of facilities on East Intercourse Island, excavating and relocating three million tonnes of rock that would be replaced by 31,000 cubic metres of reinforced concrete and 11,000 tonnes of structural steel. Ore was carried by a 2690-metre long conveyor belt from the mainland to a stockpile area on the island. Two electronically controlled stacking machines distributed the ore into eight piles, which would be shifted by conveyor to the loading wharf 214 metres offshore.[117] Completion of the facilities on East Intercourse Island in 1972 doubled Dampier's capacity. The improvements to the port of Dampier also accommodated ships with ever increasing tonnages. In 1968 Maurice Mawby predicted that within 10 years Dampier alone would be exporting iron ore tonnages equal to total exports from Melbourne and Sydney.[118] In the period up to 1976, the ship with the largest tonnage was the *Arafura Maru*, which carried 165,597 tonnes of ore. And in 1984 the *Shinho Maru* was loaded at East Intercourse Island with the first cargo greater than 200,000 tonnes to be despatched from any Australian port.[119] The improvements also allowed quicker turnaround. In 1975, for example, 412 ore-carriers visited Port Dampier, 177 berthing at East Intercourse Island and the rest at Parker Point, with a 160,000 dwt vessel taking on average 36 hours to berth, load and depart.[120]

While Dampier was the first new iron ore port to be constructed, the WA government encouraged the companies to consider Port Hedland, then an established but sleepy pastoral port carrying vessels limited to less than 5,000 ton capacity. As late as 1965, vessels entering Port Hedland had to wait until the tides of about 1.5 metres covered a rock bar five kilometres from the shore.[121] The challenge facing the WA government and the Pilbara mining companies was whether what until then was a small port with 20 foot tides could transcend its natural limitations. In 1961 Court for one had abandoned hope that Port Hedland could be transformed into a deepwater port and contemplated alternative sites such as Depuch Island, 100 kilometres west of Port Hedland, and Cape Keraudren 129 kilometres to the east.[122] In 1963, however, when MGMA received permission to export four million tons per year, they persuaded the Western Australian government to allow them to develop Port Hedland, one of whose advantages was its then already established township of about 1000 people.

MGMA took on the task of building port facilities on Finucane Island adjacent to Port Hedland and dredging the harbour and approach navigation channel to the port. One of the Goldsworthy partners, Utah Construction and Mining Company of California, was awarded the contract to construct the port facilities.[123] Utah employed a 300-ton cutter suction dredge, *Alameda*, which, when it was first built in 1958, was the most powerful dredge in the world. The *Alameda* successfully dredged a 4.5-mile channel to a depth of 29 feet and a turning circle to handle ships of 60,000 tons capacity. Dredging commenced in July 1965 but faced heavy going as the teeth of the dredge cutters were repeatedly worn down by the hardness of the rock. Despite a tropical cyclone shutting down the program, work was sufficiently completed for the 20,000-ton carrier *Harvey S Mudd* to enter the partly dredged channel on 23 May 1966. 11 days later on 3 June 1966 the ship sailed for Japan with a first shipment of 25,000 tons of iron ore from the Pilbara. After 26 months of dredging the inner harbour and approach channel, Utah completed its dredging contract for Goldsworthy in September 1967.[124]

When the Mount Newman joint venture was revived, the Bechtel Pacific Corporation was appointed construction administrator in April 1967. The Newman partners agreed to establish a ship-loading complex and industrial facilities at Port Hedland's Nelson Point. On 21 June 1967 they announced that they had awarded the biggest dredging contract ever let in Western Australia to deepen Port Hedland to accommodate ships up to 100,000 tons. The Newman partners awarded the work to a joint venture between the already tested Utah Construction Company of California and the Japan Industrial Land Development Company (Jild). The companies employed the *Alameda* and an identical Japanese dredge *Kokuei Maru*, which arrived from Japan in August. The dredges were employed to remove some 11 million cubic yards of spoil from the approach channel and harbour off Nelson Point and to use some of the spoil to reclaim 160 acres of flood land and mangrove swamp on Nelson Point in order to establish Mount Newman's industrial area to a height of 30 feet above low tide.

The Utah-Jild joint venture began work on Port Hedland harbour on 12 September 1967 and, by January 1969, the wharf and ship-loading facility at Point Nelson had been commissioned and dredging had proceeded sufficiently to accommodate 100,000-ton carriers, the first being loaded in August 1970. In April 1969 a new contract was let to further deepen the approach channel so that vessels up to 315 metres in length and 185,000 dwt could be admitted.[125] By 1971 Port Hedland had become the largest export port in Australia, an astonishing transformation, which prompted the WA government to transfer administration of the port from the Harbour and Light department based at Fremantle to the Port Hedland Port Authority, which assumed exclusive control from 15 June 1971 with Mt Newman and MGMA being nominated representatives.[126] Goldsworthy originally operated its own towage service and navigation aids. From 1969, Goldsworthy and Mt Newman jointly underwrote Adelaide Steamships' expanded tug operation.

An Australian, Harold Clough, won a bid to build the jetty at Cape

Lambert from the Cleveland Cliffs operation at Robe River.[127] Born in Western Australia in 1926, Clough won a Fulbright Scholarship to study at the University of California, Berkeley, in the 1950s. On his return to Perth in 1954 he joined J.O. Clough, the building firm begun by his father in 1919 that later evolved under his leadership into the multidisciplinary engineering and construction organisation, Clough Limited. Harold Clough's bid for the jetty was about one third of its nearest competitor – about A$13 million as opposed to A$33 million. The secret to Clough's success was his vision of the jetty. Forming an alliance with the Royal Netherlands Harbourworks Company, Clough proposed a structure supported by A-frame steel piles. While Clough's competitors were sceptical about the design, he and his Dutch partners argued that the steel piles would be sturdier than some of the land-backed structures built elsewhere. Their idea was essentially to build a longer jetty that stretched further out to sea and that was less costly to dredge for, as opposed to building a shorter jetty with higher dredging costs closer to shore.[128]

The jetty that the Clough–Netherlands Harbourworks team eventually constructed was, at more than 2.6 kilometres, the second longest in Australia. At 19.9 metres above "Indian Spring low water", it was also the tallest in Australia and indeed one of the tallest in the world. The original $13 million dredging contract extended to $25 million as the Robe River companies realised that they needed to accommodate ships of 150,000 tonnes rather than 60,000 tons. To handle such ships, the dredged water depth at the jetty had to be about 18 metres, meaning that the total height of the structure from the seabed to the deck was almost 40 metres, the equivalent of a 15-storey building. As John McIlwraith has pointed out, there were major benefits to this over-engineering. The loaders of the bigger ships that berthed at Cape Lambert could be more easily manoeuvred because of the jetty's extra height. In 1984, after a new 16-metre deep channel was opened, 207,704 tonnes were loaded into the *Hoei Maru*, a tonnage that eclipsed Port Dampier's record in 1984. And in 1993, the biggest ship then to visit an Australian port, the 322,941 dwt *Bergeland*, was loaded at Cape Lambert.[129] The technology for the large ships was

Japanese and later South Korean.

The Pilbara companies constructed not only mines, ports, and railway systems but also the manifold functions and services of developed communities such as schools, post offices, shopping centres, water supplies, roads and suburban street lighting. By the 1970s, three iron ore ports – Port Hedland, Dampier (and its township Karratha) and Cape Lambert – were operating and new townships dotted the Pilbara: Mount Newman the largest, followed by Tom Price, Paraburdoo and Pannawonica (Robe River). In the new towns a variety of clubs developed to provide relief to the claustrophobia endemic in isolated mining settlements. These clubs included outdoor activities such as archery, gemstone-collecting, horse-riding and pistol-shooting.[130] But several factors militated against the development of an older community in the towns. Marriages were often broken up by the temptations of alcohol and the ease of forming extra-marital liaisons. Educational opportunities were limited in places like Tom Price, where the high schools did not go beyond the third year and prioritised technical education. Parents often chose to leave the Pilbara because the towns did not supply sufficient social outlets for their children as they grew up.[131] There were no grandparents or family support systems.

A large percentage of the workforce in the Pilbara were migrants, as high as 37 per cent of Hamersley workers. The migrants among them were predominantly of British origin but New Zealanders and Germans were the next most numerous nationalities.[132] In common with the native-born Australians who worked there, they were motivated to earn high wages during their time in the Pilbara townships in order to buy a house or business in a capital city.[133] A single man might save $7000 per year and a man with a working wife as much as $14,000.[134] If workers could adjust to the harsh physical environment and climate, there were other advantages to working in the Pilbara. The new towns were laid out with a high degree of town-planning skills and the company housing was of a high standard.[135] The pattern in the industry was for the companies to provide accommodation at heavily subsidised rents.[136] Workers would pay a nominal rent for company-built houses and would pay no rates. Moreover, the company's maintenance employees could attend all breakdowns.[137] Still, the turnover of the workforce was high in these early years – 180 per cent at Hamersley Iron in 1967 and 140 per cent in 1968.[138] The Northwest of Western Australia was a man's country and women had to be tough both mentally and physically to endure. The effect on marriages was often disastrous.[139] Also many single people in the Pilbara had fled from marriages elsewhere in Australia.

Because the Pilbara iron ore industry was capital-intensive with a high level of fixed costs, company profits were vulnerable to any union action that delayed or prevented the shipment of ore.[140] Strikes cost the companies dearly in demurrage charges. Demurrage was incurred if f.o.b. shipments were not loaded at an agreed date and unless *force majeure* was declared. While the original intent of the *force majeure* provision was directed toward external weather events such as cyclones, it quickly morphed into being declared for industrial disputes to the extent that it became overused with the majority of Pilbara shipments in some years delivered under *force majeure*. In the Pilbara industrial relations system, the main actors on the side of management were the four major companies, which were supplied with technical assistance by the Australian Mines and Metals Association. Four major blue-collar unions in turn represented most workers. The Australian Workers' Union (AWU) and the Federated Engine Drivers' and Firemen's Union (FEDFU) covered production workers and the Amalgamated Metal Workers' and Shipwrights' Union (AMWU) and the Electrical Trades Union covered maintenance workers.[141] White-collar unions played a minor role.

From the time the industry started, various procedures for industrial relations were tried, at times included in the award by the Western Australian Industrial Commission, sometimes in negotiations by management and sometimes unilaterally imposed by the companies. These various methods could not stop a state of "near-anarchy" developing in the Pilbara as the 1970s unfolded.[142] Many strikes were overtly related to managerial policy, but most were income-related, especially the large wage increases won in 1971.[143] Other factors in the poor industrial relations environment that developed in the Pilbara were the climate and harsh physical conditions. The methods of mining had changed dramatically in half a century. Workers at Iron Knob in 1920 laboured with hand-drills and shovels after the ore had been blasted by gelignite. Then the ore was shovelled manually into rail trucks. By contrast, in the 1960s and 1970s workers in the industry were not required to do such heavy physical work. But, operating machines as they did, they still emerged as "sweat-soaked and coated in red dust as the pioneers at Iron Knob".[144] One unionist commented: "[the] company never fully appreciated the problem of working amid dust, high winds and 90 per cent humidity [in the summer]. A lot of the supervisors came from Broken Hill or big cities and expected workers to perform as they did in better climates".[145] A strike by the Hamersley workers in 1969 led to the company's first experiment with collective bargaining as negotiations commenced between the company, union officials and the Western Australian Trades and Labor Council. The upshot was a graded system of dust allowances and a safety code.[146] Unionists in the Pilbara tended to adopt more aggressive tactics than their colleagues in Perth and workers in the iron ore province tended to be less experienced than those in the cities. The result was that the Pilbara became a "frontier" industrial relations environment. In the period from the middle of 1967 to the middle of 1980 there were 1183 strikes in the Pilbara iron ore industry, rising from three in 1967 to 111 in 1972 and reaching 292 in 1980.[147]

Australia's Snowy Mountains hydro-electricity and irrigation complex, constructed between 1949 and 1974 by federal and state governments at a cost of $820 million, was a phenomenal achievement in terms of its mobilisation of capital and labour and engineering and technical complexity. Taken together, the foundational Pilbara iron ore projects, built almost contemporaneously between 1965 and 1976 at a cost of over one billion dollars, should be ranked as a comparable engineering and technological wonder of twentieth century Australia.

Lang Hancock and the Iron Ore Rush, 1966-1972

Australia began exporting iron ore in a modest way in 1963-64. By 1966-67, nine million tonnes were being shipped with exports valued at $46 million. Over the following seven years, the value of iron ore exports rose tenfold to reach nearly $500 million (84 million tonnes) in 1973-74. This figure meant that iron ore alone was reaping more export income than the Vernon Committee of Enquiry had initially predicted in 1965 for the entire minerals sector. Japan took 93 per cent of these exports in 1966-67 and an average of nearly 90 per cent over the following seven years, with most of the remainder going to the United Kingdom and the European Economic Community, where net f.o.b. returns were much larger as sales were made on a c.i.f. basis.[148] By 1966, when the WA government had concluded agreements with the pioneer Pilbara mining companies, it had also mapped and publicised the geology and iron ore occurrences in the area.[149] However, the WA government required the pioneer companies to relinquish temporary reserves held beyond the areas leased to them under agreements ratified by the WA Parliament. This gave Hanwright an opportunity to take out temporary reserves in areas where the large companies could not do so. Looking back at this state of affairs in 1971, Court reflected that "with hindsight, it is easy to see that the whole of the Pilbara region should have been placed under ministerial reserve once the original ratified agreements had been entered into".[150]

In the period after 1966 the partnership between Hancock and Wright was highly active in taking out temporary reserves partly because the pioneer mining companies could not do so and also because at that time Hancock and Wright were the "only operators prepared to believe that the market was not close to saturation".[151] In the view of the WA government, Hanwright's applications "were for the purpose of obtaining rights to ore then widely known to exist in the ground, as distinct from the intention of seeking ore in unknown areas".[152] Between 1966 and 1968 Hancock and Wright concentrated their energies on two broad geographic areas in the Pilbara. The first was Wittenoom, the site of the asbestos mine that they had earlier sold to CSR and whose surrounds contained several iron ore deposits. In December 1966, when the blue asbestos mine at Wittenoom was closed, Hancock and Wright repurchased the mine and township of Wittenoom as the first step in negotiating an agreement with the State government to establish a large industrial complex, including four iron ore pelletising plants, an oil refinery and an asbestos cement industry.[153] The Hanwright agreement was concluded in 1967 partly as a result of a highly successful media campaign mounted by the partners. To Court's irritation, Hancock named Cape Lambert as the nominated port for his proposed industrial area, setting up a conflict with Cleveland Cliffs, which also preferred that site.[154] The second area of interest for Hanwright lay 50 kilometres south of Mount Tom Price and consisted of a number of iron ore deposits between the Hardy and Ashburton Rivers. Named after the Pirraburdu Station, from an Aboriginal word meaning "feathered meat", they became known as the Paraburdoo deposits and were also part of Hanwright's 1967 agreement with the Western Australian government.[155] It quickly became obvious that Hanwright did not have the financial resources to develop the Wittenoom areas and, in their search for partners to finance the development, Hancock and Wright were unsuccessful in persuading either Daniel Ludwig or Kaiser Steel to join with them.

Failure to develop the Wittenoom complex brought other worries. Hancock and Wright were fearful about their reserves in Paraburdoo, an area rich in hematite ore that they considered held as much promise as Mount Tom Price. The partners were concerned that the Brand government might use their lack of progress in developing Wittenoom as an excuse to deprive them of their temporary reserves at Paraburdoo. Hancock's biographer, Neil Phillipson, ascribes the extraordinary animus that Hancock later developed against Robe River, a favourite of Court, to Hanwright's fear that the government would pass the Paraburdoo area to Cleveland Cliffs in compensation for the relative paucity of Robe River's iron ore reserves.[156]

When they realised that they could proceed neither by themselves nor with Ludwig or Kaiser Steel, Hancock and Wright turned to Hamersley Iron. The Hanwright partnership had tried to interest the Japanese steel mills in purchasing iron ore from Paraburdoo, but the mills were dissatisfied with the level of phosphorus in the ore from that area. This was not a problem for Hamersley Iron, which could mix the higher phosphorus Paraburdoo ore with its high-grade hematite ores from Mount Tom Price, thereby extending the economic life of both areas.[157] Hanwright therefore agreed to transfer some of its Paraburdoo reserves to Hamersley Iron.

In 1968 Hanwright approached the WA government to explain that it was unable to proceed with the terms of its 1967 agreement with the State, lacking as it did the capital, expertise and market outlets. The Brand government accepted the submission and negotiated, on Hanwright's behalf, with Hamersley Iron what became the amended Hanwright agreement.[158] Hanwright was to receive from Hamersley Iron $A3 million for the western half of Paraburdoo and a 2.5 per cent royalty on all iron ore mined. Another company jointly owned by Hamersley Iron and Hanwright, Mount Bruce Mining Pty Ltd, was set up to investigate Hanwright's 31 other reserves in the Paraburdoo area.[159] But eventually the two parties in the Mount Bruce Mining Company broke off negotiations and divided these reserves between them.

The 1968 Hanwright agreement produced a new spurt of development of the Pilbara at the peak of the Japanese-driven minerals boom in Australia. A new mine and township were to be created at Paraburdoo, a further 100 kilometres of track would be an extension of the Hamersley railway, and a second berth would be constructed at East Intercourse Island, five kilometres west of the existing facilities at Dampier to accommodate ships of up to 160,000 dwt.[160] The total cost of the expanded Hamersley project would eventually reach $350 million. Hamersley Iron had become a public company in March 1967 through a $25 million capital raising, which constituted the largest initial flotation issue made in Australia to that time. The new public company obtained a credit facility in January 1970 of $200 million worth of Eurodollars with the assistance of a consortium of North American banks to finance the development. In a further concession on Hamersley Iron's part, Russel Madigan, now its Managing Director, agreed to provide a rail link to the Hamersley railway and Port Dampier for Hanwright's proposed Wittenoom development. Much to his chagrin, Madigan learned a few days after the government's agreement to the Hanwright amendment act that Hancock had been negotiating with BHP and the Japanese, albeit unsuccessfully, to sell off part of the Wittenoom project.[161]

By 1970 Hancock was a wealthy man, earning $1.5 million per year from mining royalties although his industrial complex at Wittenoom remained undeveloped. Despite Hancock's lack of success in developing a mine of his own, his prospecting had continued and had been extraordinarily successful in the second part of the 1960s. In this prospecting, he was helped by an associate of long standing, Ken McCamey. McCamey was a weather-beaten fitter and turner, the son of a Roebourne truck driver who had been forced to leave school before matriculating. Phillipson described him thus:

> In many respects he is the sun-bronzed epitome of the popular conception of the typical "Anzac". He would look equally at home in hard hat and bowyangs, or in moleskins and carrying a stockwhip, as he does in his neatly pressed cotton slacks and shirt.[162]

McCamey entered Hancock's employ early, learning the rudiments of geology from Rio Tinto geologist Bruno Campana and becoming a pilot, although he often navigated when flying with Hancock. McCamey had a bushman's instinct for identifying from the air local manifestations of iron ore formations such as the "iron tree", a type of vegetation usually growing out of solid lump ore. Throughout the 1960s he spent many thousands of hours searching for ore from the air. One of the major discoveries attributed to McCamey's aerial prospecting lay thirty kilometres east of Mount Whaleback in an area converging on Jimblebar Creek. Describing the iron ore deposits there as "monstrous", they were for many years called after him "McCamey's Monster" (now BHP-Billiton's Jimblebar mine).

With McCamey's assistance, Hancock took out temporary reserves in other areas in quick succession: the east and west Angelas, named after Peter Wright's daughter, Rhodes Ridge, Sugar Fault, Arrowhead and Giles Mini. By 1970 Hanwright was able to persuade many interested parties that these discoveries held the key to nearly all of the iron ore deposits in the Pilbara not already covered by mining agreements. According to Court, this was far from the truth. As he later told Fairbairn's successor as Minister for National Development, Reginald Swartz:

> It is important to appreciate that the so called "findings" of iron ore by Hancock and Wright are really only delineating areas which are well known to the major companies, but which the major companies were not permitted to lay claim to at the time because of their interests in projects such as Mt Newman, Hamersley Iron, etc.[163]

By 1970 Hanwright was negotiating with American company, Texas Gulf Sulphur, for the Rhodes Ridge area, with BHP and the Goldsworthy Associates for McCamey's Monster, with CSR and Peko-Wallsend for Sugar Fault and with the giant US steel company, Armco, for the balance of his reserves. The problem that soon became apparent was that, although Hanwright might claim to have discovered new areas, it had underwritten no exploratory work. Moreover, temporary reserves had not been taken out in Hanwright's name but in those of Hanwright's associates – Don Rhodes and the brothers, J.D. and W.G. Nicholas, one of whom, James Nicholas, was Hancock's brother-in-law. Although the Hancock associates had later transferred their temporary reserves to Hanwright, this had been done without the permission of the responsible minister in contravention of the Western Australian Mining Act. Yet despite the uncertainty over its ownership of the Rhodes Ridge temporary reserves, Hanwright had secured an option agreement from Texas Gulf Sulphur to develop them.

While Hanwright was assiduously taking up temporary reserves, Court was moving in another direction. According to his policy the Pilbara's iron ore reserves should be developed by a relatively few mining companies operating under State government supervision in particular areas of influence described by their geographical or economic position.[164] In

Court's plan the State would allocate areas of influence to the established companies and would grant some other temporary reserves to steel companies undertaking to build steel operations in Eastern and Western Australia. All other iron ore areas in the Pilbara would be placed under ministerial reserve "in order to make it clear that negotiations for future developments will have to be with the Government and not people like Hanwright".[165] Court was concerned, by this policy, to avoid fragmentation and the haphazard proliferation of small companies, which he feared would concentrate on mining the premium ores at the expense of the lower quality ones.[166] Court considered that many of the temporary reserves claimed by Hanwright, notably Rhodes Ridge and McCamey's Monster, logically belonged in the Mt Newman's sphere of influence. In such cases it might have been better to give temporary reserves over such areas to the Mount Newman Mining Company with appropriate compensation paid to the actual holder of the temporary reserves. The Brand government hoped to be able to negotiate with Hanwright about its temporary reserves but was wary of antagonising as powerful a figure as Hancock, who had used the proceeds of the sale of Paraburdoo to finance a paper, the *Sunday Independent*, which regularly attacked Court and the WA government.

Hancock's fundamental objection to Court's plan was that it deprived the discoverer, namely Hanwright, of its rights and his belief that the finder of the mineral deposits should be the one who determined how they were developed. In 1969 Hancock wrote to Court asking him to:

> ...respect our ability to discover huge ore bodies (where others have failed) and to respect our persistence in refusing to be side-tracked until we have got our discoveries under way by presenting "Capital" with a clear picture on a principal to principal basis.[167]

Between January and March 1970, as Texas Gulf Sulphur started exploratory work at Rhodes Ridge, Hanwright announced the intention to float as a public company. To aid his cause, Hancock hosted Holt's successor as Prime Minister of Australia, John Grey Gorton, on a trip to the Pilbara. Shown McCamey's Monster, Rhodes Ridge and other Hanwright areas, Gorton was won over to the cause of the attempt by Australian entrepreneurs, Hancock and Wright, to develop largely Australian mining operations.[168]

But the attempt to float Hanwright as a public company faced a formidable problem. This was the State government's refusal to confirm the rights to many of the assets included in the proposed flotation. Besides a percentage of the royalty proceeds from the exports of Hamersley Iron from Mount Tom Price and Paraburdoo, Hanwright was indicating as assets 40 per cent of its Wittenoom holdings and of Rhodes Ridge, both of which projects had not yet started. Still more vague was Hanwright's intention to offer selected ventures to the public company "as they were proven", an offer which potentially extended to 50 commercial deposits in the Pilbara. One of the potential ventures most desired by Hanwright was the plan of US steelmaker Armco to set up a steel-making operation in Australia based on Western Australian iron ore.[169] Armco had at first sought to negotiate with Hamersley Iron for a joint venture based on the Mount Bruce deposits for a steel mill at Jervis Bay in New South Wales but had abandoned the plans because of unsatisfactory grades and prices of Mount Bruce ore. It had then turned its intention to the possibility of a joint venture between an Australian-owned and foreign-owned company. The Australian company would be jointly owned by Hanwright, CSR and another Australian mining company, Peko-Wallsend, while the foreign-owned company would be owned by Armco and Thyssen-Heutte of West Germany.

When it came to the allocation of Western Australian iron ore deposits to the new venture, Court reiterated his "areas of influence policy" and affirmed that Hancock and Wright did not necessarily have rights to the areas they claimed. It was up to the State government and not Hanwright to determine how the Pilbara would be developed. Armco then tried to force the issue by informing Court on 2 October 1970 that it had signed an agreement with Hanwright and that it had also offered Thyssen-Heutte a participating interest in the development of the Angelas. Armco added that it was negotiating with Hanwight over McCamey's Monster and Western Ridge.[170]

In the lead-up to state elections in Western Australia early in 1971, Hancock threw the full weight of the *Sunday Independent* behind the Opposition Labor Party led by John Tonkin who, he hoped, would offer Hanwright a better deal than Court. Hanwright also kept up his attack on Court's "areas of influence" policy by awarding options on McCamey's Monster and Western Ridge to MGMA and MIM Holdings. A former MGMA executive, David Moore, recalls being present at Perth airport early in 1971 when Court had a blazing row with MGMA representatives about having dealt with Hanwright over McCamey's Monster. Court warned the representatives: "After this weekend's event [i.e. the election], I will get you!"[171]

Publicly, Court could only declare once more that Hanwright had no legal rights to the areas:

> Each of the companies in this and similar deals understands that any value in their agreement with Hancock and Wright is entirely dependent on the decision of the Government when it finalises its review of how the Pilbara will be developed.[172]

Court's struggle against Hancock in the 1960s and 1970s resembled NSW Governor George Gipps's struggle against the squatters in the 1840s. In February 1971 Tonkin's Labor Party surprisingly defeated the Liberal and Country parties by one seat. Tonkin's campaign had been assisted by Hancock's persistent criticism of the Brand government, which had become complacent. But Tonkin and his ministers, Mines Minister, Don May, and Minister for Industrial Development, Herb Graham, quickly became embarrassed by the extent of Hanwright's temporary reserves and the number of options it had made with other companies without the knowledge of the State government. These temporary reserves extended over 2940 square kilometres in the Rhodes Ridge, Western Ridge and McCamey's Monster areas. Hanwright was also revamping a plan to develop Cape Lambert as a port in association with Armco based on the Angela deposits when that port had been confirmed for the Robe

River project. The disquiet of the Perth bureaucracy with Hanwright's plans was exemplified by the advice of Joe Lord, Director of the Western Australian Geological Survey. Lord objected that his officers and not Hancock had discovered the Angela deposits and that Stan Hilditch might even have discovered them before 1961. With Court's blessing, Tonkin cut the Gordian knot represented by a multiplicity of applications and half-negotiated agreements. In a shock announcement on 26 June 1971, the WA government confirmed Hanwright's rights to McCamey's Monster, Rhodes Ridge and Western Ridge.[173] But all other areas that they claimed except Wittenoom, which was covered by a separate agreement, would revert to the State government. This included the Angelas. The WA Supreme Court later confirmed the legality of the Tonkin government's actions. This meant that in the long term Court's "spheres of influence" plan for the Pilbara would win out against Hancock's grandiose plans.

Hancock's ambitions were defeated partly by the resistance of Court and later Tonkin but also by the slowing in the rate of growth of the Japanese steel industry in response to what the Japanese called as the "Nixon shock" of 1971, a slowing that marked the beginning of the end of first minerals boom. Harbingers of the end of the boom were Armco's abandonment of its plans in the Pilbara and Hamersley Iron's temporary mothballing its new Paraburdoo operation in 1972 when the Japanese steel mills requested a temporary cutback in iron ore imports. The definitive end of the boom would come in 1973 with the dramatic rise in the world price of oil. With Gorton's replacement by William McMahon in 1971, Hancock also lost a key political ally. Hancock and Gorton had both shared the ambition of encouraging greater Australian equity in large mining operations and a higher degree of processing of minerals in Australia. When in that year CSR fought off a campaign by Hancock to mobilise US finance to take over CSR as a basis for his operations, McMahon was motivated to introduce the *Companies (Foreign Takeovers) Act* 1972, to forestall any Hancock move against the long-established Australian company.

Conclusion

The period from 1964 to 1972 was hugely significant in mining's transformation of Australia during Australia's second minerals rush. This was the most dynamic period in the growth of the Japanese steel industry, a period in which coalmining companies in New South Wales and Queensland came to assume a dominant position in the sale of coking coal to Japan. It was also a period in which an internationally competitive iron ore industry was established in the Pilbara, two of whose companies, Hamersley Iron and Mt Newman, grew to be among the biggest in the world. The Goldsworthy operation and the Robe River Project, though not as large as the leading companies, were nonetheless substantial.

By the early 1970s Australian iron ore companies were supplying most of Japan's iron ore supplies and Australia had become one of the leading iron ore exporters in the world. By 1974 iron ore together with coal were the largest components in mining exports amounting to almost one-fifth of Australia's exports of goods and services, larger than manufacturing's proportion although still less than half of the rural sector's share. Reaching this position had been facilitated by the ability of mining companies to borrow large sums of capital from Australian and overseas sources and to overcome enormous natural challenges in constructing mines, railway systems and ports. The process of financing these large mining operations helped in part to modernise Australia's conservative banking system. In this period, both levels of government found it necessary to try to regulate the iron ore rush. The federal government's attempt to set guideline prices had proved short-lived when both the Japanese steel mills and WA governments resisted them. At the State level, Lang Hancock's huge ambitions in the west were defeated by Charles Court's plan that Western Australian iron ore deposits should be developed by relatively few companies operating, under State government supervision, in particular spheres of influence.

The Pilbara iron ore province was a huge materials handling operation in which millions of tons of ore were mined, hauled by private rail to three deepwater ports and shipped overseas. Iron ore mining in the province represented the birth of a new era of industrialisation in Australia based on natural resources, an era that had not even been recognised at the beginning of the 1960s. The creation of a miniature state in Western Australia highlights the degree of industrialisation and development that accompanied the beginning of Australia's second minerals rush as well as the new form of social and economic engineering that departed from the pattern of state-led development of the nineteenth century. Whole towns, ports and railway systems were built by private enterprise in Australia's remote northwest. The new towns and port cities attracted a mixture of foreign migrants and native-born Australians coming mainly from the large towns and drawn to the new mining operations by the prospect of higher wages in a fully employed Australia. This capital intensive industry was highly vulnerable to strike activity on the part of the relatively small but highly organised and inexperienced workforce, and the Pilbara from early on developed a highly combative form of industrial relations that only worsened in the 1970s and 1980s. The Pilbara iron ore industry, as will be seen, was also vulnerable to movements in exchange rates since all contracts were at fixed prices in US dollars and this would become a critical issue in the period from 1973 to 1975.

4

Australia's Rise to World Bauxite Dominance, 1952-1972

Up to the early 1950s Australia mined very little bauxite and produced no alumina, the chemically pure oxide that is extracted from bauxite and then smelted in electrical furnaces to produce aluminium metal. But in the early 1950s, huge quantities of bauxite were discovered in Queensland and the Northern Territory and bauxite of commercial grade was confirmed at about the same time in Western Australia. The discoveries of bauxite augmented and diversified Australia's rich resource endowment in coal and iron ore. From a position of having virtually no bauxite, Australians suddenly found themselves in possession of a major part of the world's supplies of an ore used to create a light and non-corrosive metal. From the 1960s to the present day Australia became one of the largest producers of bauxite ore, along with Brazil, India, Guinea and Jamaica. But the discovery of bauxite in Australia ushered in not only the large-scale mining of bauxite but also industrialisation on a scale not seen in Australia since 1915 when BHP launched a steel enterprise in New South Wales based on iron ore from Whyalla and coal from Newcastle. The significant difference was that the industries developing bauxite from the 1960s were geared to export markets whereas BHP steel production was serving only the domestic market.[1]

Because of the oligopolistic structure of the world aluminium industry, developing Australian bauxite necessitated involvement of a global aluminium company in each of the major Australian bauxite developments. This foreign involvement was always in partnership with

Australian capital.[2] Australian-owned or managed firms participated in joint venture arrangements with multinational companies. The result of bauxite development during the second minerals rush was that Australia quickly accounted for more than one third of the Western world's production of bauxite and about one fifth of its alumina by the 1970s. Like coal and iron ore, the development of bauxite was at first perceived as a domestic industry, but it quickly developed as one largely supplying world markets. As well as stimulating industrial development in Australia's States and territories, bauxite made Australia a vital part of the world aluminium industry and helped stimulate economic relations with Asia–Pacific countries in the second half of the twentieth century. An important by-product of bauxite mining was the development of the movement for Aboriginal land rights and the associated legal argument about native title that would ultimately overturn the previous doctrine of *terra nullius* in the 1990s.

The Global Aluminium Industry Before 1960

The history of aluminium dates back to the nineteenth century when a French company was established to work bauxite reserves in the region of Les Baux in 1855. In the 1880s the Hall–Héroult process for the electrolytic production of aluminium from alumina, the chemically pure aluminium oxide, was developed. Following this breakthrough, Alusuisse (Swiss Aluminium) was established in Switzerland in 1889 and the British Aluminium Company (BA) was set up in London five years later. Aluminium metal was prepared from bauxite in two stages. In the first, alumina was extracted from bauxite by using caustic soda usually by the Bayer process. Smelting alumina in an electric furnace then produced aluminium metal.[3] In the United States the Pittsburgh Reduction Company, founded in 1888, changed its nomenclature to the Aluminum Company of America (ALCOA), under which it would dominate the aluminium industry in the United States until World War II.[4] Alcoa also formed a Canadian subsidiary,

Aluminium Company of Canada Ltd (Alcan) in 1902. This subsidiary was incorporated as a separate company in 1928 but forced to sever its link from its American parent company in the 1950s.[5]

Historian of aluminium Mimi Sheller concludes:

> Aluminum became crucial to the making of modernity not simply as a new material out of which to make particular objects (especially those that we associate with streamlined modernism) but also as a means of innovating across the entire infrastructure of transport and communications (underlying many of the technologies that we associate with modernization). Over the course of around fifty years, from 1910 to the 1960s, aluminum came to play a crucial part in the transportation, electrical, construction, aeronautics, and ship-building industries, as well as in domestic design, architecture, technical equipment, and all kinds of banal aspects of everyday life (from packaging and fasteners, to antiperspirants and makeup, frying pans, and artificial Christmas trees).[6]

World War II saw a surge in the production and usage of aluminium and a shift in the industry's geographic centre from Europe to North America.[7] In World War II, the importance of aluminium was enhanced by its role as a strategic metal, particularly in growth of the aircraft industry. In order to maintain a high rate of aluminium production and to address Alcoa's monopoly position, the US government sold some of its surplus defence plants to two of Alcoa's competitors. The first was Reynolds Metals Ltd, originally R.J. Reynolds Tobacco Company and a long-time consumer of aluminium for cigarette packaging. The other was Henry Kaiser's Permanente Metals Corporation, a shipbuilding and heavy construction concern.[8] The latter changed its name in 1947 to the Kaiser Aluminum and Chemical Corporation (KACC).[9] After the war Kaiser Aluminum based its bauxite mining operations in Jamaica, Reynolds Metals in British Guiana and Alcoa in Suriname.[10] By 1960 the two newcomers to the American industry had taken their place in a group of six major aluminium companies that included two major European concerns. The big six companies were, in order of size, Alcoa, Alcan, Reynolds Metals, Kaiser Aluminum, the French company Pechiney, and the Swiss company

Alusuisse. These companies accounted for nearly two thirds of the non-communist world's aluminium smelting capacity in the early 1960s.[11] Not quite in the league of the six majors was the British Aluminium Company (BA). It had about $US122 million in assets in the late 1950s, including low-cost hydropower and smelting facilities in Norway and Canada.[12]

Australia on the eve of World War II mined little bauxite and so produced no alumina or aluminium metal. Three companies were engaged in the fabrication of a meagre 1300 tons of aluminium metal imported annually into Australia.[13] The most important was Australuco, a company owned one third each by the BA, Alcan and jointly the Australian companies, Electrolytic Zinc Company and Metal Manufacturers Ltd.[14] The Australian government established a Commonwealth Aircraft Corporation during the early stages of World War II. The desire to equip the Australian air force saw the wartime Menzies government announce in June 1941 plans for the creation of a small integrated aluminium industry.[15] Two years later, the Curtin Labor government introduced a bill for a government instrumentality to refine and smelt aluminium in Australia.[16] The Curtin government was convinced that an Australian alumina refinery and aluminium smelter needed to be a public enterprise. Partly this reflected the social democratic ideals of the Curtin government, but also the fact that no bauxite deposits of any consequence were known in Australia at that time.[17]

When it embarked on establishing an Australian aluminium industry, the Curtin government was motivated by strategic and economic concerns: to create an indigenous capacity to manufacture a vital wartime metal and partly to alleviate the balance of payments by not having to import aluminium. In 1944 the government revised its plan for an aluminium industry to take account of Tasmania's offer of relatively cheap hydroelectric power. In a counterpart to the Coal Industry Acts that revamped the New South Wales black coal industry, the Commonwealth and Tasmanian parliaments passed legislation in 1944 to establish the Australian Aluminium Production Commission (AAPC).[18] The AAPC was a counterpart of the

Joint Coal Board and would locate the first government-built alumina plant and aluminium refinery at Bell Bay in Tasmania.[19] In setting up the Bell Bay plant the Curtin government sought to achieve self-sufficiency in the production of aluminium for defence purposes and to ensure that Australian interests would control the aluminium industry after the war. A Copper and Bauxite Committee advised the AAPC on sources for strategic minerals. The technical secretary of the body was Consolidated Zinc's Maurice Mawby. As part of a visiting Australian mission to the United States in 1942, Mawby had established links with American companies, including the company that became Kaiser Aluminum.[20] Mawby's involvement in planning a government-funded aluminium industry and in government-initiated exploration for bauxite equipped him to play a crucial part in the most significant discovery of bauxite in post-war Australia.[21]

By 1948 the government was contemplating amending the *Aluminium Industry Act* to enable a joint venture to be established. This would be a venture between the Australian government and the British Aluminium Company to build the Tasmanian alumina refinery and smelter. But in 1949 Australian Prime Minister Ben Chifley, Curtin's successor, decided not to proceed with the joint venture. He was content with securing the BA's technical advice on how to build it. In return, the BA was given a guarantee of first right of refusal if an Australian government ever decided to sell the Bell Bay plant to private interests. Eventually, the AAPC constructed the Bell Bay operation as a purely public enterprise with a capability of producing 12,000 tons of aluminium per annum.

The Labor government did, however, decide to establish a joint venture with the British Aluminium Company for another purpose. Fearing that inadequate supplies of bauxite would impede the development of a self-sufficient Australian aluminium industry, the Chifley government promoted the search for bauxite in Australia.[22] One of the means of doing so – besides encouraging the Bureau of Mineral Resources (BMR) and AAPC to search for bauxite – was to establish a joint venture with the British Aluminium Company in a majority Commonwealth government-owned New Guinea Resources Prospecting Company (NGRP). The aim of the NGRP was to investigate the hydropower potential of Papua New Guinea and the possibility of bauxite being found in the Northern Territory. In the early 1950s the NGRP built on the exploration work of government geologists to assay a significant bauxite discovery at Gove in the Northern Territory.

In the 1950s, the alumina and aluminium plant at Bell Bay processed bauxite purchased from overseas. In 1957-58 the plant was supplying 60 per cent of Australia's aluminium metal requirements with Alcan making up the shortfall by importing Canadian metal. With the capacity to produce 12,000 tons per annum of aluminium, the Tasmanian plant was about one third the size of the smallest-size aluminium smelter in the United States. Consequently, it had to charge a price of £271 per ton of aluminium ingot, far above the price of Canadian ingot at £226 per ton. In the 1950s this made it necessary for the local producer to be protected through a combination of import licensing and compulsory allocation of local aluminium to domestic fabricators.[23]

Just after World War II any private Australian company contemplating building an integrated aluminium industry – one that mined and refined the bauxite into alumina, smelted the alumina into aluminium metal and then sold fabricated metal – faced a daunting challenge. The world aluminium industry was then concentrated in North America and Europe using bauxite located in Africa, the Caribbean and South America. In general, four tons of dried bauxite was required to produce two tons of alumina, which, in turn, equated to one ton of primary aluminium metal.[24] There were also substantial environmental costs in establishing an aluminium industry. The production of a ton of aluminium cans required four tons of bauxite ore to be strip-mined, washed and refined into alumina in a process that created four tons of caustic residue known as red mud.[25] Alumina refineries provided substantial economies of scale to companies able to build plants larger than 330,000 metric tons capacity, a size that required substantial capital investment. The then optimum size of aluminium smelters that produced the aluminium metal was smaller, about 100,000 metric tons. But aluminium smelters depended for their success on a range of other factors such as location, climate and, primarily, the price of electricity. The big producers were highly integrated and the cost of raising finance for such enterprises was a significant barrier to entry.[26] Such was the challenge faced by Australian mining companies when huge deposits of bauxite were discovered in Australia in the Northern Territory, Queensland and Western Australia in the 1950s.[27]

Comalco's Plan to Monopolise Australian Bauxite, 1952-1960

In 1931, well before the discovery of bauxite in the Northern Territory, the Australian government had proclaimed an Arnhem Land Reserve for the Aboriginal peoples. Very few white Australians visited this remote part of the Northern Territory.[28] The exception was the staff of a mission that the Methodist Church established in 1934 to provide medical services for Aboriginal people. During World War II, the Royal Australian Air Force established an airstrip at Gove. From there bauxite would have been clearly visible to the armed services personnel serving in the Northern Territory. None of them appreciated its significance. H.B. Owen, a BMR geologist, made the first reconnaissance of bauxite on the Northern Territory mainland in 1952 after AAPC geologists found the mineral in the adjacent Wessell Island Group.[29] Building on these discoveries in the early 1950s, the NGRP undertook the first detailed geological assessment of deposits at Gove in 1955 and a year later applied for a lease from the Commonwealth government to mine the bauxite.

A few years before, in June 1953, Mawby had asked his field geologists at Consolidated Zinc to look into the possibility of finding bauxite and phosphate in the Northern Territory and the Cape York Peninsula at the northern tip of Queensland. Noting the potential for bauxite at Cape York in 1953, the geologist Harry Evans returned to the area in 1955. This was the same year in which Bell Bay started producing its first aluminium and BA began investigating the bauxite at Gove. Arriving at Weipa in October, Evans took a dinghy along the Embley River to Albatross Bay.[30] Subsequent investigations confirmed that Evans had made a much larger discovery – indeed one of the great mineral finds in world history – a field containing more than 2000 million tons of recoverable ore reserves and what was to become the largest single bauxite mine in the world. Although the Weipa bauxite deposits were enormous, the Chairman of Consolidated Zinc, L.B. Robinson, cautioned Mawby:

> There is still an enormous tonnage of bauxite in the Guineas, Jamaica, Ghana, French West Africa etc. well located in relation to vast present and potential power resources, and also to the markets for the metal. This must, I think, temper any over-enthusiasm as to the possible rate of development of the Weipa deposits.[31]

After confirming the extent of the discovery, Consolidated Zinc received prospecting rights from the Queensland government over an area of 3045 square miles. By that time, the Consolidated Zinc discovery had attracted the interest of other companies. In April 1956 Alcan obtained prospecting rights over an area adjacent to Consolidated Zinc's. Moving quickly to get ahead of Alcan, Consolidated Zinc Pty registered the Commonwealth Aluminium Corporation (Comalco) in Queensland on 20 December 1955.[32] Comalco began life as a wholly owned subsidiary of Consolidated Zinc.[33] Mawby later convinced a senior Commonwealth Treasury official and fellow member of the AAPC, Donald Hibberd, to become Comalco's first managing director. Born in New South Wales in 1916, Hibberd had studied accountancy at the University of Sydney before World War II. After joining the Commonwealth Treasury in 1946, he worked on the Chifley government's plans for nationalisation of the banks. In 1953 he was promoted to first assistant secretary of the banking, trade and industry branch of the Treasury and served in that capacity as a member of the AAPC from 1953 to 1957.[34] Hibberd's involvement in Comalco exemplified the strength of the infant aluminium industry's links with government.

Lacking experience in either mining bauxite or producing alumina and aluminium, Consolidated Zinc searched for a partner in Comalco from the ranks of world aluminium companies. In favour of BA as a partner was that the Chairman of Consolidated Zinc, L.B. Robinson, was also a Director of BA. Other factors were that BA by that time had a 50 per cent interest in Australuco and a half interest in NGRP. A Memorandum of Intent signed on 11 January 1957 formalised the marriage between Consolidated Zinc and BA in Comalco. The two companies agreed that Consolidated Zinc would assign its mining rights at Weipa to Comalco; that Consolidated Zinc would also buy out the Commonwealth government interest in NGRP; and that the new Consolidated Zinc/BA partnership, Comalco, would take over NGRP's bauxite mining rights at Gove and its rights to develop the power resources of Papua New Guinea. Comalco's original concept was that hydroelectric power in Australia's external territory, Papua New Guinea, would be used to create aluminium metal from Australian bauxite.

The Queensland Labor government was tremendously excited by the bauxite discovery and the prospect of significant mining and industrial development in the State. The size of the leases requested by Comalco, however, was not readily accommodated by Queensland's mining legislation. Vince Gair, the Queensland Premier, offered Comalco instead a special mining lease to be ratified by the parliament. Gair's objective was to offer special privileges not available to other companies so that Comalco could contribute to the industrial development of the State.[35] The aims of the Gair Labor government and the Country-Liberal party government that succeeded it in 1957 were similar in respect of mining. Queensland State governments, like those in Western Australia (see Chapters Two and Three), were excited by the prospect of attracting major industrial developments generated by mining operations. Industrial development was particularly welcome in a State whose manufacturing industries were far less developed than those of New South Wales, Victoria and South Australia. Bauxite-related development would consist not only of a mining operation at Weipa but also, the Queensland government hoped, a refinery to convert the bauxite into alumina and possibly even a smelting facility.[36]

Gair's successor, Country Party Premier Frank Nicklin, and Minister for Mines, Ernie Evans, also accepted another reality. Only on the basis of giving Comalco generous treatment would it have the capacity to raise finance overseas to provide infrastructure that the State did not have. Nicklin and Evans made the leasing of 2270 square miles to Comalco more politically palatable by requiring the company progressively to relinquish portions of the lease. After exploration and testing, it would eventually reduce the area to 1000 square miles on the west of Cape York Peninsula and to 500 square miles on the eastern side of the Cape. Nicklin and Evans also reduced the normal rate of rent of £320 per annum per square mile in the Mining Acts to a rate of £2 per square mile, rising, as portions of the lease were relinquished, to £15 in 1973.[37] The Queensland government gave Comalco a tremendous advantage against competitors by granting an initial lease term of 81 years, with a right of extension by a further 21 years. Royalties at 6d per ton compared very favourably with Jamaica where mining companies were charged 1s.3d per ton at that time.[38] Admittedly, the bauxite royalties would vary according to the price of aluminium after the 21st year of the lease, but the Queensland government nonetheless afforded Comalco very generous rates for two decades.[39]

So, when the Queensland Parliament passed *The Commonwealth Aluminium Corporation Pty Ltd Agreement Act* 1957, Comalco was given very favourable conditions for accepting the considerable challenge of building an Australian aluminium industry based on Weipa's bauxite.[40] Evans's discovery had given Comalco a mine believed to be equivalent to about 25 per cent of the world's known reserves of bauxite. But this was merely the first step in a process leading to the manufacturing and marketing of aluminium. Comalco had a lease to mine bauxite but it needed to raise large sums if it wanted to refine the bauxite and manufacture aluminium. Although the agreement imposed on Comalco the obligation of building a township and harbour at Weipa, the Queensland government ended up building both.[41] There was also the issue of the impact of mining operations on Aboriginal people, a matter of increasing sensitivity by the late 1950s. The Weipa deposits were located on a reserve held by the Presbyterian Church that had run a mission there since 1898. Before beginning mining, Comalco burnt down the existing Aboriginal mission at Old Mapoon and relocated all its residents, with members of different tribes, south of Weipa.[42] In June 1967 academic Frank Stevens criticised the inequitable treatment of Aborigines in the new mining township at Weipa. He noted that "[a]gainst the $28,000 for European homes, Aborigines are placed in aluminium hot boxes costing $6,000 ... Where are the employment opportunities for the Aborigines? The company currently employs 15 Aborigines out of a total work force of 370 people".[43]

Comalco was not content with Queensland bauxite; it also wanted to add the Gove resource to its Weipa deposits. At this stage, a major dispute developed within the Australian government. The Minister for Supply, Howard Beale, had principal responsibility for the aluminium industry in the second half of the 1950s. On its creation in 1950, the functions of Beale's department included manufacture, acquisition and provision of the supply of war materiel, acquisition of materials used in producing atomic energy and the AAPC, which was concerned with a strategic metal. Beale approved of Comalco's long-term plans to establish a bauxite mining, alumina and aluminium industry based on the joint development of Weipa and Gove bauxite. On the other hand, William Spooner, the Minister for National Development, and Paul Hasluck, the Minister for Territories, disagreed.

They thought that assigning Comalco the rights to mine both Gove's bauxite as well as Weipa's would give it a monopoly. At that time the bauxite potential of Western Australia was not being considered; it was not until the end of the 1950s that a joint venture consisting of Western Mining Corporation, and two other Collins House mining companies, Broken Hill South and Broken Hill North, would come together to mine bauxite in Western Australia's Darling Ranges. One of the limitations of Comalco for Hasluck and Spooner was that the Queensland company had no intention of building an alumina refinery in the Northern Territory, and Hasluck in particular wanted to foster industrial development in the Northern Territory. Moreover, although Comalco had already committed itself to building an alumina plant in Queensland, it turned its attention away from developing the hydroelectric resources of Papua New Guinea. Instead, Comalco opted to exploit the hydroelectric potential of the south island of New Zealand for aluminium smelting and also to acquire the Australian government's aluminium smelter at Bell Bay in Tasmania in the meantime.

Prime Minister Robert Menzies sided with Beale in favouring Comalco's plan over the objections of Spooner and Hasluck. For Menzies and Beale, Comalco represented the coming together of two companies. The first was Consolidated Zinc, which, though a subsidiary of a British business, was a long-established and respected mining company based at Broken Hill. The second was the British Aluminium Company, not a member of the world's big six aluminium companies but still an important international concern. Although Comalco was then British-owned, Menzies' ministers were, as Hasluck later described them, "influenced largely by the fact that substantial control of the company was in Australian hands and the result would be the establishment of an integrated Australian aluminium industry serving export markets".[44] Menzies and Beale won the day when on 10 February 1958 Menzies wrote to Mawby granting Comalco a lease over 500 hectares on the Gove Peninsula in what became known as Special Mineral Lease Number One.

Later in 1958, American aluminium interests completed a hostile takeover of the British Aluminium Company. Although the largest American aluminium company, Alcoa, was poised to make an offer for the British company, one of Alcoa's junior rivals, Reynolds Metals, launched a pre-emptive takeover. Although BA was a market leader in the United Kingdom, it was not competitive with the major aluminium companies in North America.[45] The takeover by Reynolds Metals of BA worried Hasluck. He was concerned about the prospect that an American company, through its dominant influence in BA, would tie up the bauxite deposits at both Weipa and Gove. Hasluck's worries were reinforced when sharp differences emerged between the American company, Reynolds Metals, and the British-Australian company, Consolidated Zinc. Consolidated Zinc's plan was first to take over and expand the Australian government aluminium plant at Bell Bay. Then, in cooperation with international partners, it would build an alumina refinery in Queensland. Lastly it planned to construct aluminium smelters in Australasia with the priority being given to an international scale smelter on the southern tip of New Zealand constructed on the basis of its low-cost hydroelectric power. Later still, other smelters might be considered in Queensland or other parts of Australia.[46] The strategy of Reynolds Metals was markedly different. Financially stretched by its takeover of BA, the aim of Reynolds Metals was not to develop either the Gove or Weipa deposits in the short term. Instead, Reynolds Metals would put them in "cold storage" until it had constructed a new aluminium smelter in Canada.[47]

In the meantime, in 1959, the Menzies government decided that it wanted to sell its interests in Bell Bay to private enterprise. The now American-dominated British Aluminium Company was given first right of refusal because of the undertaking that Chifley had made to that company when it was in British hands in 1949.[48] British Aluminium duly made an offer through Comalco. But on 30 November 1959, Menzies' cabinet took Spooner's advice and rejected the offer of British Aluminium (through Comalco) to buy the plant at a price less than the total investment that the Commonwealth and Tasmanian governments had invested in the public enterprise up to that time. The Menzies government wanted to sell Bell Bay, but it also wished to avoid criticism from the Labor Opposition that it was making a concession to private interests.[49]

By April 1960, when Cabinet rejected a revised offer from Comalco, it was clear that some in the Menzies government were also harbouring doubts that Consolidated Zinc had the financial and technical capacity to build an aluminium industry.[50] Other ministers feared that Consolidated

Zinc, like the British Aluminium Company, was vulnerable to a takeover from one of the big American companies that could the tie up most of Australia's northern Australian bauxite resources in American hands.[51] The Australian managers of Consolidated Zinc shared the concerns of the Menzies government. By August 1960, tensions between the new American management of the British Aluminium Company and the Australian managers of Consolidated Zinc reached such heights that the two companies decided to dissolve their partnership in Comalco.

In the break-up Consolidated Zinc agreed to let the British Aluminium Company take over the bauxite leases at Gove from Comalco, while Consolidated Zinc retained the much larger bauxite reserves at Weipa. The Prime Minister's Department advised Menzies that it saw the split between Consolidated Zinc and British Aluminium as extraordinarily important. This was because it represented the "last opportunity for the Government to influence in any substantial way the future structure of the aluminium production industry".[52] Menzies agreed with the advice. Reversing its previous decision to support a Comalco monopoly of Gove and Weipa bauxite, the government now agreed to a duopoly in the emerging aluminium industry between BA at Gove and Consolidated Zinc at Weipa. To facilitate this new arrangement, the government agreed that Consolidated Zinc could assume the standing of BA in the purchase of Bell Bay.

Under the purchase agreement, the Menzies government exacted from Consolidated Zinc the full sum of the contributions invested by federal and State governments – nearly £11 million, an initial £2.5 million payment with the balance paid by annual instalments to 1976. In a significant sweetener, the Menzies government gave Consolidated Zinc a concessional rate of interest on the balance owing to the government each year. The company had to pay five per cent interest on the outstanding balance, but it needed to pay interest only after it had declared a dividend on the business of six and a half per cent and paid tax at three per cent. In addition, all liability to pay interest was to end when the principal was fully paid off in 1976. The Labor Opposition was incensed by the deal, arguing that the real price of the plant would be eroded through inflation and that the federal government was in fact making an interest-free loan to Consolidated Zinc. Beyond those concessions, Consolidated Zinc extracted one more advantage from the Menzies government. Comalco convinced the government to protect the plant with tariffs and import licensing during its first four years of expansion. With this assistance, Comalco was able to expand the Bell Bay plant from 32,000 tons a year capacity in 1962 to 55,000 tons a year in 1965. With its bauxite deposits at Weipa and its combined alumina refinery and aluminium smelter in Tasmania, Comalco became the market leader in the infant Australian aluminium industry in the early 1960s, just as Hamersley Iron was the leader in the emerging iron ore industry.

Comalco, Pechiney and Queensland Alumina Limited (QAL)

Around the time that Comalco was taking over Bell Bay, the Menzies government made new arrangements for mining at Gove. On 6 November 1960 Hasluck transferred the central bauxite lease at Gove, Special Mineral Lease No 1, from Comalco to BA. But he hedged his bets by allowing another company prospecting rights on the perimeter of BA's leases. That company belonged to a subsidiary of Duval Holdings. The principal, Frank Duval, was a former Australian serviceman who had established himself in Japan after World War II and then became involved in various mining ventures across Australia.[53] Through Duval's companies, retired federal Treasurer Arthur Fadden had secured for himself a position of intermediary between the Japanese steel industry and WA iron ore companies before the lifting of the iron ore embargo. Hasluck let Duval know that he would be given approval to export up to 10 million tons of bauxite from Gove's perimeter deposits and even to construct an alumina refinery in the Northern Territory if larger bauxite deposits were found. Realising that this task was beyond Duval Holdings' capacity, Duval invited Pechiney, the largest aluminium producer in Western Europe, to undertake the task. Pechiney obliged, registering an Australian subsidiary that was to acquire the mining titles at Gove for which Duval had applied. In short order, the French company made a bold undertaking to the Menzies government. It undertook to develop an alumina plant, a major industrial enterprise, in the Northern Territory with a capacity of producing 500,000 tons of alumina per annum through a consortium of aluminium producers.[54]

Meanwhile, the British Aluminium Company, with far superior bauxite leases, outlined its plans for the development of the central bauxite lease at Gove. It also sought the right to export 10 million tons of bauxite from Gove but equivocated on whether it would erect an alumina plant.[55] Spooner and Hasluck much preferred Pechiney's firm proposal to British Aluminium's. They persuaded cabinet to approve giving the French company the perimeter leases at Gove. These became known as Special Mineral Leases, 2, 3 and 4.[56] The government gave Pechiney the right to export 10 million tons of bauxite from their areas over 18 years. In return, the French company was required to develop plans within three years for the construction of an alumina plant with a capacity of 500,000 tons. The Menzies government also decided not to go ahead in granting the British Aluminium Company mining rights over the central bauxite deposit at Gove.[57] Instead, it announced on 8 May 1963 the reservation of the central bauxite lease at Gove from mining. It also invited proposals for the development of what the Australian government believed to be the only known and uncommitted deposit of bauxite left in the non-communist world.[58]

The announcement by Menzies of the decision to excise 148 square kilometres from the Arnhem Land Reserve to give to Pechiney sparked a protest by the Aboriginal inhabitants of the Northern Territory. Comalco's dispossession of the Aborigines in northern Queensland had aroused relatively little interest. But in 1963 Edgar Wells, the superintendent of the Yirrkala Methodist mission, cabled Menzies with a dramatic telegram of protest at the effects of bauxite mining on the Indigenous inhabitants of Gove. Wells's telegram sparked a major national protest, paving the way for the national campaign for "land rights" for Aborigines.[59] The telegram was picked up by Stan Davey, secretary of the Federal Council for Aboriginal Advancement, and by Aboriginal activist Barry Christophers, who detailed a statement on behalf of the Council for Aboriginal Rights:

> Aborigines have an inalienable right to security of land tenure of tribal lands which may not be superseded by any mineral rights …. The ownership of the territories concerned in Arnhem Land, including rights to all minerals on or under the land, [should] be vested in the Aborigines themselves … In circumstances where "mining interests" have already deprived Aborigine population of territory, e.g. Mapoon, Weipa etc, adequate compensation for loss of land and continuing payment of Royalties [should] be granted to the peoples concerned.[60]

Historian Bain Attwood has argued that Yirrkala transformed the Federal Council's ideas because northeast Arnhem Land was inextricably connected with "tribal Aborigines" and "Aboriginal tradition", symbols that authenticated Aboriginal land rights. Attwood points out that northeast Arnhem Land had been relatively unaffected by white colonisation and that this was unlike most other areas where pastoralism or mining had displaced Indigenous peoples.[61] Consequently, it was much easier for the Yolngu people of Gove to assert the proposition that they were the actual owners of the land. There was, moreover, little miscegenation at Gove, whose Indigenous people continued to use indigenous languages and rituals. Importantly, the Federal Council made a link between Yolngu "tribal" and "traditional" lands to all other Aboriginal reserves in Australia.[62] The Northern Territory was under federal jurisdiction and federal parliamentarians were more concerned than State parliaments about how other countries perceived Aboriginal policies. Encouraged by one such Labor parliamentarian, Kim Beazley, the Yolngu in August 1963 presented five bark petitions to parliament as a way of gaining recognition for their culture and demonstrating their land tenure and rights. One of the major effects of the minerals boom of the 1960s was thus to spark a campaign for national land rights for Aborigines, arising from the particular circumstances of bauxite mining at Gove.[63]

Around the time of the bark petition, Pechiney abandoned its plans to build an alumina plant in the Northern Territory. The prime reason for this was the unfolding of Comalco's plans for the development of Queensland's bauxite. In October 1960 Consolidated Zinc had built on contacts made by Mawby in America in World War II by taking Kaiser Aluminum as a partner in Comalco. The essence of the partnership was that the new partners in Comalco agreed to develop an alumina plant with a minimum capacity of 360,000 tons per annum in Queensland, to proceed with the expansion of the just-purchased plant at Bell Bay, and then to build a large aluminium smelter located at Invercargill in New Zealand, powered by a hydo-electric plant built on the nearby Lake Manapouri.[64] The capital required was astronomical – £45 million to construct the power system in New Zealand; £42 million for the development of the bauxite mining operation and alumina plant in Queensland, £20 million for Bell Bay, and plants at Yennora and Enfield to provide outlets for Comalco's aluminium; and £63 million for the construction of an aluminium smelter at Invercargill.[65] In 1961 efforts to raise these colossal sums failed in the United States. In the following year, Consolidated Zinc, believing that it was likely to be taken over by American Metal Climax, accepted an overture from Rio Tinto to form the Rio Tinto-Zinc Corporation (RTZ). The coming together of the Australian subsidiaries to form Conzinc Riotinto of Australia (CRA) gave Comalco greater financial resources, via its parent company RTZ, to continue with ambitious the plan to build an aluminium industry in Australia.

But in 1963, despite the establishment of CRA, Comalco was at a crossroads. Although CRA had taken over Bell Bay, it had been unable to make any progress on constructing an alumina refinery in Queensland, still less in commencing to build an international standard aluminium smelter. Mawby was convinced that if Comalco did not begin constructing an alumina plant in that year, it would lose credibility with governments and forfeit all the investment it had made in the bauxite and aluminium industry up to that time.[66] Besides the competition that Comalco faced from a new company, Alcoa of Australia, it was also under competitive threat from Pechiney and Alcan.

Alcan held contiguous bauxite deposits in Queensland as well as an interest in the Australian metal fabricator, Australuco. In 1962 Alcan was pressing the Queensland government for permission to export bauxite from Queensland to Japan.[67] It was resentful that Consolidated Zinc's purchase of Bell Bay, prior to any competitive tender, had cut the ground from under its feet in Australia. Accordingly, the Canadian company informed Queensland Premier Frank Nicklin that it was thinking of building an aluminium smelter in eastern Australia.[68] Alcan asked Nicklin whether the Queensland government would permit it to export bauxite from Cape York to Canada. If Queensland gave its permission, Alcan offered to return the bauxite as alumina for processing in an aluminium smelter that it undertook to build in Queensland. Comalco's other rival, Pechiney, was at the time the largest aluminium producer in Europe with assets of £179 million. If it were able to establish an alumina plant in the Northern Territory, perhaps in a consortium with Reynolds Metals, it might prevent Comalco from being the first company to build a large alumina refinery in northern Australia.

Comalco overcame the threats from its competitors through a number of strategies. The first was by exporting bauxite to Japan itself. The Japanese aluminium cartel at that time had a preference for spot sales of bauxite and its smelters were geared to tri-hydrate bauxite, which it purchased from Malaya, Sarawak, Bintan and India, and not the mixed mono- and tri-hydrate variety of bauxite at Weipa.[69] Comalco's Managing Director, Donald Hibberd, convinced the Japanese aluminium industry of the long-term advantages of purchasing Weipa bauxite. He persuaded the Japanese to modify their refineries from a low temperature to high temperature processing method to accommodate the Queensland variety.[70] After this,

Comalco was awarded three-year contracts with each of the Japanese aluminium producers for the sale of 600,000 tons of bauxite commencing in 1963. Coming not long after the first coal contracts to Japan from New South Wales, this bulk bauxite contract negotiated between Australia and Japan was the forerunner of longer-term contracts with two Japanese companies of three million tonnes over a 10-year period.[71] Long-term bauxite contracts and the introduction of bulk carriers led to Weipa becoming one of the largest suppliers of bauxite to non-integrated world markets.

The second of Comalco's strategies was an ingenious one. It was to co-opt its competitors into constructing with Comalco a joint alumina refinery in Queensland. By the beginning of 1962 CRA and Kaiser Aluminum were agreed on two desiderata: that an alumina plant should be built at Gladstone in Central Queensland rather than at Weipa and that, to be economical, it should have a capacity of not less than 400,000 tons of alumina per annum. The problem for Kaiser Aluminium was that it did not have an immediate outlet for all of such a plant's output. This meant that Comalco had to secure the assistance of two of the other global aluminium companies in building the refinery and taking a portion of its yearly output of alumina. At the end of 1962, Comalco resolved to approach first Pechiney and then Alcan as the most likely partners. Pechiney needed alumina for its smelter at Intalco Bellingham in Washington in the United States and Alcan required feedstock for its Kitimat smelter in British Columbia in Western Canada.[72] The problem for Comalco in 1963 was that Pechiney had plans to build an alumina plant in the Northern Territory and that Alcan was working on the option of persuading Nicklin to let it export Australian bauxite to Canada and to build an Australian aluminium smelter fed from Canadian alumina. In this manner Alcan might supplant Comalco in Australian and regional markets.

In early 1963 in an effort to outfox its rivals, Comalco invited Pechiney and Alcan to participate in an alumina plant at Gladstone. Comalco had in mind a large industrial enterprise: one of the world's largest alumina refineries. With neither competitor persuaded, Comalco deployed its growing political influence backed now by the international reach of RTZ. Val Duncan, head of RTZ, encouraged Comalco Chairman Mawby to persuade Nicklin to freeze Alcan's bauxite leases unless it gave a satisfactory undertaking to the Queensland government that it would purchase alumina from Comalco's planned refinery at Gladstone.[73] Duncan also persuaded Nicklin to press Menzies to allow Pechiney to be relieved of its obligations to process Gove bauxite in return for a commitment to join with Comalco at Gladstone.[74] Beyond that, Comalco sought to close off Alcan's option of importing cheap Canadian alumina into Australia. It encouraged the Menzies government to prohibit imports of any aluminium if there was excess capacity in Australia (from Comalco and Alcoa), a submission that was later approved.[75]

The Nicklin government, facing a State election in 1963, was determined to have a major development project launched at Gladstone as quickly as possible and was happy to oblige Comalco.[76] Indeed, the Minister for Mines, Ernie Evans, probably exceeded his authority from cabinet in his zeal to help the Queensland company. At the end of March 1963 Evans threatened Alcan that it would not be granted an extension of its prospecting authority in North Queensland unless it either began the construction of its own alumina refinery in Queensland or joined with Comalco in constructing a joint alumina refinery at Gladstone. Evans warned that failure to meet those conditions would mean that Alcan's only future rights in North Queensland would be to take up ordinary mining leases, which contained impossible conditions for a bauxite mining operation.[77]

With these inducements, Alcan agreed to join with Comalco in May 1963 while Pechiney came on board in October. Its success in persuading Alcan and Pechiney to join a consortium to construct an alumina refinery in Queensland was the decisive breakthrough for Comalco. But Comalco could not have achieved this breakthrough without the help of the

Queensland and Australian governments. The alluvial nature of the gold rush of the nineteenth century meant that individuals or small groups of miners could form viable units.[78] By contrast, the mining enterprises of the second great minerals rush were dependent for their success on relationships between large mining companies and joint ventures with governments. Like Utah, Comalco would prosper because of its strong relationships with State and federal governments. Kaiser Aluminum, Pechiney and Alcan, making up about 60 per cent of the Western world's aluminium production capacity, were thus brought together in building in Queensland what was eventually to become one of the world's largest alumina refineries. The plant was structured on the basis that Comalco sold 1.2 million long tons of Weipa bauxite under firm long-term contracts to feed bauxite into the alumina refinery. In October–November 1963 a company to operate the alumina plant was registered under the name Queensland Alumina Limited (QAL). Its shareholding consisted of Kaiser Aluminum 52 per cent, CRA 8 per cent, Alcan 20 per cent and Pechiney 20 per cent. Comalco's prospective interests were held by CRA's 8 per cent and 8 per cent of Kaiser's shareholding. Each partner in QAL was required to "take or pay" for a share of the alumina produced by the plant in proportion to each participant's shareholding in the refinery.[79]

The concept behind the contracts employed in the plant was that QAL would process each participant's bauxite into alumina and charge a toll for the cost of doing so. The toll was calculated to cover all processing costs, interest and depreciation so that the plant would have sufficient funds to repay its long-term debt. While the plant was being built, the participants gave notice that they preferred to take a higher output of alumina. The initial planned capacity of 600,000 tons per annum was increased to 1.275 million tons in 1971 and then to two million tons in 1973. This made it the largest alumina refinery in the world at that time. So successful was Comalco's consortium approach at Gladstone that it repeated it through sponsorship of the Eurallumina venture in Sardinia, Italy, a venture also based on Weipa bauxite.[80]

Mawby confided to a colleague on 4 December 1968 that "this [Eurallumina] should be a money spinner because of bauxite profits and the fact that with an alumina producer in the common market, we dodge the various duties into the European countries".[81] Its Italian venture emphasised Comalco's success in harnessing the multinational resources of the RTZ empire. All in all, Comalco's development of Weipa's bauxite was the costliest development in the history of Australian mining to that time, with nearly $300 million spent on the project in 1968 and a smelter at Bluff in New Zealand following in the early 1970s.[82] Australian restrictions on the importation of foreign metal in 1966 strengthened Alcan's resolve to build its own aluminium smelter, which was eventually opened with a 49,000 tonne capacity, not in Queensland but at Kurri Kurri in New South Wales.[83] Using alumina from Gladstone, the Alcan smelter was officially opened on 14 November 1969. At the end of 1968 Comalco also proceeded to construct its aluminium smelter in New Zealand. Building on its successful bauxite trade with Japan, Comalco constructed the smelter in association with two Japanese aluminium companies, Showa Denko K.K. and the Sumitomo Chemical Company. In 1971 the New Zealand smelter, fed with alumina from Gladstone, was opened with a capacity of producing 75,000 tons of metal each year.[84]

Alcoa of Australia's Role in the Development of WA Bauxite

Not long after the rupture of the relationship between Consolidated Zinc and the American-dominated British Aluminium Company, a new company burst into the emergent Australian aluminium industry. It was the result of the enterprising work of the Australian gold mining and exploration company, Western Mining Corporation (WMC). WMC, which will be examined in greater detail in Chapter Five, turned its attention to bauxite located in the Darling Range south of Perth in the second half of the 1950s. While the presence of bauxite in this area had long been known, mining companies had ruled out the possibility of this WA bauxite providing the basis of an aluminium industry.[85] Though patches of the bauxite carried 47 per cent alumina, they were not as rich as the 49 per cent alumina in Gove's bauxite and the 58 per cent in Weipa's, the overall resource averaged below 30 per cent alumina. The Darling Range bauxite, moreover, contained more quartz than the Weipa deposits, and many believed that would impede the process of extracting alumina.[86]

But WMC had faith in the possibilities of the Darling Range bauxite. When testing the deposits, WMC reached the surprising conclusion that the presence of silica helped rather than hindered the process of extracting alumina, as it was readily separated. In 1958 the company secured from Premier of Western Australia, A.R.G. Hawke, a temporary reserve of 6250 square miles (16,200 square kilometres) in the Darling Range – a huge area of land not far from Perth.[87] Assisting the Australian company were linkages formed with the aluminium industry in Japan, which, by the end of the 1950s, had surpassed Italy and Norway to become the sixth largest aluminium producer in the world.[88] In July 1960 WMC exported a trial shipment of 9891 tons of bauxite from the Darling Range to Japan, thus demonstrating the possibilities of trade in bauxite and alumina to a growing East Asian market.[89]

These first export sales would later lead to Mitsubishi Chemical Industries agreeing to purchase large quantities of Western Australian alumina, rising to 120,000 tons a year by 1965.[90] In order to develop its bauxite deposits, WMC set up a venture named Western Aluminium which consisted of WMC and two other Collins House companies, Broken Hill South Ltd and Broken Hill North Ltd. This venture formed the concept of building an alumina refinery at Kwinana, south of Perth. But, because of the significantly higher cost of power in Western Australia, the syndicate envisaged a large power station at the Victorian coastal town of Anglesea using Victorian brown coal to supply an aluminium smelter to be constructed at Geelong. Then a major outlet for Australia's wool and wheat exports, Geelong was an excellent port not far from a mine at Anglesea that was capable of supplying coal for an electricity generating plant. Several other major industries, such as the Ford Motor Companies plant, meant that Geelong also had a skilled manufacturing workforce.[91]

The huge capital costs involved in this enterprise, from mining bauxite to manufacturing aluminium, were beyond the capabilities of this consortium of Australian mining companies. So in 1960 it approached the American aluminium giant Alcoa, which, by taking over BA, had missed out on the opportunity of obtaining the bauxite riches of North Queensland in 1958.[92] The Australian consortium explained to Alcoa in Pittsburgh how Jarrahdale bauxite, which was previously thought to be of an inadequate grade, could be refined by simple techniques. Alcoa sent an executive, Ralph Derr, to test the Australian syndicate's claims at the end of 1960.[93] The American engineer endorsed the finding that quartz could be removed cheaply from the Jarrahdale bauxite. Within a short time he raised the estimates of the payable bauxite in the Western Australian deposits to over 100 million tons. Derr was confident that WMC's venture would be able to outsell Comalco in the Australian market, that Reynolds Metals had been right to walk out of the Weipa bauxite project, and that Consolidated Zinc's enlisting of Kaiser Aluminum into Comalco was a "desperation effort" that would also fail.[94] Derr was correct about the potential for WA bauxite, but was too pessimistic about Comalco. Comalco, the company developing Queensland bauxite, and Alcoa of Australia, through its involvement with Western Australian bauxite, would eventually flourish. Both Derr and WMC's management based their plans on a mistaken assessment that they would win out over Comalco in supplying the Australian market. As Geoffrey Blainey observed, Comalco and Alcoa of Australia would both in time depend far more on international markets for alumina and aluminium than on the tiny Australian market:

> … but it was almost impossible to predict this in the summer of 1960–61. For the Australian leaders of Western Aluminium and presumably of Comalco, Australia was a crucial market in which perhaps there could only be one winner.[95]

Within another five years, Comalco and Alcoa would shift Australia from being totally dependent on imported aluminium to "self-sufficiency in every aspect of the industry".[96]

In deciding whether to take part in the Western Australian opportunity, Alcoa of America considered the advantages of the project – low cost of mining and refining resulting from the proximity of large resources of bauxite to a large city, Perth, and a proven international port, Fremantle. Alcoa weighed these advantages against the disadvantages of the 3000-kilometre distance to an aluminium smelting site in Victoria and the looming competition represented by Comalco. Perhaps the decisive attraction for Alcoa was the proximity of the Darling Range bauxite deposits to Japan. If Alcoa did not take up the opportunity of developing them, there was a strong possibility that Japanese interests or Alcoa's American competitors might finance the project, allowing Japan to buy much of its alumina from Australia to produce cheaper aluminium metal than before.[97] In 1961 the President of Alcoa, W.H. Krome George, travelled to Australia to negotiate an agreement with the Australian companies. While George was in Melbourne, delegates from rival companies, Reynolds Metals and Kaiser, also paid court "and contending rivals chummed like foreign correspondents at the Menzies Hotel bar in Melbourne, where they drank and played bridge while the 'the Bauxite Derby' raged behind the scenes".[98]

The final result of the derby was that Alcoa agreed to team up with the Australian consortium provided that the US giant was the majority shareholder of a new company named Alcoa of Australia with 51 per cent of the shares and the right to nominate more than half of the directors, apart from the chairman who initially would be WMC's chairman, Lindesay Clark. But along with this 51 per cent American ownership came crucial benefits. Alcoa agreed to provide two-thirds of the £45 million required to construct the project, with the remaining £14.38 million subscribed by the shareholders, more than half of whom came from Alcoa of America.[99]

The announcement of the enterprise in Australia caused much excitement. Coming as it did before the iron ore ventures of the period from 1963 to 1965 and around the same time as Comalco's operations, it was heralded as one of the biggest mining and industrial ventures ever launched in Australia. In Perth, the Brand government gave its full support to the project, with Charles Court promising that the State would dredge a channel to the wharf that was to be built at the Kwinana alumina refinery and extend the State railway 26 kilometres to reach the first bauxite mine. In comparison with most of the iron ore projects that Court would assist with later in the 1960s, the bauxite ventures had the attraction of involving the much greater industrial development and jobs that came from refining bauxite into alumina.[100] For Alcoa of America, the formation of Alcoa of Australia was decisive in its emergence as a multinational company.[101] After the formation of Alcoa of Australia, Alcoa of America committed itself to a long-term strategy "based not only on projections for the Australian market, but also on the prospects for an even greater Pacific Basin market".[102]

Alcoa of Australia built its alumina refinery at the Kwinana industrial complex, the port 24 kilometres south of Perth, near British Petroleum's oil refinery and BHP's new mill for rolling steel using iron ore from Koolyanobbing. Mitsubishi Chemical regarded itself as having been instrumental in enabling Alcoa to build its Kwinana refinery by agreeing to take 120,000 tons of its annual output of alumina under long-term contracts. Indeed, for years afterwards, Mitsubishi remained resentful about WMC apparently reneging on an undertaking to facilitate Mitsubishi equity in Alcoa of Australia.[103] From the mine at Jarrahdale, mechanical shovels would load the bauxite into trucks that carried the ore to the railway. The bauxite was transported by train to the Kwinana refinery where it was then crushed into white powder, mixed with caustic soda and heated in large vessels using steam power to produce green liquor that contained nearly all the alumina. Waste was removed either as coarse sand or red mud and the purified liquid was then passed through precipitators, and reheated to about 1000 degrees centigrade to produce the alumina as a dry crystalline powder. The Kwinana alumina refinery was completed at the end of 1963 and the first ship carrying alumina to the Geelong smelter sailed on 22 February 1964. As well as the Geelong smelter, Alcoa of Australia supplied alumina to American Metal Climax, a company anxious to diversify into aluminium, as well as to a growing market in Japan. In 1965 Kwinana was shipping 65,000 tonnes of alumina to Japan; by 1970 this figure had reached 307,000 tonnes. On the other side of the continent, Prime Minister Robert Menzies formally opened Alcoa's aluminium smelter at Point Henry on the southern side of Geelong harbour on 6 May 1963.[104] By 1964 it was producing 40,000 tonnes of aluminium each year.

The new company, Alcoa of Australia, started to make profits in 1967 and expanded continuously in the second half of the 1960s, spending $35 million to increase the Kwinana refinery's capacity to 830,000 tonnes and $31 million for the steam powerhouse at Anglesea, which the Premier of Victoria, Henry Bolte, formally opened on 20 March 1969. While previously most of the finance had been raised by the parent company in the United States, in the second half of the 1960s the Australian subsidiary, Alcoa of Australia, exerted increasing independence. In February 1968 Alcoa of Australia secured a Eurodollar credit facility of $US46 million and, in 1969, two further credit facilities totalling $US130 million to finance the expansion of refining in Western Australia. Australian banks – the E.S.&A. Bank, the Commercial Bank of Australia and the Australian Resources Development Bank – supplied a portion of the funds, but most of the capital came from overseas.[105] In the four years from 1968 to 1971, at the peak of the minerals boom, Alcoa spent $247 million on construction of its aluminium empire, financed largely by overseas capital.[106]

At the beginning of the 1970s, about 500 million tonnes of bauxite had been proven in the Darling Range, and it was thought that another 500 million tonnes was likely to be confirmed after testing. By December 1970 the Kwinana refinery's capacity had been raised to 1.25 million tonnes, bringing it on a par with the QAL refinery at Gladstone and making it bigger than a soon to be completed refinery at Gove in the Northern Territory.[107] In 1971 five ships per month were berthing at Kwinana to transport alumina. A second Alcoa refinery at Pinjarra, 86 kilometres south-west of Perth, was officially opened on 3 May 1972, by which time

Alcoa of Australia had become one of the world's leading producers of alumina, while its smelter in Geelong was supplying a substantial part of Australia's aluminium needs as well as a growing share of the East Asian market.[108]

The mining boom was not only transforming Australia but also changing patterns of world trade and industry. In the case of bauxite, the rise of the new Australian mining and aluminium companies altered the structure of the world aluminium industry. Until the 1960s integrated companies dominated the world aluminium industry. In the 1960s these were joined by specialist producers, such as non-integrated aluminium smelters in Japan and the Middle East that were increasingly supplied by the new Australian mining companies.[109]

Nabalco and the Development of Bauxite in the Northern Territory

The deposits in the Northern Territory were the first large Australian bauxite deposits to be discovered. But for political and commercial reasons they took the longest time to develop. After inviting bids for the central bauxite lease at Gove, the Australian government whittled the responses from interested bidders down to two. Because by the early 1960s Comalco and Alcoa were supplying Australian domestic requirements for aluminium, any alumina produced in the Northern Territory would have to be exported. Rates of interest on industrial debentures were lower in other parts of the world than in Australia, suggesting that an Australian-financed aluminium enterprise in the Northern Territory would struggle to be competitive. Spooner and Charles Barnes, Hasluck's successor as Minister for Territories, consequently looked to proposals from proven large aluminium companies in joint ventures with Australian companies. The two preferred proposals were from Swiss Aluminium (Alusuisse) and Reynolds Metals of the United States.

At this stage of the 1960s minerals boom, State governments had not been overly concerned about the foreign ownership of the mining companies with which they dealt. Their overriding motive was to encourage development in their States and the royalties and other benefits, such as rail freight, that flowed from it. The attitude of the Australian government was more complex. It shared with the State governments a desire to encourage industrial development from mining by promoting the local processing of minerals through industrial enterprises such as iron ore pellet plants, possible steel mills, alumina refineries and aluminium smelters. But it was preoccupied with the balance of payments, a concern which had motivated its assistance to the New South Wales coal ports and the ending of the iron ore embargo in the early 1960s.

This concern led some federal ministers to seek a requirement for higher Australian equity in mining ventures in the hope of keeping more of the profits and dividends in Australia. Federal ministers and their departments were conscious by the mid-1960s that alumina and aluminium projects, though Australian-managed, were now primarily British-American in ownership. The exception was Alcoa of Australia, which, though American-owned, had 49 per cent Australian equity. In a later speech to a mining industry conference Prime Minister John Gorton graphically illustrated the benefits of local equity and local processing in the aluminium industry. He said:

> One million tons of bauxite earns us only $5 million; if it is converted into alumina it would be worth $30 million; if the alumina is converted to aluminium it would be worth $120 million; if finally it is fabricated into aluminium products it is worth $600 million.[110]

It was these stirrings of resource nationalism that saw the Menzies government seek to promote Australian equity into the bauxite-alumina venture over which the federal government had the most influence by introducing it as one of the criteria on which it would judge applications for the central Gove lease.[111] Because of the realities of the world aluminium industry, however, Menzies' ministers accepted that Australian companies would have to partner one of the majors. Only by doing so would they have a chance both of raising the finance to develop Gove and securing markets for its alumina and bauxite. In late 1964 Alusuisse firmed up its plans. The Swiss company proposed the formation of a new company, the North Australian Bauxite and Alumina Company Ltd (Nabalco), owned equally by Swiss and Australian interests. The Australian companies involved were CSR, the Australian Mutual Provident Society (AMP), Mutual Life and Citizens Assurance Company (MLC), Peko-Wallsend Investments, Commercial Banking Company, Elder Smith Goldsborough and Mount Morgan Ltd.

Nabalco offered to build a 500,000-tons-per-annum alumina refinery in the Northern Territory subject only to a favourable feasibility report and to only a five per cent return on capital.[112] In this case the Swiss were even prepared to proceed by themselves without their Australian partners. At the same time, BHP joined forces with Reynolds Metals to make an unequivocal commitment to construct a 300,000-tons-per-annum alumina plant at Gove within eight years. Charles Barnes preferred the BHP/Reynolds bid; Spooner's successor as Minister for National Development, David Fairbairn, opted for Nabalco's.[113] Cabinet decided in Fairbairn's and Nabalco's favour because the Australian partners in Nabalco collectively had more access to Australian capital than BHP, were offering to fund the infrastructure of a town at Gove and were not, as BHP and Reynolds Metal were, insisting on immediate access to Gove's perimeter leases as well as the central lease.[114] As for the perimeter leases, Pechiney's Australian subsidiary spoke to Alusuisse representatives in Zurich and proposed participating in Gove. The French company suggested either buying shares in Nabalco or entering into a tolling agreement, whereby they would convert bauxite from the perimeter leases into alumina. But eventually the perimeter leases were promised to Nabalco on condition that it agreed to build an alumina plant of larger than 500,000-ton-per-annum capacity.[115]

Nabalco set up its headquarters in three floors of the 24-storey Goldfields House in Sydney's Circular Quay. The chairman of the new company was David Griffin, Managing Director of Alusuisse's Australian subsidiary, Swiss Aluminium Mining Pty Ltd. Interned as a prisoner-of-war of the Japanese in Malaya during World War II, Griffin had practised as a solicitor in New South Wales and was an astute businessman, who would later become Lord Mayor of Sydney. Among the individuals representing the Australian partners in Nabalco, Sir James Vernon, was the most important, ably assisted by his successor as General Manager of CSR, Gordon Jackson.[116]

From its headquarters in Sydney, Nabalco undertook a one-million-dollar feasibility study that explored the bauxite at Gove, examined soil conditions, tides and the weather, and estimated profits and social aspects of the development.[117] Architects, engineers and draftsmen in Sydney turned out thousands of plans – of the town, bauxite crushing station, alumina plant and port. The final plan for the town emerged as a mixture of Swiss taste and Australian practicality. It provided for a town centre, built about six miles from the alumina plant, including a shopping mall, hotel-motel, community hall, library, town administration offices and primary school. Residential areas rose north and south of the town centre, mixing low-set family homes and double-storey bachelor units.[118]

The feasibility study, conducted by Nabalco over about two years, indicated that an alumina plant with a 500,000 ton annual capacity would cost $A212 million to construct or $A424 per ton of capacity. The immensity of these capital sums is illustrated by the fact that $A200 million in 1968 is equivalent to about $A2.3 billion in current dollars. This was a great problem for the Australian investment partners, not so much for the considerable amount of money that had to be raised as in the meagre profits generated to service the agreement. One of the reasons for the poor economics of the project was the need for Nabalco to provide the cost of community infrastructure such as township, power supply, water supply and port in an undeveloped and remote part of Australia. This contrasted with the position of other enterprises being built in established areas, such as Alcoa's plant at Kwinana and the QAL refinery at Gladstone.

Before the feasibility study was formally submitted to the government, representatives of Nabalco divulged the alarming conclusions to Barnes in December 1967.[119] Having a 500,000-ton-per-annum plant, they informed him, was a "hopeless" proposition and government assistance would be required to see even a moderate profit.[120] The cost estimates relieved the Australian partners of any formal commitment to the project under their original 1964 agreement with Alusuisse. There were, however, two areas of strength in the report. The first was Nabalco's verification of larger bauxite deposits in the central mineral lease, at least 162 million tons. The second was the capacity of Alusuisse to absorb a much higher quantity of the bauxite and alumina output. Nabalco indicated that it could surmount its problems by building a much larger alumina plant in return for the government permitting some bauxite to be exported from the Northern Territory. The company sought early government approval of bauxite sales from Gove since Alusuisse would have to modify its European alumina plants to accommodate the different quality bauxite. In addition, Nabalco suggested that the Commonwealth government should increase its financial contribution to the costs of building the township from $A3 million to $A39 million.[121] Company representatives concluded by informing Barnes that Nabalco would go ahead at Gove with at least a 500,000-ton-per-annum alumina plant. But it needed to make a decision by 31 May 1968 on whether the Australian partners in the company would still be participating.

The prior notification to Barnes of the impending bad news on Gove saw detailed consideration by the Gorton government – particularly the departments of National Development, Territories and the Treasury – of how it should respond to the requests for assistance.[122] Barnes would later advise his cabinet colleagues: "C.S.R. are evidently coming to suspect that the Australian partners have been used as a 'stalking horse' to obtain the leases for Nabalco and now that this end is in sight, the Swiss are no longer concerned with the continued participation of the Australian Partners".[123] Without further Commonwealth action, the Australian partners would withdraw, but the Australian government had to decide whether it was worth encouraging Australian participation at Gove or whether it should simply let the contemplated Australian investment lapse. This was a political problem rather than simply a question of economic calculation. The Australian Labor Party Opposition subsequently criticised the mere reduction of Australian equity in the project as a "policy of sell-out by default".[124] Because of the clouds hanging over the project, CSR was most reluctant that Nabalco should sign an agreement with the Commonwealth. But the Gove development had been delayed so long that the Gorton government insistently pressed for a decision.

So it was that in February 1968 the Secretary of the Department of Territories, George Warwick Smith, "more or less held a gun at Vernon's head and insisted on the Agreement being signed forthwith".[125] Vernon submitted and on 23 February 1968 the agreement was signed.[126] It committed Nabalco to build an alumina plant of not less than 500,000 tons per annum capacity by the end of 1971 and to retain the existing 50 per cent Australian equity unless its feasibility study showed an expected after tax rate of return of less than 7.5 per cent. If Nabalco's drilling of the perimeter leases proved favourable and the Commonwealth agreed to its request for the extra leases, it would be required to build a plant of not less than 750,000 tons per annum. The company undertook to pay royalties of 30 cents per ton on untreated bauxite and 20 cents per ton on bauxite treated in Australia. Under Hasluck's influence, Commonwealth policy towards Aboriginals in the Northern Territory had been evolving from protection to assimilation in the 1950s and 1960s. In 1953, Hasluck had persuaded the Northern Territory Legislative Council to amend the *Aboriginals Ordinance* and *Mining Ordinance* to allow for mining on Aboriginal reserves. This was on the reasoning that the "national" interest in encouraging the development of industries based on bauxite mining equated to the interest of the Aboriginal peoples who would earn royalties from the development.[127] Under the terms of the agreement with the federal government, royalties from bauxite mining at Gove were directed to the welfare of the Aboriginal inhabitants of the Northern Territory in fulfilment of the Menzies government's belief that mining benefited all Australians, including Aborigines.[128]

Despite the agreement having been signed, it was nonetheless not a bankable document to secure the Australian share of the financing of the project. The biggest obstacle was a fundamental difference of objectives between the Australian and Swiss partners in Nabalco. This had always been latent and it came to the surface in meetings in Zurich in 1968 before and after the signing of the agreement.[129] The fundamental objective for the Swiss giant was to secure for its international operations a large and secure supply of alumina. For the Australian group of companies, on the other hand, it was necessary that the investment in the Gove project should meet their minimum rate of return. The Australian companies insisted that they would need to enjoy reasonable profitability and to have sufficient cash flow to cover the costs of interest payments on borrowed capital.

Following an inaugural meeting in Zurich in February 1968 the Australian participants in the Gove venture regrouped in Sydney. They agreed that that they would try to stay in the project by negotiating a different arrangement with the Swiss.[130] At this meeting James Vernon confirmed his opinion that the Australian partners were entitled to see a nine per cent return after tax on their investments during the first three years and thereafter to see a growth in their prospective return. CSR representatives visited Zurich again in May. They conveyed to Alusuisse their thoughts about the high costs of the project and also their opinion that the world price of alumina in Australian dollars per metric ton was higher than the price of $A51 per metric ton that the Swiss wanted to pay for Gove alumina.[131] The Swiss countered with a firm insistence that $51 was the "going price". The Swiss also warned that they had options other than Gove to obtain bauxite and alumina such as through participation with Comalco at Eurallumina.[132] Alusuisse would not agree to paying their Australian partners nine per cent return on a 500,000-ton plant and nor would they negotiate an agreement for a larger plant with the sword of Damocles (a commitment to pay the Australians a guaranteed return) hanging over their head. In these circumstances the Swiss indicated that they would have to fly to Canberra and inform the Australian government that they would not be spending any more money on the Gove project.[133]

In support of CSR and the Australian partners, the Australian government tried to intervene later in May 1968 by pressing the Swiss on the alumina price. By this time the Department of Territories had been split into a Department of External Territories and a Department of the Interior, the latter responsible for the Northern Territory. Peter Nixon, the Minister for the Interior, wrote to Nabalco's David Griffin on 28 May suggesting that the alumina price of $51 per metric ton was significantly lower than the Commonwealth government anticipated and would divert a substantial share of the profits from the development of the bauxite away from Australia.[134] Griffin countered by referring to *realpolitik* in the world aluminium industry.[135] He pointed out that the quantity of alumina produced from an enlarged plant at Gove would be one of the largest from a single alumina plant anywhere in the world. There were, he added, only four aluminium companies in the world producing sufficient aluminium metal to justify their purchasing one million tons of alumina and these companies were fully integrated. In these circumstances, it would not be possible for a vendor to offer such a quantity of alumina for sale (in excess of $1,000 million worth over 20 years) "if the price were not within the general levels paid within the industry for such quantity by fully integrated competitors".[136] Further, he argued that it would be inconceivable for any aluminium company to commit itself to purchase the whole of its requirements for alumina from a single source for so long a period and bind itself to paying more than what its competitors were prepared to pay – that would expose it to "extinction as a viable member of the industry". The price of $51 per metric ton, Griffin concluded, was not only reasonable but "would be the highest possible price that Nabalco could expect to receive for it".[137]

After the debacle of their earlier meetings, CSR and Alusuisse set about negotiating an entirely new agreement for a larger alumina plant.[138] Alusuisse and CSR hammered out a compromise in a five-page "Heads of Agreement". In doing so, CSR and its other Australian partners set aside their previous aspiration to receiving a nine per cent return on their investment. The two sides further agreed that, instead of having a single company, Nabalco, there would be a joint venture arrangement whereby ownership was held by two separate companies as tenants in common.[139] The revised arrangements were then submitted to the Australian government. The Gorton government did not like reducing Australian equity in the project. But insisting on a higher price for the alumina would see the Swiss revert to a smaller size alumina plant and the Australians partners' withdrawal.[140] The government therefore agreed to the price of $51 per metric ton of Gove alumina, bauxite being exported by the Australian partners, and a slightly higher Commonwealth contribution to the town, now known by its Aboriginal name, Nhulunbuy.

Later in 1969, local Aboriginal peoples attempted to halt the mining on the grounds that it was against the interests of the local owners of the land. In 1963, as we have seen, the Yolngu people at Yirrkala had sent a petition, painted on framed sheets of bark, to the House of Representatives, demanding rights over their land against the encroachments of mining. A parliamentary inquiry recommended that compensation be paid but not restoration of the land being excised from their reserve for bauxite mining. In 1968 the Yolngu people took the decisive step of submitting their case to the Northern Territory Supreme Court.[141] The Commonwealth government took the lead from Nabalco in fighting the claim that was built around the first articulation of the concept of native title in Australian courts.[142] Justice Richard Blackburn ruled that the land was *terra nullius* (land belonging to no one) at common law. Accordingly, he found that the Commonwealth had validly granted a mining lease to Nabalco. After Justice Blackburn's decision against the plaintiffs in 1971, the mine went ahead and Prime Minister William McMahon formally opened it on 1

July 1972.[143] The Gove alumina refinery, when later completed, was the Northern Territory's largest industrial development.

Conclusion

By the mid-1970s, Australia's known reserves of bauxite had increased from practically nothing in 1950 to what were believed to be the largest in the world. Reserves at Weipa were estimated to be in excess of 3000 million tons; Gove was thought to contain at least 250 million tons; and reserves of economic grade in the Darling Range were thought to be about 1000 million tons spread over several locations. Another significant deposit of over 200 million tons of bauxite had been proved but not developed in the Mitchell Plateau of the Kimberley District in Western Australia.

Today Australia's reserves of bauxite are second only to those of Guinea, Australia is the largest producer of bauxite in the world and produces about 20 per cent of the world's alumina each year.[144] Mining Australian bauxite from the mid-1950s led not only to major bauxite mining operations at Weipa, Gove and the Darling Range but also to significant industrial development in processing the mineral across Australia – in Queensland, Western Australia, Victoria, the Northern Territory and New South Wales. Industrial development of the scale involved in processing bauxite into alumina and aluminium had not been seen in Australia since 1915 when BHP launched its iron and steel enterprise based on coal from Newcastle and iron ore from South Australia. The major difference was that in 1915 BHP concentrated on making steel for the domestic market. By contrast, Comalco, Alcoa and Nabalco were geared for competitive export markets. As Max Griffiths has noted:

> This had enormous implications for the Australian economy and the development of the aluminium industry marked the beginning of the redressing of Australia's serious imbalance of trade. For the first time since World War II, the mining industry looked set to become a major contributor to Australia's export income.[145]

By the middle of the 1970s Alcoa's Pinjarra alumina refinery already had a capacity of two million tonnes per annum to which was added Kwinana's capacity of 1.4 million tonnes. Together, Alcoa's refineries produced one seventh of the non-communist world's alumina. Adding the QAL refinery at Gladstone, the Gove refinery in the Northern Territory and Bell Bay in Tasmania, Australian alumina refineries were supplying nearly 30 per cent of the Western world's alumina and Australia was overtaking North America as the main producer of alumina.

Mining bauxite and the massive industrial development and inflow of foreign capital that went with it were one of the most significant dimensions of the minerals boom of the 1960s and 1970s. Comalco and Alcoa were protected in their initial stages, but one of the striking characteristics of the new industrial empire based on bauxite was that, like the coal and iron ore industries, it was outward looking and, unlike most of Australia's manufacturing industry, it did not depend on protection by government. As a result of the mining and development of bauxite from the 1950s on, Australia became self-sufficient in one of the metals of modernity and a major source of the world's bauxite and alumina. Bauxite mining at Gove sparked a campaign for national Aboriginal rights that clashed directly with the interests of mining companies. The Aboriginal defence against the legal arguments of Nabalco and the federal government articulated for the first time a concept of native title for land that progressively resonated throughout Australia. As we will see, mining companies would resist the efforts of federal governments to establish a national land rights regime in the 1970s and 1980s. Eventually, however, the concept of native title would be accepted in Australian law and the mining industry would come to terms with and actually welcome the new legal regime for promoting better relations between the mining industry and Aborigines.

(Above) Bell Bay Aluminium Works, Tasmania, 1961. NAA: A1200, L39493

(left) Bauxite mining at Weipa, Cape York, Comalco. NAA: B942, ALUMINIUM [1]

5

The Boom in Other Minerals, 1966-1972

The first four chapters have described the first stage of the second Australian minerals rush as it related to coal, iron ore and bauxite. The minerals boom, as we have seen, had its origins in the late 1950s and early 1960s and would extend until 1973. This chapter, covering the years from 1966 to 1972, focuses on the discovery of resources other than coal, bauxite and iron ore, including petroleum, natural gas, nickel and uranium. It also examines several ores that had been mined from the nineteenth century – manganese, copper, lead and zinc – and their increased exports during the minerals boom due to rising overseas demand and new discoveries. The rest of the chapter discusses the political and economic impact of the minerals boom. The spectacular rise and fall of some mining stocks in the period from 1969 to 1971 accentuated public concern over the mounting foreign ownership of Australia's mineral industries. This helped pave the way for the election of a federal Labor government in 1972 with a radical and nationalist agenda for Australia's mining industries. The minerals boom left a lasting legacy – mining industries that became the backbone of Australia's exports. Mining also wrought an astonishing reversal of fortunes for a country dependent on agricultural exports and beset by chronic balance of payments problems from the 1930s to the early 1960s. In the early 1960s, Australia was deeply worried by the threat to its agricultural exports posed by Britain's application to join the European Economic Community (EEC). By the time Britain actually entered the EEC in 1973, Australia's mining exports had helped Japan overtake the United Kingdom as Australia's main trading partner.

Petroleum and Natural Gas

Exploration for oil was unfruitful in the first half of the twentieth century. It began in the Coorong in South Australia in 1881, but that well was dry, as were most of the wells drilled up to the 1950s.[1] It was not until 1961 that the first small commercial discovery of oil was made at the Moonie field 320 kilometres west of Brisbane.[2] A few years later in 1964 Western Australian Petroleum Limited (WAPET) found larger amounts of oil at Barrow Island, 50 kilometres off the Pilbara coast.[3] But in 1967, when WAPET started extracting oil from that source, Australia was still producing less than 10 per cent of its domestic oil requirements.[4] By the early 1960s, the failure to find oil had led some to conclude there was no prospect of significant discoveries in Australia.[5] Others, less pessimistic, thought that the search was faulty. Too much emphasis, they said, was being placed on the absence of surface indications of oil and funding for exploration had been inadequate.[6]

Deterred by the failure to find oil on land, some British interests believed that oil might be present in the Sydney Basin area off the NSW coast. Ian McLennan, then a senior general manager in BHP, decided to pre-empt British interests by securing exploration leases in the area ahead of the Australian company's overseas competitors.[7] In the 1950s, BHP's subsidiary, Australian Iron and Steel, explored the southern part of the Sydney Basin, drilling in likely areas. But their efforts were in vain.[8]

A few geologists were convinced that the Sydney Basin was not the right place to search. Nicholas Boutakoff, who worked for the Victorian Mines Department, suggested that exploration should be undertaken in Bass Strait.[9] An American geologist named Lewis Weeks shared Boutakoff's instincts. Weeks was born on a farm in Wisconsin on 22 May 1893, graduated with a degree in geology and then searched for oil across the world before retiring from Standard Oil of New Jersey in 1958.[10] Employed as a consultant to BHP in 1960, Weeks argued that Australia's southern waters should be explored. To conduct such a search, Weeks indicated that he would need £350,000 for a magnetometer survey and another £1 million for seismic surveys. As BHP's net profit for 1959-60 was £9.4 million, these were very considerable sums even for Australia's largest company.[11]

In deciding whether to search for oil in Bass Strait, BHP was greatly assisted by the Menzies government's *Petroleum Search Subsidy Act* 1957, which provided direct financial assistance for approved projects.[12] The government was highly motivated to help companies find oil. By the early 1960s Australia had the world's fourteenth highest oil consumption due to the rapid spread of car ownership since World War II.[13] In 1966 imports of crude petroleum and petroleum products cost the country over $200 million, a major dent in Australia's balance of payments.[14] BHP's chairman, Colin Syme, commented: "It was like putting your money on a rank outsider. Had the government incentives not been there, we would not have taken it on".[15] BHP's aeromagnetic and seismic surveys indicated the possibility of several large anticlines where oil could be trapped in Bass Strait. But even with government assistance, BHP did not possess the technical expertise to conduct more intensive searches on its own. Accordingly, BHP sought a partner from the ranks of the world's major oil companies, asking its potential partners to pay for the costs of exploration and assume all the risk. But BHP would still own half of any discoveries and only have to contribute 50 per cent of the capital costs when a production program began. Despite these strict conditions, BHP convinced Standard Oil, trading in Australia as Esso Standard Oil (Australia) Limited, as its partner. In 1964 negotiations were finally concluded for a 50-50 partnership between BHP and Esso.

In that same year, Esso exploration vessel, *Glomar III*, struck gas in a well on the East Gippsland Shelf. Renamed Barracouta 1, this well was about 24 kilometres offshore and was subsequently estimated to contain 52,600 cubic metres of natural gas.[16] This gas would provide the basis for significant industrial development in Australia. McLennan recalled:

> I'll never forget the message I received that day. The gas had just blown them out of the water. We were looking for oil, and I guess

> we were just about as surprised as anyone when we found gas. We were at a bit of a loss what to do with it.[17]

In 1965 the BHP–Esso partnership made a second major finding of gas in the Marlin field, about 100 kilometres off the Victorian coast. Then, in 1967, oil was discovered at Kingfish, 88 kilometres from the coast. Initial estimates were that Kingfish held recoverable reserves of 1130 million barrels of oil. In July of that year more oil was found at Halibut, 72 kilometres off the coast. Oil fields discovered later were the Cobia Field in 1972 and the Kingfish, Mackerel and Fortescue Fields in 1978. Oil was transported from the wellhead to oil refineries by sea or by pipeline. The most important pipeline ran from Bass Strait to Melbourne and Geelong.[18]

The Commonwealth helped the BHP-Esso partnership by setting an industry-wide price of oil at $US3.47 per barrel. This was considerably above the price of Middle Eastern oil, which was then less than $US2 per barrel. For decades the world price of oil had lagged behind inflation. But in 1960 the major oil producing countries – Iran, Iraq, Kuwait, Saudi Arabia and Venezuela – formed the Organization of the Petroleum Exporting Countries (OPEC). Its aim was to raise the royalty and tax rates private oil companies paid to host countries. From 1960 onward Western economies became increasingly dependent on oil from the Middle East, the prices of which were increasingly influenced by the policies of the OPEC cartel. With an anticipated flow from Bass Strait of 350,000 barrels per day, the Bass Strait discoveries converted Australia from almost total dependence on foreign crude-oil supplies to around 70 per cent self-sufficiency. This development would prove particularly beneficial to Australia following the OPEC-engineered oil price rises in the 1970s. By 1978 Bass Strait accounted for 93 per cent of total oil production from three Australian oil fields – Bass Strait, Barrow Island and the Moonie–Alton field in Queensland.[19] Recoverable reserves amounting to 2100 million barrels of oil saved Australia more than $1.5 billion per year in foreign exchange by the mid-1970s.[20] By virtue of the Bass Strait discoveries, BHP the steelmaker was now half-owner of significant underground reservoirs of oil and natural gas. Ian McLennan recorded in 1979:

> There are three major dates in the history of the [BHP] company. The first was when Broken Hill was found; the second was when the decision was made to start in steel; and the third was when it decided to go into the oil business. The entry into oil and gas had a most major effect on the company. It's made it possible to develop in all sorts of other ways. It's kept it as the largest company in Australia.[21]

Bass Strait natural gas was found as a by-product of the search for oil. The situation was similar in South Australia when a company established to find petroleum ended up finding gas. Although the South Australian Department of Mines had been adamant that South Australia had no potential for hydro-carbons, a dissenter was Reginald Sprigg, one of the department's own geologists. Sprigg set himself up as a consultant to a new company established by a member of the well-known Bonython family to search for oil – South Australian and Northern Territory Oil Search (Santos). Sprigg encouraged Santos to focus on the Cooper Basin, an area of about 100,000 square kilometres in the north-west corner of South Australia.[22]

Sprigg's reports on the geological structure of the Cooper Basin received such favourable attention in the United States that several American companies vied with each other to partner with Santos. The capacity of Santos to explore the Cooper Basin was greatly reinforced when ultimately it invited Delhi Corporation of Texas to become a partner.[23] In 1963 the Santos-Delhi partnership identified a large reservoir of gas at Gidgealpa in the north-west corner of South Australia. Two and a half years later, another much larger reservoir of gas was found a short distance to the south of Gidgealpa at Moomba. Following these discoveries, the South Australian Gas Company signed a 20-year contract for Santos to supply gas to Adelaide. The distance from Moomba to Adelaide was 800 kilometres, not much less than the distance from Sydney to Melbourne. A pipeline was constructed over this distance that by 1969 was supplying natural gas from the Cooper Basin to Adelaide. The Australian Gaslight Company later signed a 30-year contract for Santos to supply gas from Moomba to Sydney after construction of a 1360 kilometre pipeline, the largest then constructed in Australia.

By 1969 natural gas was being supplied to customers in Brisbane, Melbourne and Adelaide and became available in Perth at the end of 1971. It would not be until 1976 and the completion of the pipeline from Moomba to Sydney that Australia's most populous city would be connected to natural gas. Brisbane was supplied from a small field at Roma; the major regional centres in Victoria were supplied from Bass Strait; and Perth received its gas from a field at Dongara, 351 kilometres north-west of Perth. Adelaide and Sydney would both draw their supplies of gas from the Cooper Basin gas fields.[24] Government instrumentalities owned the pipelines between gas fields and major population centres: the Gas and Fuel Corporation of Victoria had one between Bass Strait and Melbourne; the Pipeline Authority of South Australia owned another from Moomba to Adelaide; and the Commonwealth Government, through the National Pipelines Authority, would eventually take over the Moomba to Sydney pipeline in the 1970s. The discovery and extraction of natural gas brought with it significant changes to Australian industries and households.[25] In time natural gas, like coal, would become a significant export industry after world-ranking deposits of natural gas were discovered off the North West Shelf off Western Australia. The significant development of the North West Shelf will be dealt with in Chapter Seven.

Manganese, Lead, Zinc and Copper

The most dynamic of the mining industries in the Australian minerals boom were the new exports, iron ore, coal and alumina. The boom, however, also extended to ores that had been mined and exported since the nineteenth century. Chief among them were copper, manganese, lead and zinc. Australia had exported non-ferrous metal ores – that is, metal ores that did not contain iron – from the nineteenth century. These constituted only 3.5 per cent of all exports in the early 1960s.[26]

From 1840 to 1880 more than one thousand copper deposits were found in Australia. Although many of them were worked, annual production of copper did not reach 20,000 tons until 1899.[27] Between 1899 and 1920 production exceeded 20,000 tons but after 1919, as prices plunged, it fell significantly and fell below 10,000 tons in 1926. In the mid-nineteenth century Australia's copper production came from mines in South Australia such as Burra and Moonta. A mine at Cobar in New South Wales assumed importance in the 1870s and the Lake George mine at Captain's Flat, south of Queanbeyan in New South Wales, began production in 1882. In 1896 copper mining commenced at Mount Lyell in western Tasmania.[28] In 1939 production of copper in Australia was 21,310 tons but consumption throughout the war ran at 30,000 tons annually. The post-war resurgence of copper began at Mt Isa in western Queensland. Mount Isa Mines had realised in the 1930s that the copper there was at least as important as the lead-zinc-silver lodes.[29] The reason for the delay in mining copper until the war was lack of money. But in the more prosperous early 1950s, when a new copper smelter became operational, Mt. Isa became the largest copper mine in Australia.

Other significant producers of copper after World War II were Peko-Wallsend Investments and Cobar Mines Pty Ltd. Peko-Wallsend was formed in 1961 from a merger between Peko Mines N.L. and Wallsend Holding and Investment Co. Ltd.[30] The company had interests in coal at Cessnock and rutile and zircon at Newcastle. Rutile occurred in beach mineral sands and is used in the manufacture of pigment and for the production of titanium metal. Resistant to corrosion and heat, zircon is used in engines, electronics, spacecraft and the ceramics industry. Peko-Wallsend also mined copper and gold at Tennant Creek in the Northern Territory. Cobar Mines was formed in 1957 from a partnership between Broken Hill South and CRA. But the most significant discovery of copper in the 1960s occurred not on mainland Australia but in the Australian external territory of Papua New Guinea (PNG). CRA had begun searching for copper in eastern Queensland in 1960. But a company geologist, Ken Phillips, persuaded CRA that Papua New Guinea offered a more favourable area to explore for ore-bodies.[31] In 1963 CRA obtained a Special Prospecting Authority to search in the area of Panguna on Bougainville Island, part of PNG.[32] By applying traditional exploration techniques including stream sampling, geochemical study of soil and diamond drilling, CRA geologists discovered a major deposit well known overseas as "porphyry copper" on the western side of the island 12 miles from Kieta.[33]

By late 1966 ore reserves of 900 million tons had been identified in Bougainville. In the following year CRA proposed a large-scale copper mining operation of the same order of magnitude as for the Hamersley and Mount Newman iron ore projects – with a capital investment of $135 million.[34] The project offered tantalising benefits to the Territory: $10 million per annum revenue when fully operational, export earnings of about $50 million per annum, and an opportunity for the Territory government of 20 per cent equity. The project promised roads, power, a port for general use, and employment and training for a substantial number of locals.[35]

Bougainville Copper Pty Ltd (BCPL), a subsidiary of CRA, negotiated long-term contracts with smelters in Japan, Germany and Spain for sales of copper concentrates with a total copper content of nearly two million tons over 15 years. To run this operation CRA and RTZ set about borrowing $US250 million from a consortium of 27 banks in eight countries. The consortium made two separate loans: the first an intermediate loan of $US154 million directly to BCPL; and the second a long-term loan of $92 million that would be provided to the Commonwealth Trading Bank of Australia, which would then pass on the equivalent in Australian dollars to BCPL.[36] The Bougainville copper mine became for a time the largest in the world and was one of CRA's best performing assets in the 1970s.

A discovery in 1960 in the Gulf of Carpentaria put an end to another deficiency in Australia's mineral economy. Manganese is an essential element in the production of steel and has other important uses in the manufacture of dry cells and batteries and the chemical industry.[37] Manganese ores had been mined in Australia since 1882, but until the 1960s annual production had been relatively small. Before World War II production was less than 900 tons per year.[38] Manganese ores are relatively common, jet-black and conspicuous. The largest deposits of manganese discovered in Australia were along a coastal plain on the west side of Groote Eylandt, an island in the Gulf of Carpentaria of about the same size as the Australian Capital Territory. In November 1960 two geologists, P.R. Dunn of the BMR and P.W. Crohn, then a Senior Resident Geologist of the Northern Territory Administration, arranged to assay the Groote Eylandt deposits. The results of the testing were so encouraging that in 1963 BHP became interested and organised a larger program.[39] In October of that year, Sir Ian McLennan announced that a manganese mine would be developed and in 1964 a wholly owned subsidiary, Groote Eylandt Mining Company (GEMCO), was set up. GEMCO was granted leases over about 34 square kilometres in return for a commitment to construct facilities for producing up to 70,000 tons of manganese ore annually by 1966. In return for the extension of the leases BHP established a ferro-manganese furnace at Bell Bay in Tasmania for supply to the steel industry.[40] GEMCO loaded the first shipment of manganese ore for smelting at Bell Bay in March 1966. By the end of 1966, 130,000 tonnes of ore had been shipped, of which 32,000 tonnes went to Japan.[41] Within a decade BHP had invested $70 million in Groote Eylandt and had become the world's third largest exporter of manganese, selling 75 per cent of its output to 13 countries. Japan was the largest consumer, taking 850,000 tonnes per year.[42] Well before such enlightened attitudes were fashionable in the Northern Territory, GEMCO enforced a radically inclusive policy on Indigenous labour. Aborigines were employed from the outset, at first as guides and surveyors' assistants. In due course they operated trucks, loaders and rail haulage trains and helped in the maintenance of towns. From the beginning, Aboriginal workers received equal award wages to those of Europeans performing the same work and they lived with white workers in the single men's barracks.[43] If they wished, they were entitled to all other regular benefits such as annual leave and travel allowances. White workers who made offensive remarks about Aboriginal workers were instantly dismissed. Historian Alan Trengove concluded that "[n]o policy of indiscrimination has ever been more stringently applied" than at GEMCO.[44]

Until the minerals boom of the 1960s and early 1970s, mid-twentieth century Australia was known in the minerals world mainly for its production of lead and zinc. In the 1960s Australia was the largest lead producer in the non-Communist world and the third largest producer of zinc after the United States and Canada (fourth if the Soviet Union was included).[45] From the late nineteenth century, Broken Hill had been the major contributor to Australia's output of lead and zinc with combined production reaching 182,703 tons in 1910. The next largest was Mt. Isa Mines Ltd at 64,000 tons.[46] The Tasmanian Read–Rosebury mines of Electrolytic Zinc of Australia on Tasmania's west coast were a major supply source. In the 1950s, lead and zinc were Australia's second largest earners of US dollars. But international competition for lead and zinc markets was strong. Towards the end of the 1950s the Australian lead and zinc industries were troubled when the United Kingdom stockpiled lead and dumped its surplus supplies onto world markets, and again when the United States threatened to impose a protective tariff on both commodities. To prevent this action, Mt Isa Mines agreed voluntarily to reduce its output of lead and zinc by one quarter in 1959.[47]

During the course of the minerals boom iron ore became by far the largest ore exported. But ores as a group, including the new iron ore export industry and the old lead and zinc industries, increased as a proportion of all exports from less than 10 per cent in 1968-69 to 14 per cent in 1970-71 and averaged more than 12 per cent in the first three years of the 1970s. As the economic historian Brian Pinkstone has observed, this made ores as important in Australia's overall export mix as wool.[48]

Nickel and Uranium

Nickel, which had not previously been mined in Australia, became front-page news in Australia in April 1966. In that month the Australian gold-mining company, Western Mining Corporation, announced that it had drilled through a high-grade deposit of nickel at Kambalda, south of Kalgoorlie in Western Australia.[49] At the time of the discovery, three places dominated world nickel production – Canada, New Caledonia and the Soviet Union. Canada was the largest producer in the world and the International Nickel Corporation (INCO) was Canada's largest nickel miner, accounting for 80 per cent of Canadian production with Falconbridge and Sherritt Gordon supplying the rest. The Soviet Union was the second largest nickel-producer followed by New Caledonia, which supplied ore for the French company Société le Nickel. Canada and New Caledonia between them held more than half of the world's known nickel reserves and nearly 70 per cent of the Western world's.

In the 1960s technological changes led to rising world demand for nickel, a metal that was mainly used in the production of stainless steel but also in alloys and nickel-plating and the aerospace and munitions industries. Nickel also had important strategic applications at a time when the United States was fighting in Vietnam in the period. In the 1960s a surge in the worldwide demand for nickel led to shortages and a 52 per cent price rise. At the same time as demand was increasing, supply was tightened by a major strike of INCO's workforce in 1966 and an even bigger one in 1969 that Falconbridge miners joined. The price of nickel was $US992 per ton in 1950; and it had risen to $US1940 in 1967.[50]

For many years INCO dominated the technology of nickel refining, which was undertaken in three steps: mining; smelting to create a sulphide matte; and refining to produce nickel powder or briquettes. In the late 1960s INCO started exploring for more reserves in Indonesia, Guatemala, New Caledonia and Australia, using aircraft with an electro-magnetic sensor that detected changes in the conductivity of the earth. Other nickel companies also intensified exploration. Falconbridge began searching in the Dominican Republic. But Freeport Minerals Corporation of the United States, which had entered the nickel industry in the 1930s, took a minority interest in Metals Exploration Ltd and collaborated with it in a joint venture that gave it control of the Greenvale nickel deposits in Queensland.

Western Mining Corporation (WMC), as we have seen, was the company that helped develop an aluminium industry based on WA bauxite. The company had brought gold mines into production in Kalgoorlie in Western Australia in 1934 and established a subsidiary, Central Norseman Gold Corporation N.L., in 1935. WMC's gold production played a significant part in helping Australia recover from the Great Depression. But by the 1950s the gold industry had declined because its price remained fixed while production costs rose. By that time WMC had become mainly Australian-owned as a result of laws that exempted Australian gold mining companies from taxation. In 1973 Arvi Parbo, the managing director of WMC, argued that tax concessions for the gold industry designed by Labor governments in the 1940s had greatly helped in "buying back the farm":

> … in the period between the close of the Second World War and the commencement of the recent mining boom. In the case of Western Mining Corporation Limited there was a change from almost complete foreign ownership in 1948 to a substantial majority of Australian ownership by 1966 because of taxation provisions which were more favourable to Australian shareholders than to overseas shareholders.[51]

From the 1950s onward WMC embarked on a strategic program of diversification, one of the results of which was its interest in bauxite. The company's exploration philosophy stressed the need for detailed mapping. As senior WMC geologist Ray Woodall, who became a legend because of his later Olympic Dam discovery, explained:

> An ore body is a most abnormal concentration of metal or mineral. Such concentrations are not only microscopic in size by comparison with the vastness of the Earth's surface, but are also extremely rare. If the search for concealed ore is ever to become something more

> than a losing gamble for all except those heavily endowed with luck, methods must be found which will reduce the size of the target area to manageable limits without the expenditure of vast sums of money.[52]

By 1962 the company was involved in the detailed geological mapping of Kalgoorlie. Two years later Woodall wrote a memorandum to the company's superintendent in Western Australia, Laurence Brodie-Hall, pointing out that the Eastern Goldfields of Western Australia displayed all the geological features thought to be essential for the presence of base metals. In the meantime, Woodall was approached by a Western Australian farmer, George Cowcill. In the process of prospecting for uranium near Kambalda, Cowcill had found metalliferous samples which proved to be nickel.[53]

At that time the nearest economic deposits were in New Caledonia. Excited by Cowcill's discovery, Woodall persuaded WMC to give its official approval to exploration for base metals in Western Australia and to finance the geological mapping of Red Hill near Kambalda. In 1965 WMC received approval from the WA government for extensive reserves covering the potential for nickel areas extending south and south-east of Kambalda.[54] In the meantime, Arvi Parbo, then deputy general superintendent of WMC, began investigating the economics of mining and transporting the ore and to determine which types and grades might be commercially attractive.

Destined to become chairman of WMC and later of BHP, Arvi Parbo was born at Tallinn in Estonia in 1926. Fleeing the advancing Russian army in 1944, he was separated from his parents and put in a displaced persons camp after which he completed his schooling in Germany and studied at the Clausthal Mining Academy. The young man migrated to Australia in 1949, doing a variety of jobs before winning a Commonwealth Scholarship to Adelaide University. After graduating as a mining engineer in 1956, he began working for WMC.[55]

Drilling began in January 1966 at Red Hill and produced sulphides that contained 8.3 per cent nickel. The leaders of WMC at the time were veteran miner Sir Gordon Lindesay Clark, Laurence Brodie-Hall and the English-born mining engineer Bill Morgan, the managing director. By the 1960s Clark was one of Australian mining's wise men with a great faith in scientific exploration. Laurence Brodie-Hall had come to Australia at the age of 14 and took a range of menial jobs before training as a mining engineer.[56] Bill Morgan had prospected for gold in Bougainville and worked for the State Electricity Commission in Victoria before joining WMC. He was also one of the first Australians to appreciate Japan's potential as a market for Australia.

Clark, Parbo, Morgan and Brodie-Hall knew that, while there might be a temporary world shortage of nickel, it would not be long before the established producers increased their output to meet the demand. They had to decide whether to risk developing the Kambalda site, which might not prove as substantial as the company anticipated. So, when the initial drilling results were positive, the company faced two choices. The first was to hope there was significant nickel below the surface and get into production as quickly as possible, establish WMC in the world nickel market, and generate an immediate cash flow. The other option was to wait for further drilling before sinking a prospecting shaft. Clark and the WMC board agreed in taking the risk of the first option. Construction of a shaft began on the Lunnon Shoot in July 1966 in a project that proved highly lucrative for WMC. Bill Morgan helped his company by negotiating long-term sales contracts, which were then a rarity in world nickel markets. As a result of his efforts, WMC announced on 18 May 1967 that it had signed a contract with Japan's Sumitomo Metal Mining Co. Ltd for the sales of concentrates with a total nickel content of about 40,000 tons over 10 years.[57]

The decision to embark on an intensive program of development made it necessary for WMC to secure finance. Clark believed that loan finance would expose the company to unacceptable risk and took the view that a share issue was the correct way to proceed. The world market in nickel was very competitive and Clark considered that WMC was able to enter it only because of a temporary shortage in supply. Accordingly, in September 1966 WMC made a share issue of 2.5 million 50-cent shares in the proportion of three for ten at a premium of $2.50. Of the $7.5 million raised in this manner, $5 million was required for the opening of the nickel mine at Kambalda, a plant to produce nickel concentrate, and to continue exploration and prospecting.[58] By August 1967, infrastructure had been established at Kambalda. A powerhouse had begun generating electricity and, by the end of the year, houses, a supermarket, a post office, temporary school-rooms and medical facilities had been constructed. WMC shipped its first nickel concentrates to Canada and Japan in August 1967 in addition to 10,000 tons of concentrates to the Sherritt Gordon refinery in Saskatchewan for refining. WMC's first nickel contract in Australia was to supply the Royal Mint with the metal for alloying in their coins, and in September 1967 WA Premier David Brand opened the Kambalda project, the first commercial nickel mining operation in Australia.

WMC's timing was fortunate. Its nickel operation came into production when consumers, particularly in Japan and Europe, were having difficulty securing supplies. In January 1968 WMC raised $46 million through another share issue. This enabled it to build a nickel refinery at Kwinana to produce nickel powder and briquettes, as well as by-products of ammonium sulphate, copper sulphide and nickel-cobalt sulphide. WMC chose Kwinana as the site for the refinery because of its proximity to Perth, infrastructure and access to vital raw materials and water supply. Bechtel Pacific Corporation began construction of the refinery in 1968 and completed it in 1970. WMC then began building a nickel smelter that was brought into operation in December 1972 with a capacity of processing 200,000 tons of concentrates per annum.

By the early 1970s nickel had taken gold's place as WMC's principal product. The first discovery of nickel at Kambalda resulted in a string of other nickel discoveries from Norseman to Wiluna and by 1970 Western Australia's nickel output was $A87.4 million. Discoveries of nickel along with other minerals, such as iron ore and bauxite, saw the market value of mining in Western Australia surge to $A579.4 million in 1970. The bonanza wrought dramatic changes in Western Australia's economy. In 1963–64 mining and quarrying accounted for 6.2 per cent of all recorded production in Western Australia and 11.6 per cent of primary production. By 1974-75 these figures were an astonishing 30.9 per cent and 48.8 per cent respectively.[59]

The additional search for uranium in Australia to replace the former sources at Rum Jungle and Mary Kathleen was hampered by an embargo on exports imposed in 1967 by David Fairbairn, the federal Minister for National Development. But about 60 companies were exploring for, or about to explore for, uranium in Australia. By this time, advanced geophysical techniques had displaced the Geiger counters or hand-held scintillometers used by prospectors in the 1950s. In this new round of exploration, uranium was discovered first at Lake Frome, 500 kilometres north of Adelaide. Then, after drilling around the original area of the Mary Kathleen mine in Queensland, CRA decided to re-open the mine in 1970. In September of that year, a uranium discovery was made in Australia that paralleled the world-ranking discoveries of bauxite in the 1950s and iron ore in the early 1960s. The Australian company Queensland Mines, after only six months of prospecting, found "the world's richest uranium strike" on an Aboriginal Reserve in Western Arnhem Land in the Northern Territory.[60] By comparison with most of the world's uranium deposits, which averaged between two and 10 pounds of uranium oxide per ton, Queensland Mines announced that its deposits contained 55,000 tons averaging 540 pounds per ton, with an estimated value of between $600 and $700 million.[61] Queensland Mines named the deposit "Narbalek" after the Aboriginal word for the small kangaroos that lived in the area.

At about the same time two other Australian companies working together, Peko Mines NL (later Peko-Wallsend) and the Electrolytic Zinc Company of Australasia (EZ), announced the discovery of a 70,000ton deposit of uranium at Ranger, 50 kilometres south of Narbalek.[62] This deposit proved to be part of a series of deposits containing about 100,000 tons in all. Six months later, Noranda Australia, a subsidiary of a Canadian copper company, discovered a smaller 21,500-ton deposit 20 kilometres south-west of Ranger at a place that became known as Koongarra.[63] Finally after exploring throughout 1971 and 1972 Pancontinental Mining and Getty Oil found about 24,000 tons of uranium, averaging 8 pounds per ton, at Jabiluka, 20 kilometres to the north of Ranger. Taken together, a group of mainly Australian-owned companies had confirmed that the Alligator Rivers region of the Northern Territory was a uranium province of world importance.

Economic Change and the Australian Mining Industry Council

Writing in 1971 in a book he called *The New Australia*, Colin Simpson argued:

> Australia has been virtually rediscovered in the past ten years. The new exploration of the continent ... has uncovered vast hordes of mineral wealth, along with reservoirs of oil. It is no overstatement to say that the country that has stepped affluently into the seventies is a New Australia.[64]

In the decade leading up to 1974-75, the contribution of the manufacturing sector to Australia's gross domestic product (GDP) fell from 26.7 per cent to 23.3 per cent. In part this reflected the phenomenon named after Australian National University economist, Bob Gregory – the "Gregory effect". Gregory's 1976 analysis of the minerals boom of the 1960s was that increased investment in mining drew money from other sectors of the economy, the chief victim being manufacturing for export.[65] The share of the rural sector in Australia's gross domestic product also fell from 12.4 per cent in 1964-65 to 6.5 cent in 1974-75. Over the same period the mining industry's share of GDP rose from 1.8 per cent to 4.4 per cent.[66] This increase underestimated the significance of the minerals sector.[67] Between 1964 and 1969 the value of mineral exports, other than petroleum products, grew from $A220 million to

$A695 million. As a percentage of merchandise exports the growth was from eight per cent to 21.5 per cent.[68] In 1968-69 these exports were equal in value to the combined totals for wheat, flour, meat, sugar and butter. Almost 70 per cent of the rise in mineral exports between 1963-64 and 1969 came from three sources: iron ore, coal and bauxite, the industries that laid the foundation for the minerals boom. Growth also occurred in other minerals such as lead, zinc, copper and mineral sands. In 1972 one quarter of Australia's exports of minerals and fuels consisted of non-ferrous metal ores and concentrates; two-fifths were of iron ore; 6.7 per cent comprised petroleum products; and about another one-fifth was made up of coal and coke.[69] The production of minerals in the 1960s and 1970s was highly capital-intensive, requiring much more than the pan and shovel of the gold prospectors in the nineteenth century. Neither was there any large influx of immigrants doubling the existing population as had happened in the 1850s. But as Ian McLean has argued:

> the indirect and longer term effects on the labour market were qualitatively similar. The high levels of profitability in mining resulted in wages in that industry becoming the pace-setters. The indirect demand for labour (infrastructure development, new towns to be constructed in remote mining locations, other inputs and supplies sourced from within Australia) ensured the spread to other sectors of higher wages and hence prices.[70]

Another effect of the minerals boom was to accentuate the fundamental change in trading patterns that had been developing in the 1950s and 1960s. The Japanese-inspired minerals boom of the 1960s and early 1970s saw a significant transformation in the composition of Australian exports. In 1954-55 the rural sector accounted for 80 per cent of Australian exports. In 1972-73 the rural sector still accounted for the majority of exports, at just over 50 per cent. But iron ore, the leading mineral export, made up 7.1 per cent, coal 4.5 per cent and alumina 2.5 per cent.[71]

Taken as a proportion of export of goods and services, the rural sector accounted for 43.6 per cent in 1972, manufactures 19.5 per cent and services 16.3 per cent. The share of mining in Australia's exports had risen to 18.9 per cent, still less than half of rural exports but only slightly less than manufactures.[72] Manufacturing exports increased from 7 per cent of merchandise exports in 1950 to 19.4 per cent of exports of goods and services in 1975.[73] Many of these manufactured exports were in the category of "simply transformed manufactures" – resources that required little processing such as pig iron, paper, clay, bricks and plaster.[74] From 1975 the share of manufacturing in Australia's merchandise exports would decline while mining's share would rise.[75] The lasting result of the first stage of Australia's second minerals rush, namely the boom from about 1960 to 1973, was that the minerals industries established themselves again as one of the vital export sectors of the Australian economy.[76] The growth of mining came, moreover, at an opportune time when many of Australia's traditional rural industries faced considerable difficulty. Wool had to contend with increased competition from synthetics and rising costs. From dominating Australia's merchandise exports in the mid-1950s, its share had fallen to less than one fifth in 1972.[77] World markets for wheat were glutted and held out prospects of lower returns, while dairy products and sugar had to compete with competition from highly protected overseas markets and the loss of British markets as the United Kingdom moved into the European Economic Community.[78]

The mining sector's share of foreign exports would grow even larger in the 1970s as resource projects that had been initiated in the previous decade reached their full potential. By 1978 mining's share of total exports of goods and services would reach 29 per cent as the rural sector's share dropped to 37.7 and manufactured exports to 15.8 per cent.[79] The mining boom of the 1960s and early 1970s also had an extraordinary effect on Australia's balance of trade. Australia had a trade deficit of $218 million in 1967-68. But as the value of mineral exports increased, Australia earned surpluses of more than $400 million in 1969-70 and 1970-71, $931 million in 1971-72 and $2138 million in 1972-73.[80] Without the export income from the minerals industries, Australia would have been much less prosperous in the 1960s and 1970s, especially when the import savings from the production of Australian crude oil are also considered.[81] The subject of mining did not appear in the index of Donald Horne's analysis of Australia in 1964, *The Lucky Country*.[82] Horne's classic ended on the pessimistic note that Australia's luck was running out and that "if things go on as they are Australia will slip down the per capita national income scale".[83] After the ensuing six years of minerals boom, however, Horne wrote about a country saved by six years of minerals boom:

> if the minerals rush hadn't happened, Australia would by 1970 have been a country in economic crisis. There would have been no compensation for the sluggishness of farm exports. The point would have been reached which I forecast for the end of the 1960s in *The Lucky Country* – the beginning of a serious fall down the international prosperity ladder.[84]

In those six years, capital inflows had increased markedly. There were important differences between the late 1960s and early 1970s and other periods when Australia had borrowed heavily, such as the 1880s and 1920s. In those earlier periods governments had been responsible for much of the borrowing. In the late 1960s and into the early 1970s, however, the borrowing was by private companies, now corporations. The Australian Bureau of Statistics (ABS) estimated that between 1963-64 and 1972-73 there had been $2062 million of foreign direct investment in the mineral sector (including smelting, extracting and refining). It also recorded that foreign portfolio and institutional investment into Australia over the same period amounted to $2847 million. The economist R.B. McKern estimated that as much as 70 per cent of this portfolio investment was directed towards the minerals industry. An increasing proportion of foreign investment in the 1960s and early 1970s was by US companies; mining was a major agent of Americanisation in Australia. In 1963-64, private capital expenditure in mining represented only 11 per cent of total manufacturing investment; by 1968-69 its share had jumped to 50 per cent.[85] A corollary of this increased foreign investment was higher foreign ownership of the mining industry. The ABS estimated that foreign ownership of the mining industry increased from 27.3 per cent to 51.8 per cent between 1963 and

1974-75. US interests trebled their ownership from 9.2 per cent in 1963 to 28.7 per cent in 1974-75. By 1971-72 investment in coal and iron ore accounted for 40 per cent of foreign direct investment and 27 per cent of total private capital inflow.[86]

The minerals boom of the 1960s and 1970s also had the effect of loosening the economic and political ties between Australia and the United Kingdom and strengthening the linkages between Australia and Japan. In the period from World War II through to the 1960s the Anglo-Australian economic nexus had been maintained by Australia's membership of the sterling area – a group of countries that pegged their currency to the pound sterling. Thus when the British government devalued the pound sterling in 1949, Australia had followed suit and loyally imposed import restrictions on dollar imports to conserve the sterling area's reserves. Australia's membership of the sterling area was not purely sentimental; it was backed up by strong trade ties because for 70 years after Federation the United Kingdom had been Australia's biggest trading partner. In the early 1950s Western Europe took 50 per cent of Australia's exports and the United Kingdom, on its own, took 33 per cent.[87] Australia's dependence on the British market explains the degree of Australian anger and dismay when in 1961 the Macmillan government sought membership for the United Kingdom of the European Economic Community (EEC). The Menzies government reacted strongly to the threat both of lost markets for rural products in Britain and the danger posed by the EEC's highly protectionist Common Agricultural Policy. In 1961, however, the government could not foresee how completely the minerals boom would transform Australia's economy. By the early 1970s Japan had displaced the United Kingdom as Australia's main export market and Australia's economy was immeasurably stronger because of the dynamism of the mineral export industries. Whereas in 1950-51 Britain took a third of Australia's exports and Japan six per cent, in 1973-74, when Britain finally joined the EEC, Japan took 31 per cent and Britain a meagre seven per cent.[88]

The British economy weakened in the 1960s as a result of the British government's forlorn efforts to maintain sterling as an international currency and to retain many of its international commitments abroad. At the same time, Australia's economy strengthened on the back of the minerals boom. In 1966 the British government introduced a system of voluntary restraints on British companies such as RTZ in order to try to slow down the flow of British capital overseas.[89] The British government argued that the strain on the United Kingdom's balance of payments was so severe that even Australia's large international reserves of sterling were no counterweight.[90] But British notice to Australia of the curbs prompted vociferous complaints from Australian Prime Minister Harold Holt and his Treasurer William McMahon. McMahon complained that the British restraints on investment represented the "final tolling of the bell for the Sterling Area" and asserted that they amounted to the severing of "that last substantial link which has held the Sterling area together".[91] As Charles Johnstone, the British High Commissioner in Canberra, explained:

> Although of every £10 invested in Australia £9 are raised on the Australian capital market, the remaining £1 is crucial in that it presents long-term risk capital with an initial low dividend yield which could hitherto only be raised overseas. Our own voluntary programme, coupled with restraints on the outflow of United States capital, is therefore regarded by Australians as an obstacle in the way of the development of their mineral resources from which they hope to derive a substantial part of their export income in the future ....British restraints on direct investment, if prolonged indefinitely, would bear heavily on Australia's capacity to expand existing industries and establish new ones. Any move which we might make towards the resumption of the inflow of British capital into Australia would compensate in some measure for the damage which British entry into the EEC is bound to inflict on Australian overseas trade.[92]

Australian complaints led to the British introduction of a milder version of capital restrictions. Johnstone noted around the middle of 1967 that Australia's search for new markets had brought about significant changes in the pattern of Australia's foreign relations. Johnstone added:

> The publicity given to the iron ore developments in the North-West has concentrated attention on new and broader horizons of foreign trade. In 1961–62 iron ore shipments did not figure at all in the trade returns of Australian exports; this year, even though shipments only began within the last year or so, iron ore accounts for more than 1 per cent of total Australian exports, and this percentage will climb steeply as the mining concerns fulfill contracts for the export of some 300 million tons over the next 16 years. Meanwhile, the percentage share of exports held by coal has almost doubled with increased shipments to Japan.[93]

The mining boom's impact on Australia was reflected in the Holt government's reaction to the British government's decision to devalue sterling by nearly 15 per cent on 19 November 1967. The British initiative was of worldwide significance and confronted all major trading nations with a choice; they could either adjust their currency downwards in line with sterling or retain their existing exchange rates.[94] The Country Party and John McEwen's Department of Trade and Industry wanted Australia to devalue in line with the pound to support rural primary producers who were hard pressed both by falling commodity prices and the imminent prospect of Britain's entry into the EEC. If the Australian government were to follow its traditional path, it would have emulated Chifley's Labor government, which had devalued the Australian pound when Britain had last devalued in 1949. But this was not the path recommended by the Treasury and McMahon. In contrast to the situation in 1948-49 when Australia and 65 per cent of world trade devalued along with the British pound, none of the major Western economies – the United States, the members of the European Economic Community, Canada and Japan – devalued. McMahon convinced his cabinet colleagues that devaluing was not in Australia's interests because it would add to the cost of Australian imports from countries other than Britain. As Holt explained it, only 13 per cent of Australian exports went to Britain in 1966 and, all told, the countries that had devalued took only 20 per cent of Australia's exports.

The other strong argument against devaluation in 1967 was the favourable balance of payments outlook brought about by the mining boom.[95] Holt explained that Australia's reserves in gold and foreign exchange amounted to $1070 million and its drawing rights with the IMF were $630 million, making Australia's reserves $1700 million.[96] Holt and McMahon argued essentially that mineral exports, import savings from the discovery of indigenous oil and capital inflow meant that Australia's balance of payments would remain healthy and that Australia need not devalue. It was significant that one of Australia's largest mining companies, CRA, supported Holt in 1967 in not devaluing the currency against sterling.[97]

At the time of the sterling devaluation in 1967, Australia's mining industries came together to form the Australian Mining Industry Council (AMIC). The establishment of the organisation had much to do with conflicts between the new iron ore companies in the 1960s and their need to present a more harmonious front to the federal government.[98] In 1972 McMahon, by then Prime Minister, acknowledged the utility of the new organisation:

> For our part we have for a considerable time now conducted regular consultations at Cabinet level with the organisations representing the major groups in the economy. Your Council has become a more important part of these regular economic consultations in recent years, reflecting the dramatic increase in its contribution to the national economy. We consider this direct consultation to be an important two-way stream between the Government and other component parts of the economy. We want it to continue and develop and I am sure you do too.[99]

In December 1971 the United States devalued its dollar by 8.6 per cent as part of an agreement with Japan under which the latter increased the value of the yen by 17 per cent against the US dollar.[100] These currency changes were part of the international crisis that had led in August 1971 to the ending of the Bretton Woods system of fixed exchange rates and of convertibility of gold with the US dollar. In these circumstances the Australian mining industry and AMIC waged a strong campaign to oppose revaluation of the Australian dollar at a time when two-thirds of

Australia's mineral exports were contracted in US dollars. An increase in the Australian dollar's value against the greenback would have undermined the international competitiveness of the mining sector. In the end the smaller party in the federal coalition government, the Country Party, aided the cause of the mining industry by refusing to accept the Australian dollar's revaluation by more than six per cent, although this was against the advice of the Treasury and the Reserve Bank.[101] The consequence of the decision was mounting inflationary pressure in the Australian economy, fuelled by the minerals boom, which a federal Labor government was forced to tackle later with more drastic measures.

The Stock Market Bubbles, 1969-71

The stock market boom and bust in the years 1969 and 1970 were primarily related in particular to nickel and uranium with other minerals following their wake. In 1966 a Sydney entrepreneur, Boris Ganke, became the largest shareholder of a destitute Adelaide-based mining company named Poseidon NL. Ganke took up an option on nickel deposits at Bindi Bindi, 193 kilometres north of Perth.[102] These were owned by a syndicate headed by mining engineer Norman Shierlaw, who subsequently became the largest shareholder of Poseidon NL. Shierlaw hired Ken Shirley, a full-time prospector who explored in the area of Mount Windarra, 24 kilometres north-west of Laverton in Western Australia. Shirley noted the caps of a nickel sulphide lode in a banded iron ore formation and pegged 10,800 acres of land that proved to contain one of the best nickel deposits found in Australia.[103] Windarra would prove to be a substantial nickel mine in the long term. But the neophyte nickel miner, Poseidon, exaggerated the richness of Windarra's nickel and underestimated the cost of extraction.[104] In the beginning of September 1969, few Australians had heard of Poseidon or Windarra, but towards the end of the month the results of drilling at Windarra led to a flurry on Australia's stock exchanges. Poseidon stocks, which closed at $1.15 in Sydney on 24 September, rose to

$1.48 in Sydney, $1.53 in Adelaide and $1.55 in Perth.[105]

Shierlaw's own company, Shierlaw and Associates, began heavy buying of Poseidon stock on 25 September, leading to a further increase in the price of the stock. When, Poseidon directors reported on 29 September 1969 that a second drill hole at Windarra had struck nickel and copper sulphides, speculators pushed the stock higher. On the following day Poseidon shares closed at $7, bringing the stock of big companies like CRA and BHP up with it. The journalist and historian Trevor Sykes has remarked on the irony that a small nickel company, which had not achieved anything but some exploration results, added $65 million to the capitalisation of Australia's largest company, BHP.[106]

As the price of its shares exploded, Poseidon became a household name. Poseidon's announcement on 1 October 1969 of assay results, suggesting 10 times more nickel than there really was, sparked a share boom even bigger than that of the 1880s in Australia. In the month of national elections in Australia, Poseidon stock rose, reaching $20.10 on 6 October. Although professional share traders dominated the trading spree, they were soon joined by small investors. In August 1969 the value of shares traded in minerals and oils was $28 million, but in October that figure climbed to $168 million and to nearly $200 million in each of the months of November and December.[107] In November 1969, when Poseidon announced the imminence of another assay report, its shares leapt to $51 and in the following month, on 23 December, they rose from $130 in Sydney to $185 before closing at $175.[108] After Christmas, the stock broke through the $200 barrier, reaching $214 before closing at $210 on New Year's Eve.

At the end of 1969, $1000 million had been traded in mining and oil stocks in Australia, while industrial turnover had been barely half that figure. In London, the *Economist* remarked:

> Australia more than any other country is being carried along by Japan's explosive economic growth. The spectacular history of the Poseidon shares rather exaggerates the overall mining boom in

> Australia. But there is little doubt that historians will see the late 1960s as comparable with the legendary gold rushes of the 1850s. Australia is expected to become the world's largest exporter of a whole range of minerals – certainly iron ore, coal and alumina, possibly copper, nickel, pig iron and aluminium.[109]

Poseidon's spectacular downfall in the following year had to do with a range of factors. One was the fall in the world demand for nickel that was accompanied by a rise in production in other parts of the world such as the Dominican Republic, South Africa, the Philippines, Greece, Indonesia and Rhodesia. With demand stagnant and production in oversupply, the European free market price of nickel fell from $US3.90 to $US2.00 in 1969. WMC had been able to sell all of its nickel output from Kambalda in its first three years of operation under more favourable prices, thus generating cash flow. Then, in advance of the free market prices falling, WMC switched to selling under long-term contracts. In mid-1970 it was too late for Poseidon to follow this strategy.[110]

The second factor militating against Poseidon's success was the rise in inflation in the early 1970s. This meant that the costs of operating a nickel mine in the 1970s escalated at a far greater rate than in the 1960s. The third factor was that the grade of the Windarra deposits proved to be less than Poseidon initially estimated. On 1 October 1969 Poseidon announced 3.56 per cent nickel. But published data for the years 1975 and 1976, after Windarra had been taken over by WMC, indicated that the richer sections of the mine provided an average mill grade between 1.85 and 1.97 per cent. Unable to raise finance to develop its mine in 1970, Poseidon's stock fell to almost nothing by the end of that year. The project was rescued in October 1972 when WMC arranged for temporary finance through the Australian Industry Development Corporation to allow the Windarra mine to proceed. On 3 January 1973 WMC took a 50 per cent stake in the mine and agreed to finance its further development.

The mining of the uranium deposits at Narbalek is a similar story to the extraction of nickel at Windarra. Roy Hudson, the director of Queensland Mines, was a veteran of the Australian mining industry whose discovery of iron ore at Savage River in Tasmania had given him an annual income from royalties of over $100,000. Hudson was instrumental in forming Kathleen Investments in 1958 as a vehicle to retain an Australian stake in the Queensland uranium mine, Mary Kathleen Uranium (MKU), which had been established in the 1950s and whose major owner in 1970 was CRA. Kathleen Investments acquired the shareholding in Queensland Mines in 1965 and it was floated as a public company in 1967. Hudson announced on 31 August 1970 that the uranium deposits at Narbalek contained 55,000 tons of uranium oxide at 540 pounds per ton. One estimate was that Narbalek alone had increased the world's known uranium deposits by 10 per cent. Hudson's announcement saw a surge in the shares of Queensland Mines and its parent company, Kathleen Investments. After the announcement, shares in Kathleen Investments soared from $4.80 on 31 August to $17.50 in October and shares in Queensland Mines increased from $7.20 in August to $46 in early October.[111] Such was the scale of buying that Hudson persuaded Gorton to foreclose the possibility of foreigners buying up the deposit. Because Kathleen Investments was a company registered in the Australian Capital Territory, the economically nationalist Liberal Prime Minister was able to introduce an ordinance to the effect that foreign investment in Kathleen Investments and Queensland Mines could be no greater than 15 per cent.[112]

The movement in the share prices of Kathleen Investments and Queensland Mines attracted the attention of share trading company Mineral Securities Australia Ltd (Minsec). By December 1970 Minsec had spent $11 million purchasing shares in Kathleen Investments and $16.6 million buying shares in Queensland Mines, all on the basis of the false estimation that Queensland Mines had 55,000 tons of ore with 540 pounds of uranium oxide to the ton of ore.[113] On 12 August 1971 the truth was revealed that the Narbalek deposits amounted to 8932 tons of which some was at 16 pounds of uranium per ton and some at 240 pounds per ton. The collapse in Queensland Mines and Kathleen Investments, after the downgrading of its Narbalek deposits, was instrumental in an even bigger collapse, that of Mineral Securities.[114]

In the first five years of its existence after 1965 "Minsec had achieved the most remarkable success of any company in Australia's history".[115] It was conceived by Australian entrepreneur Ken McMahon. McMahon had studied geology at Sydney University in the 1940s and worked as a mining surveyor for CSR at Wittenoom in the 1950s, then for the Sydney-based investor in mining shares Commonwealth Mining Investment Australia Ltd, and later for Consolidated Gold Fields of Australia. Using his experience as a geologist and mining share trader, McMahon formed Mineral Securities Australia Ltd in 1965 as a share trading and investment company with initial capital of $300,000. By the end of 1970 it would have assets of more than $100 million.[116]

Minsec earned early profits by offsetting gains from extensive trading in shares against investment in new mining stock.[117] In 1968-69 Minsec bought $31.4 million worth of shares and sold $16.5 million, and in 1970-71 the company purchased $73.8 million worth of shares and offloaded $49.7 million. In the first seven months of 1970-71, it had purchased $75 million worth of shares, including the $29 million spent on Kathleen Investments and Queensland Mines. The Senate Committee under Senator Peter Rae that inquired into the share market collapses of 1969 and 1970 made an instructive comparison of Minsec with Australia's largest life insurance office, the Australian Mutual Provident Society (AMP). In the early 1970s AMP had about one third of the total assets of all life offices in Australia. The life office informed the Rae Committee that its annual rate of share purchases in the early 1970s was $40 million and its sales about $8 million. By contrast, in the calendar year 1970, Minsec bought $107 million worth of shares and sold $47 million.[118] During 1969-70, when Australia's stock exchanges were dominated by the Poseidon boom, the Sydney Stock Exchange recorded business of $1900 million; the turnover of Minsec alone in this period was $122 million.[119] Moreover, Sykes has estimated that the turnover of Minsec and the group of companies associated with it or controlled by it approximated 10 per cent of the Sydney Stock Exchange's turnover in that year.[120]

When the fall came it was spectacular. On 5 February 1971 Mineral Securities Limited crashed into insolvency.[121] Over a period of six years it had developed from a small company into an apparent major mining and investment empire – but it had been one built on the basis of being able to earn large profits from share trading indefinitely. This philosophy justified borrowing from abroad and on the short-term money market and paying high interest on funds that could be recalled at short notice. Sykes concluded that Minsec's principals had:

> overstated the bull market and were losing large sums on trading at the finish. Its largest ever market manoeuvre breached the fundamental commonsense of trading. In the long run it aided foreign take-over of the mining industry because nearly all the mines it bought had previously been in Australian hands, whereas after Minsec crashed many of them passed into overseas control.[122]

At the height of the stock market boom, Comalco had established itself as the largest, most vertically integrated and most profitable of Australia's aluminium producers. The company was, however, increasingly criticised because the greater part of its profits went as dividends to its overseas owners, RTZ and Kaiser. To counter such complaints, Comalco announced in 1970 that 13 million shares, or 10 per cent of its capital, would be offered to the Australian and New Zealand public. The biggest allotments went to financial institutions such as the Australian Mutual Provident Society (AMP). Politicians, including the Victorian Premier Henry Bolte, the WA Premier, David Brand, the wives of the Queensland and NSW Premiers and the Queensland Treasurer, Gordon Chalk, took up a smaller allotment of shares. The Comalco floatation may never have aroused political controversy had not Gorton revealed in May 1970 that he had declined the share offer and instructed his cabinet ministers to do likewise.[123]

The ensuing controversy over the share issue prompted academic Frank Stevens to reiterate his criticism of Comalco's treatment of Aborigines and for a few days there was significant public criticism of Comalco and its share offer. Eventually, however, the story went off the front pages, prompting historian Richard West to muse that, if a similar situation had occurred in Europe or the United States, "the sale of shares at par by a foreign company to politicians and civil servants would cause a furore, and could bring down a Government".[124] While the Comalco flotation did not bring down a government, it highlighted how foreign ownership of mining had become such a political issue that mining companies were responding by offering Australians equity in their operations and politicians like Gorton and McEwen were creating institutions, like the Australian Industry Development Corporation (AIDC), established in 1970, to help finance Australian mining operations.

Conclusion

Central to the first stage of Australia's second great minerals rush – the minerals boom from the early 1960s to 1973 – were new export industries based on coal, iron ore and bauxite. The boom extended to petroleum, natural gas, nickel and uranium as well as to long-established mining industries – such as manganese, copper, lead and zinc – that also profited from rising international demand for Australian resources and discoveries of new deposits such as at Groote Eylandt and Bougainville. The permanent legacy of the minerals boom was mining industries that would become the backbone of Australia's exports and a pillar of the Australian economy. The boom transformed and diversified a large and unexplored country dependent on agricultural exports that were subject to price fluctuations leading to balance of payments instability. In contrast to New Zealand (which had no mining sector) and Britain, Australia embarked on a period of relative affluence. The boom also triggered a massive inflow of foreign capital that, while strengthening the balance of payments, led to higher foreign ownership of Australia's mining industries and contributed to some Americanisation of Australia.

In 1964 Donald Horne had concluded *The Lucky Country* on a

pessimistic note, predicting the coming economic decline of Australia. The minerals boom that gathered pace in the years after the book's publication engendered a mood of extraordinary optimism and nationalism, especially during the terms of the Gorton and Whitlam governments. In the late 1960s and early 1970s Australia's restrictive immigration policy, the White Australia policy, began to be dismantled, Australian passports replaced British passports and an Australian national anthem superseded the royal anthem. Decimal currency was introduced in 1966, the Department of "External" Affairs" was replaced by the Department of "Foreign Affairs" in 1970, Britain began to be treated as just another foreign country and Australia set Papua New Guinea on a path to independence. Writing in 1980 Horne asked: "How could one imagine Gorton without the mining boom? His nationalism was mounted on oil derricks, cliffs of bauxite and mountains of ore".[125] Similarly, "Whitlam was borne aloft and into office with a $2 billion growth in overseas reserves in one single year".[126] The prosperity brought by the mining boom gave Australians the confidence to elect the reformist Whitlam government in 1972 after Labor's two decades in the wilderness at the federal level. At the same time, the stock market bubbles from 1969 to 1971 and mounting public concern with the increasing level of foreign ownership of mining led to pressure on all Australia's political parties to introduce policies to promote increased Australian ownership of Australia's mineral resources. Whitlam's political campaign in 1972 benefited from this change of mood, as we will see in Chapter Six.

6

"Buying Back the Farm": the Whitlam Government and Mining, 1972-1975

At the end of 1972 Labor returned to power federally with Gough Whitlam as prime minister. In one sense the timing was fortunate as the bounty earned from mineral exports in the 1960s and early 1970s had strengthened the economy enough to enable Whitlam to implement his ambitious agenda of social and economic reforms. The sharp rise in the price of oil in 1973, however, threatened Whitlam's agenda and the government's very existence. The oil price hike stalled the minerals boom and ushered in years of higher inflation, higher unemployment and lower economic growth, but the end of the minerals boom of the 1960s and early 1970s was only a hiatus in Australia's twentieth and early twenty-first century minerals rush, its second rush. During the period of the Whitlam government from 1972 to 1975, Australia's major minerals industries, coal and iron ore, improved as new production entered the market. Moreover, not long after the Whitlam government's demise, a new "resources boom" was triggered in a delayed response to the oil price increases of 1973 and 1974. During the period of economic transition of the Whitlam government between the end of the minerals boom in 1973 and the start of what came to be called the "resources boom" in 1977 a key aspect of Labor's policy agenda was to "buy back the farm", a phrase coined by McEwen in the 1960s when he warned of the dangers of uninhibited foreign investment.[1]

After the 1972 general election, Whitlam appointed Reginald Francis Xavier (Rex) Connor as Minister for Minerals and Energy. Rex Connor was born in 1907 in Wollongong.[2] He joined the Australian Labor Party in 1950, narrowly won the State seat of Wollongong–Kembla and was a member of the NSW Parliament until 1963 when he left State politics to enter the House of Representatives as member for Cunningham.[3] Although on the left of the Australian Labor Party, Connor aligned himself closely with Whitlam when the latter was Deputy Leader and later Leader of the Opposition. This cost Connor his position on the ALP executive in the late 1960s when the left of his party failed to support him in a caucus ballot. Although Connor lived in the political shadows between 1969 and 1972, Whitlam never forgot his support.[4] He eventually rewarded Connor's long-standing interest in mining by giving him the newly created portfolio of Minerals and Energy, which absorbed many of the functions of the old Department of National Development. But it would become more powerful than the old department.[5]

Many of Connor's statements in Opposition had signalled the trend of his thinking – insisting that Australian coal and other minerals be sold at world parity price; championing construction of a national natural gas pipeline; promising to assert national authority over the resources in the continental shelf; stopping "principalities" being granted to companies like BHP and Burmah-Woodside, a British–Australian consortium that became active in the North West Shelf (see Chapter Seven); and fundamentally opposing further foreign control of Australian industry, the mining industry in particular.[6] Connor immediately strengthened his department by appointing Lennox Hewitt, formerly Secretary of the Prime Minister's Department under Gorton, as his permanent secretary. Connor was clear about Labor policy on mining:

> There shall be at least majority Australian control over both equity and policy in resources development, and we will devise and implement an integrated and coordinated national fuel and energy policy. In particular we will regulate exploration, development and transportation, marketing, and use of oil, natural gas, and all related hydrocarbons. We will also prevent the pollution of the ecology and environment by fuel extraction.[7]

Connor failed to achieve the centrepiece of his policy, the setting up of a Petroleum and Minerals Authority; and the Whitlam government fell in November 1975 partly because of an aborted attempt to borrow petro-dollars, a notional unit of currency earned from the export of petroleum, to "Australianise" the mining sector. But the national control of mineral exports that Connor introduced helped Australia win higher prices for coal and iron ore. They also gave the government leverage to try to persuade mining companies to set up Australian industry associations to negotiate with the Japanese steel mills. Aided by these higher prices and new industry associations, the coal and iron ore industries earned significantly higher export income despite the difficulties faced by other mining industries. The second minerals rush gathered momentum in the Whitlam years, particularly in Queensland's coal industry. One sign of the success of Whitlam's policies was that the coalition government led by Malcolm Fraser would retain federal control over mineral exports and strengthen the policy of fostering Australian equity in new mining projects by establishing the Foreign Investment Review Board (FIRB) in 1976.

Revaluation, Australian Equity Guidelines and Control of Mineral Exports

After the Whitlam government was elected in 1972, it was forced to act immediately against the inflationary spiral that had gathered pace during the minerals boom. Both the Country Party and the mining industry had resisted revaluation during William McMahon's prime ministership. On 23 December 1972 Whitlam appreciated the Australian dollar unilaterally by seven per cent. In announcing this revaluation, Whitlam also gave notice that overseas-owned mining companies and oil explorers in Australia would not be able to bring investment funds into the country for their activities unless they first deposited 25 per cent of their capital with the Reserve Bank of Australia.[8] On 4 February 1973 Whitlam further appreciated the currency when he decided not to follow the US government in devaluing the dollar. The aim of these revaluations was to restrict capital inflow, boost imports and check exports in an effort to bring inflation under control.[9] But appreciation of the Australian dollar meant that exporters received less in the local currency for a given amount of exports.

The mining industry's first reaction to Rex Connor was negative.[10] Connor did not hide his scorn for mining executives responsible for writing mineral contracts in US dollars. As a result, miners were among the major casualties of the realignment of the Australian currency in 1972 and 1973. The Pilbara iron ore companies, for example, shared contracts totalling $6 billion, more than $5 billion of which was denominated in US dollars.[11] Connor also displayed little sympathy for foreign-owned mining companies.[12] Moreover, in February 1973 he vented his anger when learning of the McMahon government's eleventh hour approval of uranium contracts worth $100 million. One of these had placed 10 per cent equity in mining Australian uranium in the hands of the Italian corporation *Ente Nazionale Idocarburi* (ENI).[13] Connor believed a world energy crisis to be imminent and questioned why the McMahon government had hastened to sell uranium cheaply when prices for energy raw materials were about to rise steeply.

Despite some acerbic criticisms of Connor, the mining industry was not uniformly opposed to his policies on mining. Early in 1973, parts of the coal mining industry actually lobbied the Whitlam government for strong federal intervention. One of the most significant voices urging action was B.W. Hartnell, Chairman of the Joint Coal Board. For many years, Hartnell had criticised the failure of Australian coal mining companies to present a united front when negotiating with a cartelised Japanese steel industry. He had also castigated foreign-dominated Queensland coal companies for undercutting the price of NSW firms.[14] Hartnell and the NSW Coal Proprietors lobbied the Whitlam government to intervene to regulate the coal mining industry nationally and suggested two possible options to Connor – to create a Commonwealth agency responsible for the marketing of all Australian coal overseas; or to leave individual contracts in the hands of private companies but to subject them to Commonwealth control.[15]

Hartnell's advice was supported by Connor's own advisers in the Department of Minerals and Energy and Connor decided to act. He made two submissions to cabinet in January 1973.[16] Both were based on the ALP's 1971 Platform, which stated that the ALP would permit only exports of minerals sold at full market price and not allow the depletion of fuel and energy resources needed for Australia's own national development. Connor argued that the rapid development of Queensland's open-cut coal mining posed the risk that Australia could deplete its open cut coal, leaving the more costly underground coal, resulting in "significant effects on future costs of Australian industry and long-term export capacity".[17] He argued that the Japanese steel industry operated as a "monolithic' buyer and was adept at playing one company off against another. Japanese steel companies based their purchasing policy, Connor argued, on the formula of "cost plus a reasonable profit" and "sought to profit by the competition created through excess capacity, which has been created, with their connivance, in the older sectors of the country, with supply at lower prices emerging elsewhere in Australia".[18] The NSW Combined Colliery Proprietors' Association, the Miners' Federation, the Joint Coal Board and the Department of Minerals and Energy all supported federal export controls to achieve the orderly marketing of Australia's coal resources. Cabinet approved Connor's recommendation that the government should achieve the "balanced development" of Australia's black coal resources so that the production of coal for export was in the best interests of Australia. Cabinet also agreed that new coal contracts, or expansion of existing mines, could be approved only if Connor was satisfied that comparable coal at reasonable prices was not available from existing mines.[19]

Connor's submission on control of the coal industry was put forward in tandem with another for Commonwealth control of mineral exports generally.[20] At that stage there were already export controls on some minerals – though the embargo had been lifted in the 1960s, iron ore contracts nonetheless still had to be approved by the Australian government;

copper was controlled to ensure that Australian industry had adequate supplies; and natural gas was controlled to make reasonable provision for Australia's domestic needs. But there were no controls in place for exports of bauxite, alumina, nickel and coal. Connor argued that Australia's prices for mineral exports were being affected by overseas buyers presenting a united front to competing Australian sellers, particularly in the coal and iron ore industries. He also pointed to companies, such as Queensland Alumina Limited, that were exporting at the cost of production or on a similarly non-commercial basis. Connor also complained that the Commonwealth had no systematic knowledge of commercial arrangements for the sale of Australia's mineral exports other than for iron ore. Embarrassingly, the Commonwealth government had to glean its intelligence from public or semi-public sources, State governments, private companies and the press. To help achieve the government's policy objectives, Connor obtained cabinet's approval to apply export controls to all minerals except aluminium metal and refined lead and zinc. The federal controls on mineral exports would be retained by the Fraser government from 1975 to 1983 and by the Hawke Labor government until the second half of the 1980s.

The Australian Industry Development Corporation, the Fitzgerald Review and the Petroleum and Minerals Authority

To encourage a higher degree of Australian ownership of mineral projects, the Whitlam government began by using existing machinery. This included the McMahon Government's *Foreign Takeovers Act* 1972, which allowed the Treasurer to veto foreign takeovers not in the national interest, and the Australian Industry Development Corporation (AIDC). The AIDC, as we have seen, was a government corporation established by Gorton and McEwen in 1970 to marshal funds, mainly from foreign sources, to provide assistance to Australian companies that requested AIDC investment. Whitlam and Connor wanted the AIDC to be able to raise risk capital to allow local companies to invest in one of the sectors that most needed it – Australian mining.

In March 1973 Whitlam announced a vision for an enhanced AIDC. He sought to remove restraints on the corporation's activities and to establish a National Investment Fund to raise capital within Australia for its work.[21] Under Whitlam's plan, the AIDC would be required to focus on improving Australian ownership of resources and their development. In August 1973, the Labor government introduced legislation to expand the AIDC's functions and enhance its money-raising powers in the Australian capital market. The amending legislation irritated business groups and the Opposition-controlled Senate, who suspected that the AIDC initiative disguised government attempts to control private industry. As a result the amending legislation languished in the Senate.[22]

In the meantime, Whitlam's ministers disagreed about how the AIDC should actually work. On the one hand, Connor argued for the prohibition of future equity investment by foreign corporations in new Australian mining projects.[23] He also pressed Whitlam to notify the State premiers "that the States must not continue to usurp the rightful functions of the Australian Government by their indiscriminate allocation of huge tracts of land as mining leases without regard to national planning".[24] On the other hand, the Minister for Overseas Trade, Jim Cairns, advocated partnerships between overseas interests and the Treasury strongly opposed a complete prohibition on foreign investment in new mining ventures.[25]

On a visit to Japan in October 1973, Whitlam sought to end the uncertainty within the government by issuing a comprehensive statement on foreign investment policy. Whitlam announced that the government required equity in new energy projects – coal, oil, gas and uranium – to be in Australian hands while still looking for foreign participation "through access to technology, loans and long-term contracts". As for non-energy resources, the Whitlam government's policy was more flexible; the Prime Minister announced that "[w]e desire partnership between Australian and foreign equity capital. I want to make it quite clear that there is no proscription of foreign equity participation in mining".[26] Under the new system announced by Whitlam, all foreign investment in Australia would be screened and would require approval by the Treasury.[27]

By the time Whitlam made this statement in Tokyo, planning was under way for another corporation, much more powerful than the AIDC, to help the government in the field of minerals and energy. The Petroleum and Minerals Authority (PMA) was Rex Connor's brainchild. In March 1973 he had sketched an idea for a new government corporation to perform three functions. It would be a government oil company capable of undertaking the full range of activities from exploration to distribution of petroleum and natural gas; a government mining company empowered to perform the full range of mining activities from exploration to refining metals; and an entity to help Australian companies engaged in any of the activities under the PMA's sphere of operation. In April 1973 Connor advised Whitlam that the enterprise he had in mind could best be thought of in terms of the Italian venturer ENI, which had recently purchased Australian uranium.[28] In 1953 Italy established ENI as a state-run energy corporation which over the next 20 years would set up 180 subsidiary and associated companies operating in about 30 countries, employing 80,000 people and with gross sales of $2.5 billion in 1973. ENI acted as a state holding company for its wide-ranging subsidiaries that were supposed to act as private companies under Italian law.[29] The idea for a Petroleum and Minerals Authority (PMA) harked back to earlier government involvement in corporations such as the Commonwealth Oil Refineries.

Connor's idea was that the legislation to establish the PMA would also proscribe new foreign equity participation in mining exploration and development by using the Commonwealth's corporations power, which the High Court had construed generously in the *Concrete Pipes* case of 1971. In the absence of foreign investment, the PMA would fill the void by investing in private Australian mining companies or engaging in exploration and mining in its own right. As Connor argued to cabinet in June 1973, "we would generally not grant exploration permits other than to Australian interests and we would ask the States not to grant exploration permits other than to Australian interests pending the introduction of the new legislation".[30] However, the new authority might allow foreign exploration and exploitation of some areas that the PMA might discard, or in areas such as deep sea drilling, which required foreign expertise.[31]

On 16 April 1973 the government decided to discontinue federal subsidies for companies searching for petroleum and to discontinue tax concessions for the acquisition of shares in mineral and petroleum exploration companies. It also authorised "legislation to be prepared on the same principles as those approved for the National Pipeline Authority, to establish a Petroleum and Minerals Authority responsible for exploration, production, transport and refining of minerals".[32] Cabinet left the responsibility for drafting the PMA legislation with Connor, who later explained that, in framing the legislation:

> I conceived of the Authority both as a vehicle for increased Australian ownership and control of our resources and also as a revenue producer for Australia. A major national objective of the Authority is the creation and exploitation of opportunities for participation in future discoveries which will lead to financial returns for Australia as well as to increased ownership of the industries based on these discoveries.[33]

Cabinet intervened in the drafting of the legislation in November 1973 to ask Connor to remove sections from the draft bill intending to void mining rights granted to companies under State laws unless those rights were approved in applications made to the PMA. Whitlam, the Treasurer, Frank Crean, and the Trade Minister, Jim Cairns, were all concerned that including such a provision would lead to a confrontation with the States on matters they regarded as their domain.[34]

When Connor introduced the PMA bill into parliament in 1973, the mining industry was strongly opposed to the legislation. While it recognised that the Whitlam government had a mandate to create the PMA, the Australian Mining Industry Council (AMIC) argued that such a body should compete on equal terms with private mining companies and be subject, as private companies were, to State mining legislation.[35] AMIC, the peak mining body established in the late 1960s, argued that the bill gave the PMA "authority, against the wishes of the occupier, to explore, occupy and mine anywhere in Australia including exploration areas under leases already held and worked by mining companies". The PMA, AMIC feared, would become the "overriding Mines Department of Australia with the power to act, as it saw fit, over any land regardless of prior rights".[36] In February 1974 AMIC, working through the Country Party, pressed for a complete rewriting of the bill.[37]

The main Opposition parties in the Senate, the Liberal and Country (later National) parties, were adamantly opposed to the bill, as were the non-Labor States, Queensland and Victoria. Moreover, even the Labor States held reservations about the foreign investment policies associated with the PMA legislation. The Premier of South Australia, Don Dunstan, was one dissenter. He explained to Whitlam that minerals exploration in South Australia had been sustained by foreign companies associated with such discoveries as natural gas in the Cooper Basin and uranium at Lake Frome.[38] He pointed out that SA governments had granted mining tenements to foreign companies without discrimination and that, under South Australian law, the holders of these tenements had the right to exploit any discoveries. Similarly, the Labor Premier of Western Australia, John Tonkin, complained to Whitlam on 17 May 1973 about the deleterious consequences of the federal government's mining policies for a State whose economy was predominantly geared to the mining and processing of mineral resources. In particular Tonkin criticised Whitlam's decision to require foreign companies to lodge 25 per cent of their investment as an interest-free deposit with the Reserve Bank of Australia. This decision, Tonkin considered, had adversely affected the planning of an American–Australian consortium proposing to develop iron ore deposits at Rhodes Ridge. Tonkin did not favour 100 per cent development of mineral deposits by overseas companies. But he argued that, where the capital cost was in the vicinity of $300 million as was the case with Rhodes Ridge, "we should not 'shut the gate' to a proportion of foreign capital".[39]

Resistance to the PMA bill continued in 1974 until the Senate's failure to pass it became one of the grounds on which Whitlam advised the Governor-General, Sir Paul Hasluck, to dissolve and hold elections for both houses of parliament. After Labor won the ensuing election on 18 May 1974, an historic joint sitting of both houses of parliament eventually passed the PMA legislation. Speaking in favour of the bill, Connor defended the fact that the PMA would not be subject to State laws on the ground that its powers flowed from the Constitution and that it would be a denial of the Constitution to make the exercise of its powers subject to any State. He argued that the creation of the PMA would not result in an impractical and massive takeover of the minerals industry. It would not be the end of private enterprise, Connor insisted, but the beginning of a partnership between government and private enterprise that would lead to a greater expansion of the petroleum and minerals industry under the aegis of the federal government in the national interest. Connor's arguments on the need for public involvement in mining were bolstered by the conclusions of Tom Fitzgerald, whom Connor had commissioned to inquire into the Australian mining industry and taxation. Fitzgerald controversially reported that between 1967 and 1972 federal government assistance to the mining sector exceeded tax payments from mining companies by $40 million.[40]

Despite the passing of the PMA Act by the joint sitting of the House of Representatives and the Senate, the High Court later invalidated the legislation on a technicality.[41] To be sure, if the legislation had not been struck down on a technicality, the States would have challenged its substantive provisions. But if the legislation had survived this challenge and if the Whitlam government had lasted beyond 1975, the history of mining in Australia might have been different. The PMA would have been a much more powerful tool to marshal capital to promote Australian ownership of mineral resources. Connor also had in mind that the PMA would play a large role in developing the North West Shelf natural gas reservoir in the same way that Statoil had helped to develop the petroleum and natural gas resources of the Norwegian Shelf. Because of the degree of opposition to the PMA in the 1970s and because the natural gas in the North West Shelf was developed by a private joint venture in the late 1970s and early 1980s, the Australian Labor Party decided not to reintroduce the measure when it returned to power in 1983.[42]

Connor and the Coal and Iron Ore Industries

Although the Whitlam government was thwarted in its attempt to establish the PMA, other aspects of its mining policies were more successful. The biggest success came with its management of mineral export controls. The controls introduced in 1973 gave Connor powerful leverage to persuade sectors of the mining industry to present a united front against Japanese buyers. The controls met with strong resistance from the Japanese government, which claimed they were causing "uncertainty and anxiety" and would compel it to reconsider its position in undertaking further mining investments in Australia.[43] In 1967, as we have seen, the Japanese steel mills had persuaded the Holt government to withdraw its guideline prices on iron ore exports by carrying out its threat to give further iron ore business to Australia's competitors in South America. After the Whitlam government's introduction of export controls on all minerals in 1973, Japanese steel firms tried the tactic again by threatening to source their coal imports from the Soviet Union, South Africa and Canada.[44] By 1973, however, Australian miners of coking coal and iron ore had established themselves as the lowest cost and largest suppliers of the Japanese steel industry. Consequently, the Whitlam government was able to stare down the threat of trade reprisals.[45]

In 1973, after Connor persuaded Australian iron ore companies to present a united front in negotiations with Japan, they secured a 17.5 per cent price increase to compensate for revaluation and a further 20 per cent price increase in 1974 to make up for inflation and rising costs. Iron ore miners thus earned a 40 per cent price increase from the depressed price levels of 1973.[46] In the early 1970s iron ore was Australia's most important mineral export. The volumes exported by that time had established Australia as the world's third largest producer and second largest exporter of iron ore.[47] At $439 million, iron ore accounted for 7.5 per cent of total exports in 1972-73. Aided by the new mineral export controls, unit prices rose 25 per cent to $8.28 per tonne in 1974-75. Since tonnages had also increased by almost 30 per cent to 85 million tonnes in 1974-75, export receipts from the sale of iron ore rose to $706 million and 8.4 per cent of merchandise exports in that year.[48]

In 1973 the export controls helped Connor and the Joint Coal Board achieve another long-held goal. In July of that year, after meetings between the NSW Coal Proprietors' Association and the Queensland Coal Owners' Association, the two State organisations agreed to promote a national coal industry through the vehicle of the Australian Coal Association (ACA). The ACA represented NSW and Queensland producers. Its chairman, Edward Ryan of Bellambi Coal, appointed a delegation, including representatives from mining companies Coalex, Utah and Thiess Bros, to work out a timetable for discussions in Japan.[49] Most of the contracts between the NSW coal producers and the Japanese were due for renewal in 1974 and negotiations for them would commence as early as the second half of 1973.[50] In persuading coal producers to unite in an Australian Coal Association, Connor was contributing to his goal of "Australianisation" by having industry organisations arguing for Australian positions with the Japanese steel industry.

Thus in 1974 producers of hard coking coal in New South Wales and Queensland negotiated with the Japanese as representatives of a unified Australian coal industry working under federal oversight. Though coal negotiations were conducted separately by coal producers in 1974, they were nonetheless conducted under guidelines set by the Department of Minerals and Energy and approved by Connor. The first approach was in early 1974 by ACA chair Edward Ryan, who left for Japan to negotiate on behalf of producers from the NSW south coast – Bellambi, Kembla Coal and Coke and Clutha Development.[51] On 28 March 1974 the Australian Broadcasting Commission (ABC) reported that government intervention had compelled the Japanese steel mills to offer the south coast producers spectacular increases for Australian coking coal, reportedly after Connor had threatened to ban the export of coking coal to Japan if prices were not improved.[52] The prices of Bellambi and Kembla Coal and Coke coals improved from a base price of $US20 per tonne with escalation allowing for wage increases to $US33.04 per tonne in 1974. Similarly, Clutha's Wollondilly coal price improved from $US19.06 per tonne to $US31 per tonne.[853]

The Queensland producers, who were next to negotiate, benefited from the price rises achieved for hard coking coal from the NSW south coast, which was of similar quality to the coking coal mined by open cut methods in Queensland.[54] Indeed, Utah achieved larger price increases than the NSW south coast companies. The price for Utah's Peak Downs and Goonyella coals doubled, from $US15.05 per tonne to just under $US30 per tonne after escalation. Thiess Peabody Mitsui (TPM) went to Japan for negotiations at the same time as Utah and achieved a price increase for its Moura coal from $US17.54 to $US25.75 per tonne. TPM improved the price to $US28.60 per tonne. After the success achieved by the producers from the NSW south coast in March 1974, the Department of Minerals and Energy advised Australian mining company Thiess Brothers to go back and renegotiate the price of its South Blackwater coal, which it did, improving its price per tonne to $US31.40. The price increases secured by NSW and Queensland coal producers in 1974 improved the value of Australian coal contracts from $A290 million in the year ending in March 1974 to well over $A500 million a year later.[55]

Connor's success in helping the mining industry achieve higher prices for mineral exports earned him praise from unlikely quarters. The *Australian Financial Review* observed in 1974 that Connor had transformed minerals and energy, previously a "rather dull, administrative area of little consequence", into a "top policy portfolio".[56] The paper further assessed that:

> [b]ecause he was obdurate on energy exports and because he was proved right by the dramatic turn of events in the international scene, Mr Connor presided over a radical change in public acceptance of government involvement in energy policy ... Anybody who considers that Australian fuel policy should return to the laissez faire days of a few years ago is sadly misreading the changed relationship any government anywhere in the world must have to the energy industry.[57]

The negotiations on coal prices in 1975 for the Japanese fiscal year 1975-76 were critical because the Japanese mills were now dealing with a unified national industry backed by a powerful federal department and minister. The negotiations turned into a trial of strength between the unified Japanese steel mills and the Australian Coal Association backed by the Department of Minerals and Energy. On 10 March 1975 Connor and Hewitt, his departmental secretary, met with representatives of the ACA in Sydney to discuss negotiating tactics. There were two schools of thought within the new association. One view voiced by Bellambi and Utah was that it was realistic to aim at prices of between $US50 and $US52 per tonne for Australia's high-quality hard coking coal.[58] The other more optimistic view, pressed by Kembla Coal and Coke and Clutha, was that that the Australian coal exporters should push for a $20 per tonne increase so that Australian high-quality coal could reach a price of about $US57 per tonne, near the level recently secured by US coal exporters. Connor chose the higher price of $US57 as the guideline price Australia should seek for the high-quality coking coal from Queensland and the NSW south coast and $US40 per tonne for the soft coking coal from the Newcastle area. But the Japanese steel industry, which was facing a decline in steel production in the forthcoming year, adopted a tough approach with the Australian negotiators. By that time, Australia was the leading exporter of metallurgical coal to Japan, having contracted to export 32 million tonnes of coking coal to Japan in the Japanese fiscal year 1975-76, amounting to 45 per cent of Japanese imports of foreign coking coal. The Japanese steel industry was determined to keep the price of Australian coking coal down, aiming at a price of around $US45 per tonne for south coast hard coking coals and around $US32 per tonne for the Newcastle area soft coking coals.[59]

As in 1974, Edward Ryan, the ACA President, opened the negotiations in March 1975 in association with Clutha Development and Bellambi. Bellambi and Clutha negotiated for NSW hard coking coals, which were expected to set the benchmark for the entire Australian coal industry. While Clutha pressed for a price increase amounting to $US20 per tonne, the Japanese mills indicated that they would go no higher than $US42 per tonne on the grounds that Japanese steel production was depressed. The Japanese argued that the Australians should not have the benefit of both long-term contracts and high market prices but should accept the formula of "costs plus a reasonable profit".[60]

Ryan countered that the low price which Japan had habitually paid for Australian coals had helped the Japanese steel industry become the third biggest in the world. Ryan informed Hewitt that the stand taken by the mills reflected that they knew about the agreement between Connor and the ACA on 10 March 1975 on prices and were reacting harshly to what they regarded as "unwarranted interference by Government in normal commercial relationship between buyer and seller".[61] Under these circumstances, Ryan recommended that Connor let the industry settle for lower price increases than Connor had stipulated on 10 March 1975. But after speaking to Frank Duval, the experienced Australian businessman resident in Japan who had involved himself in developing Gove bauxite, Hewitt advised Connor to let the Australian negotiators return to Canberra. Having asked for a $US20 per tonne increase and only being offered a $US8 per tonne increase, the Australians, Duval thought, should certainly press for a better outcome. In addition, Duval criticised the coal proprietors for telling the mills that Connor was behind their initial bid for a $US20 per tonne increase.[62]

The price negotiations remained at stalemate until Connor made a visit to Japan in June 1975 to discuss a range of matters with government and industry, including coal liquefaction and uranium enrichment. In discussions in Tokyo, Saburo Tanabe, the Managing Director of Nippon Steel, explained to Connor that American coal exporters were putting the Japanese steel industry under pressure by demanding excessive spot prices for their coal and that Australia should not attempt to link its coal prices, under long-term contracts, to these American prices. Connor replied to Tanabe that the Australian government did not want to intervene in the coal negotiations but that protracted negotiations were not good for either Australia or Japan.

In the end, the two men brokered a deal whereby Connor agreed to allow the Australian coal exporters to settle for prices of around $US50 per tonne for hard coking coal and $US38 for soft coking coal. These prices were less than Connor had previously wanted but, in return for this concession, Connor received an assurance from the Japanese steel industry on its intent to increase its coal purchases from Australia in the longer term. The assurance came in the form of a letter to Connor from Yoshihiro Inayama, Chairman of Nippon Steel, to the effect that Japan intended to increase its coking coal purchases from 27-29 million tonnes in the mid-1970s to 44-49 million tonnes by 1980, an increase of some 20 million tonnes per annum. This had been based on a formal assurance from the Australian government, through Connor, that Australia could meet the Japanese requirements for Australian coking coal.

Connor hailed the Inayama letter and the associated price increases, which were up 71 per cent on the previous year, announcing that the value of Australia's coal exports should increase from $A400 million in the current year to as much as $A1700 million per year by 1980. He described the arrangements – which would result in earnings to Australia of about $A7 billion over five years – as "the highest commercial value agreement in Australian history"; they "gave the guarantee of stability to the Australian coal industry for production and employment throughout a period of world economic uncertainty".[63]

In 1973-74 the average unit price for coal was $12.50 per tonne and exports were worth $327 million. By 1975-76 price increases negotiated on long-term contracts had allowed the average unit price for coal to rise to $35 per tonne. The value of total exports of coal products then increased to $1023 million or 11 per cent of Australian exports. In this period coal overtook iron ore as Australia's most valuable mineral export.[64]

During the course of 1974 the Japanese steel industry had become anxious to secure another 20 million tonnes of coking coal annually from Australia by the end of the 1970s. Although affected by the economic adversity of the early 1970s, it maintained an appetite for low-cost Australian coals. It had its sights on three particular projects which were under development – Utah's planned mine at Norwich Park, Thiess Peabody Mitsui's project at Nebo, a mine initially earmarked by William Donovan in the 1950s, and a consortium of mainly Australian producers wishing to mine at Hail Creek in Central Queensland. Connor agreed on the need to increase Australia's coal exports and aimed to raise them by 28 million tonnes annually within five years.[65] However, he delayed approving any of the three new projects on the basis that existing mines in Queensland and New South Wales could provide the extra tonnages Japan desired.

By the middle of 1975 the Queensland Premier Joh Bjelke-Petersen and his Treasurer, Sir Gordon Chalk, were voicing such strong criticisms of the Whitlam government's coal policy that Bill Hayden, the Queenslander who had replaced Jim Cairns as Treasurer in June 1975, wrote to Connor and Whitlam urging them to place Norwich Park and Nebo before the Whitlam government's investment advisory body, the Foreign Investment Review Committee.[66] At the heart of Connor's opposition was a deep dislike of multinational corporations in general and Utah in particular. When Warwick Parer, an Australian senior executive in Utah, first met Connor with another executive, Ralph Long, Parer recalled:

> We went to Canberra, Ralph Long and I, Ralph Long was this gentle man. At the first meeting we had with Rex Connor, he said: "I am pleased to meet you Mr Long. If I had been here ten years ago you wouldn't be here".[67]

Utah had approval from the Queensland government to export 300 million tonnes of coal from the North Bowen Basin and 100 million tonnes from its Blackwater lease in the south of the region. In 1975 Utah was asking the Queensland government to waive the limit of 300 million tonnes on its exports from the North Bowen Basin and also to grant permission for Utah's fifth open cut mine, at Norwich Park. Moreover, the company was seeking permission to export up to 30 per cent of its mineable reserves in the North Bowen Basin, which the company had now estimated to contain some six billion tonnes of coal. In a press release, Connor objected:

> Despite its reserves from four existing mines, [Utah] was pressing to open up another shallow open cut at Norwich Park ... without any further concessions or increased Australian shareholding. It is claiming the right to rip out and export 1600 million additional tonnes of Australia's best coking coal, valued at current prices at around $A60,000 million. Utah will gain over 53 per cent of Australia's coal export profits during the forthcoming year.[68]

Connor was strongly predisposed against approving the Norwich Park development. He was also inclined to delay a decision on Thiess Peabody Mitsui's Nebo project. He hoped that a legal battle in the United States would result in Peabody's owner, Kennecott, divesting itself of coal properties in the United States and then divesting itself of Peabody's stake in TPM, which could be taken up by Australian interests. Connor was most favourable to Hail Creek, a deposit 96 kilometres south-west of Mackay in the same general area but nearer the coast than Utah's Goonyella, Peak Downs and Saraji fields. An Australian consortium, Associated Australian Oilfields and Interstate Oil, had discovered the field in 1969 with indications that it contained 800 million tonnes of coal. In mid-1971 a syndicate known as Hail Creek Associates had formed to develop the deposits. The consortium consisted of Australian Associated Oilfields and Interstate Oil with 60 per cent equity, Western Mining Corporation with 25 per cent, and the residual 15 per cent taken up by Marubeni Corporation and Sumitomo Shoji Kaisha.[69] But because the $A227 million project was premised on sharing infrastructure with Nebo, it was held up by Connor's delay on approval of that project.

Connor's opposition throughout to any change in the Whitlam government's foreign investment policy bolstered his case against supporting new coal projects. But in September 1975 Connor was beaten in cabinet on the issue of foreign investment in mining, after which Whitlam announced a more "open-door foreign investment policy".[70] The government now set a 50 per cent maximum ceiling on foreign ownership and control for new development projects and merged the Foreign Investment Committee and the Committee on Foreign Takeovers into the Foreign Investment Advisory Committee. In the following month Whitlam announced that the government had given approval for the planned projects at Hail Creek, Nebo and Norwich Park in Queensland. The reason for the change in policy was to give the Japanese steel industry the security of an additional 15 to 20 million tonnes of coal annually and to promote additional iron ore business from Japan that might be dependent on the new coal contracts being approved.[71] Despite the Whitlam government's defeat in the general elections of December 1975, the modified September 1975 foreign investment policy continued to be implemented by Fraser's Liberal-National coalition and carried on in the early years of the Hawke Labor government.

It was under this bipartisan foreign investment policy that Utah's Norwich Park mine was eventually approved by the Fraser government in March 1976. While the new Minister for Trade, Doug Anthony, insisted on 50 per cent Australian equity in Norwich Park, he acceded to a request from Utah and the Queensland government that this equity could be spread across all of the leases of Utah's Central Queensland Coal Associates (CQCA) coal mines. A 50 per cent Australian equity in Norwich Park translated into an increase in Australian equity from 10 to 20 per cent across the four CQCA mines and the port of Hay Point.[72]

Uranium and the Loans Affair

In the emerging uranium industry Connor had what he thought was the perfect vehicle for the full implementation of the Whitlam government's policies on minerals and energy.[73] Since most of Australia's uranium deposits were in the Northern Territory, they were a federal rather than a State responsibility and the Menzies-era *Atomic Energy Act* 1953 vested all the Northern Territory's uranium in the Commonwealth. The industry was mainly Australian-owned and it was in an earlier stage of development than industries such as coal, iron ore and aluminium. Connor refused to agree to further uranium contracts besides those approved at the eleventh hour by the McMahon government, and, with Whitlam's approval, he ensured in September 1974 that the Commonwealth would market all sales of Australian uranium.[74] In the same month he predicted to the House of Representatives that Australia would be providing the Japanese market with 300,000 short tons of uranium by 1980. Connor estimated that merely exporting uranium as yellow cake would earn $7 billion while selling enriched uranium would earn Australia $35-40 billion.

Frustrated by the High Court's striking down of the PMA Act, Connor persuaded Whitlam and other senior ministers at the end of 1974 to approve the government's initiative to borrow $4 billion Arab petrodollars to provide the "necessary infrastructure for the emergency development of those resources based on the energy crisis".[75] When the Nixon administration decided in August 1971 to abandon the Bretton Woods system of fixed exchange rates, the weakening US dollar was allowed to float freely. Depreciation of the American dollar eroded the profits of oil-producing countries whose revenues were denominated in US dollars. This gave a strong motive to the oil cartel, the Organization of the Petroleum Exporting Countries (OPEC), to announce a 70 per cent increase in the price of crude oil in October 1973. OPEC had enormous market power, controlling 40 per cent of world oil supplies. As a result of monthly cutbacks in oil production the price of oil quadrupled from around $US3 to $US12 dollars per barrel. The oil price shock of 1973 adversely affected Western economies, by making everything that required oil more expensive and triggering inflation. For the Whitlam government, the OPEC oil price increases were catastrophic. The profitability of companies and the terms of trade fell, unemployment increased and the economy went into recession. The saving grace for Australia was the 1960s discoveries of oil in the Bass Strait, which had made Australia 70 per cent self-sufficient in crude oil.[76] Although Whitlam had come to power on the platform of "buying back the farm", the OPEC oil price increase made the task harder by increasing the cost of importing energy. But in the OPEC oil crisis Connor saw an opportunity – to borrow petro-dollars earned from Middle Eastern oil sales to buy back Australia's own energy resources. Connor knew that the French had made an agreement with the Shah of Iran to enrich uranium under which the Shah would lend the French $US1 billion for a 10 per cent share in the output of uranium from an enrichment plant in the Rhone Valley.[77]

In a meeting of the Executive Council on 13-14 December 1974 Whitlam, Cairns (Deputy Prime Minister and Treasurer), Connor and Lionel Murphy (Attorney-General) agreed to borrow US$4 billion. This borrowing followed an Australian Atomic Energy Commission (AAEC) estimate that investment in uranium enrichment would cost $A3 billion. The initial list of items requiring finance included pipelines, rail electrification, upgrading of coal exporting harbours and $225 million for "three uranium mining and milling plants".[78] Development of uranium enrichment plants did not feature in this initial list. But as historian Wayne Reynolds has shown, uranium enrichment became central to the loans affair. On 12 December 1974 Whitlam approved a proposal for the AAEC to borrow $US4 billion. On Attorney-General Lionel Murphy's advice, the government was persuaded that these funds could be borrowed for "temporary purposes", thus not requiring approval from the Loan Council, the joint Commonwealth–State body set up under the Constitution to regulate borrowings of the Commonwealth and the States.[79] The ensuing attempt to borrow petro-dollars was one of the factors that precipitated the Whitlam Government's, and Connor's, downfall.

After the OPEC-engineered spike in oil prices and a visit to Japan by Minerals and Energy secretary Hewitt in September 1973, the Japanese had agreed to explore the possibility of developing a joint program to enrich uranium.[80] Although unable to formalise this cooperation with the Japanese, Connor pressed ahead with his plan to establish a uranium enrichment plant in Australia in 1975. After Connor had decided on the site of a plant at Lake Phillipson in South Australia, the AAEC persuaded him that the site needed to be in Central Australia to support the Ranger and Mary Kathleen uranium deposits.[81] On 28 May 1975 Connor announced to the House of Representatives that Australia had made substantial progress on uranium enrichment and that it might be possible for Australia to "go it alone" without the help of American technology by using the centrifuge process, which used a large number of rotating cylinders. In the meantime, however, the US government had ruled out any US support for Connor's program and the Japanese government appeared to succumb to US diplomatic pressure by telling Connor when he visited Japan in 1975 it would be "premature" to go ahead with a joint project on enriched uranium.[82] When Connor's role in the attempt to borrow petro-dollars was leaked to the press in 1975, Opposition Leader Malcolm Fraser found in it the "reprehensible circumstance" on which to base the Senate's refusal of supply to the government. The Senate's refusal to pass supply in turn led to the Governor-General, Sir John Kerr, dismissing the Whitlam government, installing Fraser as caretaker prime minister and dissolving parliament. After Whitlam was defeated in the subsequent election in December 1975, no future Australian government would be successful in fostering the enrichment of Australia's uranium. Commenting on this failure, a taskforce advising the Howard government in 2006 stated that "for Australia to possess such a large proportion of the world's uranium resources – approximately 40 per cent of the global total – and not to have taken up opportunities over the past 35 years to develop uranium enrichment industries is highly regrettable".[83]

Conclusion

The Whitlam government's attempt to "buy back the farm" is most commonly associated with failure: particularly the striking down of the PMA legislation, the failure of Connor's long-term plan to pipe natural gas from the still undeveloped North West Shelf to the eastern States and the role of the loans affair in the fall of the government. Another way to look at the policies, however, is to examine how the successor Liberal-National coalition government dealt with the legacy left by Whitlam and Connor. Not long after taking office, the Fraser government outlined a policy to "restart" the mining industry, promising a lesser role for government and a greater role for private enterprise. But despite its free-enterprise philosophy, the Fraser government "elected to formalise, codify – and in some areas advance – the nationalist agenda for resources during the late 1970s and early 1980s".[84]

The Fraser government retained the Whitlam government's 50 per cent local ownership guideline for new mining projects and also set up an independent review agency, the Foreign Investment Review Board (FIRB), as a successor to Whitlam's Foreign Investment Advisory Committee to advise the Treasurer on the *Foreign Acquisitions and Takeovers Act* 1975. The FIRB was given the role of assessing foreign investment applications against a set of "national interest" criteria, including whether the proposed project would produce a "net economic benefit"; whether the project would serve "Australia's best interests"; whether "sufficient" Australian ownership had been achieved; and whether greater scope for Australian ownership existed. In other words, for more than a decade after Connor's resignation in 1975 bipartisan support remained for a policy of 50 per cent Australian ownership and Australian control of "key areas", including minerals and energy. Bipartisan support for a Foreign Investment Review Board has continued to the present day. The 50 per cent Australian equity rule for new mines would finally be ablolished by the Keating government in 1992.

The establishment of the FIRB allowed the Fraser government to institute a cooperative process to enable foreign investors to bargain over the degree of Australian ownership. A higher degree of foreign ownership could be bartered for other benefits such as infrastructure spending or local minerals processing. In 1974 and 1975 the Whitlam government seriously contemplated "buying back the farm" by purchasing Australia's largest mining house, Conzinc Riotinto of Australia (CRA), from its British parent Rio Tinto Zinc (RTZ). The initiative was overtaken by the Governor-General's dismissal of Whitlam, but the "Australianisation" of CRA was later achieved by other means. In 1978 the Fraser government announced a policy under which foreign-owned firms would be exempted from screening requirements for new projects if they committed to increasing Australian ownership of their existing operations to 50 per cent according to a timetable negotiated with the FIRB. Under this policy CRA began the naturalising process in 1978 and achieved 50 per cent Australian ownership in 1986. By that time, following the recession of 1981 and 1982 and more troubled times for the global coal trade, BHP had acquired Utah's coal operations in Queensland and the Australian diversified company CSR had acquired the coal operations of Thiess Brothers, thus considerably adding to Australian ownership of the mining industry. In iron ore, CRA bought out the American Kaiser Steel's position in Hamersley Iron and Australian miner Peko-Wallsend took over the management of the Robe River iron ore project from Cleveland Cliffs. In this way, large parts of the Australian mining industry were "Australianised" in the 1980s with the assistance of by policies initially implemented by Whitlam and Connor.

The Fraser government also retained Connor's mineral export controls, despite vigorous efforts by the Japanese steel mills – and by State governments such as Bjelke-Petersen's in Queensland and Court's in Western Australia – to have them dismantled. Although the Fraser government was forced by pressure from the Court Government and the Japanese steel mills to modify the controls on iron ore in 1979, federal controls on coal were retained by the Fraser and Hawke governments in the 1980s to ensure that Australia as a whole received reasonable prices for its mineral resources.

Whitlam's period in government spanned the end of the 1960s and early 1970s minerals boom and the OPEC oil price increase that induced a recession in 1974-75. While it was a period of difficulty for the mining industry, particularly as it adjusted to the revaluation of the Australian dollar, in other respects Australia's second minerals rush gained momentum. In 1972 mining's share of Australia's exports of goods and services was 18.9 per cent; in 1976 it was 27.3 per cent, largely because of increases in volumes and prices of coal and iron ore.

7

The Resources Boom, 1976-1982

The years of the Fraser government, from late 1975 to 1983, continued to be the era of "stagflation". They were years of higher inflation and unemployment and lower rates of economic growth than during the minerals boom from the early 1960s to 1973.[1] Australians had annual rates of growth of 4.7 per cent from 1961 to 1974, but they would experience only 2.2 per cent in the years from 1975 to 1983. Annual per capita growth in incomes fell by almost two-thirds, from 2.5 per cent in the 1960s and early 1970s to under one per cent between 1975 and 1983.[2] That there was a slowing of growth in the 1970s but no precipitous plunge into depression, such as there had been in the 1890s and the 1930s, was partly due to the strength of Australia's mining industries. Ian McLean has convincingly argued that two factors helped insulate Australia from what might have become a much more dramatic economic slowdown in the mid- to late-1970s and early 1980s. Both related to the mining industry and its exports. One was the oil price increase of the 1970s. This spike had initially driven down real incomes, but later the price increase stimulated demand for substitutes for oil like steaming (thermal) coal, natural gas and uranium. The other factor that warded off a more dramatic slow-down was that Australia's changing export mix in favour of minerals and energy exports had induced a shift away from a historic dependence on Britain as an export market towards reliance on other markets, particularly Japan. According to McLean:

> In a process that has been sustained to the present, a diversification

> of Australian export markets was occurring toward the emerging economies of East Asia, which were recording growth rates above those of the mature industrial economies in Europe and North America. This diversification partially insulated Australia from what otherwise may have been a more dramatic slowdown in the 1970s.[3]

Economists have described the years from about 1977 to 1982 as a "resources boom" because they were times of large-scale investment in the export of energy raw materials like steaming coal, uranium and natural gas and in low-cost coal-fired power stations built to expand production of energy-intensive commodities such as aluminium metal.

In addition to stimulating a demand for energy raw materials, the higher price of oil reduced the competitiveness of most developed economies. This led to slower growth in steel than had been hoped for in the late 1960s. But the rate of expansion of the Japanese steel industry in the late 1970s and early 1980s was still enough to justify the Japanese steel mills pushing for several more coking coal projects in Queensland. At the same time, higher oil prices stimulated a demand for steaming coal (coal for generating electricity) in New South Wales and Queensland. Australia was able to meet the new demand because of the investment in transport and infrastructure that had already been made for coking coal and because of increased investment by State governments in ports and railways during the resources boom. In Western Australia no new iron ore mine was established, but the existing Pilbara operations expanded. The reduced competitiveness of Japan and the United States encouraged aluminium producers to set up smelters in Australia on the basis of relatively cheap coal-powered electricity in Queensland, New South Wales and Victoria and the easy availability of bauxite. Investment in metals increased sharply in 1979-80 and 1980-81, mainly due to investment in aluminium smelters, but the resources boom peaked in 1982-83 due to the onset of a brief global recession.

The resources boom was thus of shorter duration than the Japanese-inspired minerals boom which had lasted from the early 1960s to 1973. The coming to an end of the first minerals boom had been foreshadowed by the Poseidon stock market collapse, the "Nixon shock" of 1971, and cutbacks in the Japanese steel industry in the early 1970s. But following these shocks, the Whitlam years from 1973 to 1975 had been characterised by increasing exports of already-operating coal and iron ore mines. The minerals boom of the 1960s and 1970s had been driven by strong and growing Japanese demand for Australia's mineral products, while the resources boom was driven by increased energy prices. The resources boom from 1977 to 1982 brought a large increase in production capacity, which resulted in an oversupply of coking coal in the first half of the 1980s and a slower-than-expected growth in the demand for steaming coal due to delay in the construction of overseas power stations. The upshot was that the resources boom did not result in the same rapid growth of mineral exports as had accompanied the Japanese-inspired minerals boom.

Nonetheless, at the end of the resources boom in 1983, the share of mining in Australia's total goods exports had risen to nearly 32.6 per cent, slightly more than the value of rural products. Coal exports alone were earning over $3 billion or 14.8 per cent of merchandise exports in 1982-83. At double the share of wool, wheat and iron ore, this meant that coal had become "by far the single most important Australian export" at the end of the resources boom.[4] Moreover, the mining industries collectively had become Australia's dominant export sector for the first time in the twentieth century.[5] Because of the momentum gained during the resources boom, coal would continue as the top commodity export in the 1980s and 1990s.[6] By the beginning of the 1980s, too, the trading relationship between Japan and Australia had grown to be the seventh largest in the world, with Australia supplying 50 per cent of Japan's iron ore and 40 per cent of its coking coal.

The New Black Gold

The first post-World War II coal boom from the late 1950s to the mid-1970s had been driven by the needs of the Japanese steel industry, but the energy crisis of 1973 was important in creating a structural shift in the demand for coal. Developed nations realised that oil had become overpriced and that they were vulnerable to OPEC control of basic industrial fuel. Consequently, they sought to diversify their energy supplies and develop a new reliance on coal as an alternative source of fuel for power stations and heavy industry. The question was whether "Australia could develop an overall ability to export coal at a rate which would meet the demands of the whole of industrial manufacturing nations".[7]

Australia began to export steaming coal after the first OPEC oil crisis in 1973 when NSW producers entered into contracts for the sale of eight million tonnes of steaming coal to Japan commencing in the years 1974 and 1975. In 1977 the US President, Jimmy Carter, announced a national energy plan. Among other things, it aimed to increase the production of American coal by two-thirds.[8] The Organisation for Economic Co-operation and Development (OECD) issued a report in January 1977 that examined ways in which industrial countries might develop strategies to cope with higher oil prices by diversifying their energy supplies. Between 1975 and 1990 the output of coal-generated electricity doubled in the United States.[9]

In June 1977 Japan's MITI published an "Interim Outlook for Long Term Energy Supply and Demand" that was much more categorical about the likely effect of what was being called the "energy crisis" on Japan's future economic growth and the coal trade. The report predicted six per cent economic growth in Japan from 1975 to 1985 and five per cent from 1985 to 1990. It also discussed ways in which Japan could reduce the costs of energy. One of the ways MITI suggested was for Japan to increase imports of coal from 62 million tonnes in 1975 to 102 million tonnes in 1985 and then to 144 million tonnes in 1990. These projections were based on Japan's developing a number of coal-fired power stations to be fuelled by Australian steaming coal. Later, in May 1979, the OECD's International Energy Agency (IEA) adopted "Principles for IEA Action on Coal" and a "Decision on Procedures for Review of IEA Countries' Coal Policies". The objectives of these reports were to increase greatly the supply of coal from countries like Australia, to encourage the flow of investment into coal projects, and to remove impediments to the production and international flow of coal.[10]

From the mid-1970s onwards, international oil companies began to pinpoint the black coal reserves of New South Wales and Queensland as among the most significant sources of fuel in the world. Australian black coal deposits were estimated to be less than five per cent of the world's total reserves of coal, but their importance was enhanced by their proximity to Japan and the relative ease with which they could be mined and exported. The linkage made between Japan's energy needs and Australian energy resources gave the cue to international oil companies to rush to buy into the Australian coal industry, beginning in 1976.[11] By that year, there was already significant foreign ownership of the Australian coal industry. The American-owned Utah Development Corporation was then shipping about 15 million tonnes of Australian coal per year, about half of Australia's coal exports, and Clutha Development, owned until 1976 by American businessman Daniel Ludwig, was the second-largest Australian coal exporter.[12]

Much of the overseas rush to buy into the Australian coal industry during the resources boom occurred in New South Wales. The OPEC energy crisis catapulted the soft coking coal and steaming coal of New South Wales – previously regarded as a poor cousin to Queensland's hard coking coal – into world demand. New South Wales had supplied the bulk of coal exports to Japan in the 1960s before being overtaken by Queensland in 1972-73. The two States were to draw level in 1978-79, largely because of the onset of the OPEC-inspired boom in steaming coal.[13] In 1976 British Petroleum (BP) purchased 50 per cent of Clutha Development and the remainder in 1978.[14] Shell took a 37 per cent interest in New South

Wales company Austen and Butta and 16 per cent of Thiess Brothers.[15] Austen and Butta was already exporting steaming coal to Europe and had extensive reserves of that variety of coal, including the Hartley project near Lithgow, while Thiess Brothers owned some of the largest and best steaming coal projects on the eastern seaboard, including at Callide in Queensland and the Drayton deposits in the Hunter Valley.[16]

In 1977 CRA was still majority owned by the Rio Tinto-Zinc Corporation. Now looking to develop further by taking over remunerative coal operations, it launched a determined campaign for an $87 million takeover of Coal and Allied Industries.[17] To satisfy the requirements of the Fraser government about Australian equity, CRA joined forces in its takeover bid with the Australian coal and shipping company Howard Smith Limited. The federal government agreed to a CRA–Howard Smith takeover of Coal and Allied Industries, but the NSW Labor government, led by Neville Wran, stymied the manoeuvre.

By the time that the Wran government was elected in New South Wales in 1976, Australia's most industrialised State had been particularly affected by the economic downturn of the 1970s. It affected manufacturing industry more than mining. In the 1950s and 1960s the flow of foreign capital into the manufacturing sector, high immigration and an expanding domestic market had produced strong growth in New South Wales. But falling rates of profits, stronger competition from countries like Japan and slower immigration in the 1970s had seen a large decline in manufacturing employment.[18]

The Labor Party under Wran had only narrowly defeated the Liberal and National parties in 1976 and on coming to power it had few instruments to influence economic restructuring because the Commonwealth government controlled the tariff, investment allowances and export incentives. The big exception was resource development. The State government had power over mining leases, the provision of infrastructure for natural resources, and electricity charges for large users like aluminium smelters. With these instruments the Wran government sought to organise:

> the transition of the state economy from an industrial base predominantly serving the domestic market, to a resource-based economy predominantly exporting natural resources and manufacturing commodities with a high energy component.[19]

The second minerals rush was taking the State of New South Wales in the direction of greater openness before the economic reforms of the federal Hawke Labor government in the 1980s.

Heading a Labor government, Wran shared the Whitlam government's ambitions to Australianise coal mining, or at the least to try to keep the NSW coal industry as much in Australian hands as was possible. Wran's Minister for Mines, Pat Hills, commented: "The thing is everyone talks as if energy policy is a federal thing, but they forget that control over coal rests with the States. It's my responsibility to protect these – our God-given right – and not have it taken over by some multinational".[20] CRA would have to wait for almost a decade, and the relaxation of foreign investment rules in the 1980s, before it would secure in 1989 a 70 per cent interest in Coal and Allied Industries.[21]

CRA had determined to secure the prize of recently discovered deposits that Coal and Allied had started exploring for in the Upper Hunter in New South Wales in the 1970s. These were the approximately 1000 million tonnes of soft coking and steaming coal in the Warkworth area in the Upper Hunter west of Singleton. They were believed by some to be equivalent to half of Utah's reserves of coal in Queensland. Coal and Allied Industries had been looking to export from that area 3.5 million tonnes of coal per year, nearly half of which would be steaming coal, starting in the 1980s.[22] One of the ways in which Wran sought to fight the battle against foreign-owned companies and to preserve Australian equity in the NSW coal industry was to allow the State Electricity Commission (SEC), one of the large producers of coal in the State, to enter the export market.

In the late 1940s the electricity system in New South Wales had been chaotic. It had been provided by a multiplicity of small county councils and private companies. In 1950 the NSW government had established the SEC to "expedite a better supply of electricity to the public" and to "promote and encourage the use of electricity and especially the use thereof for industrial and manufacturing purposes and for the purposes of primary production".[23] The SEC was structured, as the JCB was, around a core of experts maximising the supply of energy. Over time it won some autonomy from elected governments. In 1977, with Wran's approval, the SEC entered into a joint venture with the Japanese Taiheiyo Mining Company to export steaming coal from Newnes near Lithgow.[24] The Taiheiyo Mining Company, owned by the Mitsui group and with close links to Japanese power utilities, was one of Japan's largest miners of steaming coal. In 1977 the Wran government forestalled CRA's takeover of Coal and Allied Industries by controversially granting some of Coal and Allied's Warkworth exploration area to the SEC. In the midst of this takeover controversy, Coal and Allied Industries had secured an agreement for a steaming coal contract over 20 years estimated to be worth $500 million. This was a long-term contract with the Japanese conglomerate Ube Industries to increase the annual tonnage sent to Japan from 400,000 tonnes to one million tonnes after 1980-81. J.R. Bennett, the chief executive of Coal and Allied Industries, described the agreement as "the most significant coal supply arrangement in respect of period, tonnage and the type of coal ever entered into between an Australian coal producer and an individual Japanese company".[25]

After CRA's attempted takeover of Coal and Allied Industries had been aborted, the NSW government reversed an earlier decision. In February 1978 it awarded that company, still in majority Australian ownership, two separate leases in the Warkworth region: one comprising 56 million tonnes north of the Hunter River that would be known as Hunter Valley Number One; and another containing 350 million tonnes south of the river, known as Hunter Valley Number Two.[26] From this time onward, Coal and Allied Industries moved much more heavily into open cut mining in the Upper Hunter area of New South Wales in a decisive break with its more than one hundred years history of underground mining.[27] Coal and Allied borrowed more than $60 million, including $30 million from the Australian Resources Development Bank, $15 million from the Australian Industry Development Corporation, $10 million from the Bank of New South Wales and $10 million from the State Superannuation Board. With this money the company's new Upper Hunter mines commenced operations in the late 1970s and exported coal not only to the Japanese electricity companies and steel industries but also to power utilities in West Germany, Israel, South Korea and Hong Kong.[28]

The resources boom was not just a period of development in mining steaming coal. Despite the difficulties facing the Japanese steel industry in the 1970s, that country's steel mills supported the initiation of several new coking coal mines. In August 1977 BHP signed a contract worth $1300 million with the Japanese steel mills for the sale of 27 million tonnes of coking coal over 15 years from its Gregory Open Cut mine in Central Queensland.[29] The project involved BHP spending $200 million on a 73-kilometre spur line from its mine to the existing railway at Blackwater and on a new coal-loading facility at Clinton Estate upstream from Gladstone.[30] In the same year, BHP paid $100 million to buy out Peabody's stake, and thereby obtain a controlling interest, in the Queensland mining group Thiess Peabody Mitsui (TPM).[31] A few years later in 1981, BHP signed a contract to supply 47 million tonnes of coking coal worth $2600 million over 14 years from a mine at Riverside. The first shipments were to start in 1983 through Hay Point. TPM had been trying to negotiate a contract for the project, originally known as Nebo, in 1972 and had reached a basic agreement in 1975. However, the OPEC-induced slump in the Japanese steel industry had seen the development postponed. Under BHP's leadership, the group, now Thiess Dampier Mitsui (TDM), would secure in 1981 a contract described by the mine's manager as "in terms of volume and length of contract … one of the biggest in mining history".[32]

Besides BHP's mines, several other major new developments began in Queensland in the second half of the 1970s. In 1978 the Queensland government gave American oil giant Houston Oil authority to proceed with developing coal deposits at Oaky Creek in the Bowen Basin on condition that the American company spent $80 million on developing the mine in the years up to 1980.[33] Houston Oil, however, had difficulties in putting together a consortium that would satisfy the Fraser government's requirements for a satisfactory degree of Australian equity in the project.[34] Frustrated by the task, the American oil company withdrew from the project in 1980 after selling its stake in the project to half-Australian owned MIM holdings. MIM Holdings had been established in 1970 as the parent company of Mount Isa Mines Limited and was also involving itself in the iron ore industry through part-ownership of the Goldsworthy mine in 1977. MIM developed Oaky Creek with minority equity partners and began operations in 1982.[35] By 1990 MIM Holdings would be one of the largest mineral-based companies in Australia.[36] A few years earlier, in 1978, the Queensland government had approved another Central Queensland coal project at German Creek. This was a $400-million project financed by a consortium of Australian, British and German companies, including the British National Coal Board, the largest coal miner in Britain, Rührkohl, Europe's largest coal mining operation, and Austen and Butta. In a breakthrough for Queensland, the German Creek mine began operating in 1981 with its output going mainly to Europe and Britain.[37]

Another even bigger project was the development of a mine in Queensland to export steaming coal from Blair Athol. In the 1960s the Blair Athol mine was almost moribund, having a large ore body but a restricted Australian market.[38] CRA had taken a 62 per cent stake in this long-established Queensland mine in 1968 before the OPEC oil price rises of the 1970s spurred an international demand for steaming coal. A mine that was meeting the relatively small local demand for steaming coal suddenly became a prized asset in the mid-1970s with the rise in the price of oil. In the late 1970s, a Daniel Ludwig subsidiary, S. & M. Fox Pty Ltd, which owned the residual 38 per cent of Blair Athol, decided to sell its shares. CRA wanted to buy all the minority interest so as to own the mine outright. Then it planned to sell a minority stake to Japanese power companies as an inducement for them to agree to a long-term contract to purchase steaming coal from Blair Athol. CRA feared that it would not be able to secure long-term coal contracts if it did not offer the Japanese equity participation.[39] There was, however, another potential buyer of the residual shares in view – the American oil company Atlantic Richfield (ARCO).[40] Malcolm Fraser's Treasurer, John Howard, refused to intervene in CRA's favour and, as ARCO had the first purchasing option, the American company secured the shares, joining the long list of international oil companies which had by then bought into the Australian coal industry.[41]

In August 1979 the Blair Athol Mining Company (a consortium now consisting of CRA and ARCO) succeeded in signing a letter of intent with the Electric Power Development Company of Japan (EPDC). This gave EPDC, representing nine Japanese power utilities, an option to purchase a 19 per cent stake in the Blair Athol Mining Company. In June 1978 the Australian government had granted CRA "Australianising" status, meaning that the Australian government regarded CRA as Australian-owned, even though local equity in CRA was only 38.9 per cent, because of a commitment that CRA had made to lift its Australian equity over time to more than 50 per cent. The Blair Athol Mining Company sought to give the Japanese Electric Power Development Company equity in the Blair Athol mine by reducing the CRA and ARCO share. But Howard refused the application in April 1980 on the grounds that there would not be enough Australian equity left in the Blair Athol venture.[42] Opposition spokesman for Minerals and Energy, Paul Keating, was strongly critical of the Fraser government's policies towards Blair Athol, first in allowing ARCO to buy the minority interest in Blair Athol and then in putting obstacles in the way of Japanese equity participation. "The result", he complained, "is that an American oil company without any previous interest or connection with the project was awarded equity over and above that of a Japanese group which could provide both equity and contracts for the development of the mine".[43]

Despite the setback, CRA managed to arrange the Blair Athol consortium so as to satisfy the Treasury and complete a $500 million development with new infrastructure, including the construction of 116 kilometres of railway and a second coal-loading facility at Hay Point.[44] The expanded mine was officially launched in 1984. In 1981, as the Blair Athol mine was being expanded, the Department of Industry and Commerce reported that investment of more than $10 billion had been earmarked for new coalmines across Australia during the resources boom. This represented about 31 per cent of projected mining and manufacturing investment and covered 49 projects in New South Wales and 18 in Queensland, including CSR's new $200-million Drayton mine in the Hunter Valley and its $30-million Callide expansion; BHP's $305-million Saxonvale project; and MIM Holdings' $500-million Newlands Steaming Coal project and $300-million Oaky Creek mine.[45]

Under the management of Vernon's successor, Gordon Jackson, CSR had further diversified into minerals and energy in the 1970s and 1980s after its highly successful iron ore and bauxite ventures in the 1960s. At Mitsui's suggestion CSR had purchased the Hunter Valley soft coal producer Buchanan Borehole Colliery in the early 1970s. In the latter half of the decade it had secured a stake in a major coking coal project at Hail Creek in Central Queensland. But Jackson's most ambitious coal acquisition was the takeover in 1980 of Thiess Brothers. In 1979 Jackson launched a hostile $460-million takeover bid, the biggest up to that time in Australian history.[46] Bolstered particularly by its coal purchases in New South Wales and Queensland, CSR developed into a large diversified industrial group with gross sales of more than $3 billion in 1982 and five divisions – sugar, building materials, aluminium, minerals, and coal and petroleum – and more than half its export sales going to the Pacific Basin.[47] CSR's takeover of Thiess Brothers gave it particularly valuable coal properties at Callide in Queensland and Drayton in the Hunter Valley.

In the battle between Queensland and New South Wales for coal markets during the resources boom of the late 1970s and early 1980s, one critical area of competition was in development of infrastructure. There are similarities between the Wran government's surprise at the intensity of the coal rush during the resources boom and the realisation of previous State Labor governments that they would have to act decisively to develop the NSW coal ports in the 1960s. One commentator remarked in 1981:

> To an extent, the Wran government appears to have been caught napping by the suddenness and size of the upturn in the NSW coal industry's fortunes, and, struggling to get its act together, to be grasping awkwardly at hurried ways of servicing the boom.[48]

As a result of the Commonwealth–State program to upgrade the New South Wales ports in the 1960s, in the early 1970s New South Wales had three coal loaders with a total exporting capacity of 19 million tonnes per year – Newcastle with 10 million, Balmain three million and the new inner harbour of Port Kembla six million.[49] By 1974 it had become obvious that capacity at Newcastle was inadequate to meet demand.[50] So grave was the situation that in 1974 the Joint Coal Board issued Order Number 30, requiring the JCB's approval for any vessel loading at Newcastle. In March 1974 the JCB extended the order to Balmain and Port Kembla and it would stay in operation until 1977. To meet the increasing demands on Newcastle as a coal port in June 1974, mining companies financed the construction of a second privately-built coal loader at Port Waratah that was opened on 25 May 1977. The new $89-million facility boosted the annual loading capacity of Newcastle to 20 million tonnes per annum and was designed to handle 120,000 dwt ships, which would be able to enter the harbour once a $70-million project to deepen Newcastle harbour was completed in 1981.

Dredging the harbour involved removing 1.7 million cubic metres of rock from the harbour's entrance and deepening the Steel Work Channel to a depth of 13 metres. The ultimate plan to deepen the harbour to 15.2 metres would make it the third deepest port in Australia.[51] Even this expansion, however, was soon seen as insufficient. In March 1979 the Wran government established a task force that recommended that a third coal loader be developed at Kooragang Island. It would be completed in 1984 with a loading capacity of 15 million tonnes per year. By this time, Newcastle's loading capacity was raised to 42 million tonnes per year and Newcastle was on the path to becoming Australia's largest coal port and one of the largest in the world.[52]

Plans to upgrade coal-loading facilities on the south coast of New South Wales were less successful than at Newcastle during the resources boom. The downturn in the Japanese steel industry in the early 1970s and widespread political and environmental criticisms in New South Wales forced Clutha Development to abandon the idea of a privately built coal loader on the south coast. As with Aboriginal land rights, the impact of mining on the environment became an issue for Australians in the 1960s and 1970s.[53] The environmental impact of mining had been apparent in earlier times, notably in the barren and sickly landscape produced by the Mount Lyell copper mine on Tasmania's beautiful western coast in the period following World War I. But in the 1960s, with the advent of the second minerals rush, a national consciousness about mining and the environment began to develop. This consciousness was based on an aggregation of local concerns such as lead poisoning in Broken Hill and Mount Isa, the deadly effects of asbestos at Wittenoom in Western Australia, dust pollution in the new Pilbara towns and the effect of open-cut coal mining on the rural landscape in Queensland and New South Wales.

Mining affected the environment in different ways. Exploration produced large holes and excavation pits; operating mines produced emissions and waste products such as sulphuric acid and heavy metals released in the process of extracting minerals. Underground mines rendered land unusable; open-cut mines stripped whole ecologies away; and sluicing and dredging caused irreversible damage by diverting and polluting watercourses.[54] The campaign against Clutha's proposed private railway and ports was one of the early major environmental campaigns in New South Wales in the 1960s and 1970s.

Later, Clutha became part of a consortium of collieries from the southern and western area of the Sydney basin proposing the development of a coal loader at Botany Bay. As lower-grade coking coal and steaming coal from the area around Lithgow became more profitable during the resources boom, it became critical for collieries in that region to have access to coal loaders that could admit ships of up to 200,000 tonnes.[55] The collieries developed a plan for a coal loader at Botany Bay that would begin operating in 1979 with a loading capacity of seven million tonnes per annum and a capability of expansion to 25 million tonnes. The Askin coalition government had given approval in principle for the construction of the Botany Bay coal loader. But the Wran government responded to environmental and political criticisms of the proposed loader by announcing in June 1977 that it would not be built at Botany Bay. Instead, the Wran government announced that a new coal loader would be built at Port Kembla; that the coal loader at Balmain would be upgraded to assist in shipping coal from the west of New South Wales; and that a new rail link would be built to transport southern and western coal to Port Kembla. Industrial strife, however, meant that the new coal loader at Port Kembla would not begin to operate until late 1982. Due to port congestion in Wollongong, the Wran government had to ask the JCB to administer a scheme to divert coal from Port Kembla and Balmain to Newcastle in the period from May 1978 until the end of 1981. In the meantime, companies like Clutha Development had to use a combination of railway and road to get their coal to the existing coal loader at Port Kembla.[56]

To help coordinate the development of infrastructure, the Wran government established a Coal Resource Development Committee with representatives from trade unions, collieries and government authorities. The committee's *NSW Coal Strategy 1981* was mainly concerned with government provision of infrastructure for coal and the means of financing it. Public investment was as important as private investment in mining during the resources boom. A 1982 report of the New South Wales Department of Industrial Development and Decentralisation reported that most of the $5563 million forecast for public investment in New South Wales was concentrated in resource infrastructure, especially in railways and coal-fired electricity generation. This was equivalent to the total amount of private sector investment at the time.[57]

As the Wran government grappled with expanding the infrastructure for the coal industry, a symptom of the problem of increasing demand for coal and bottlenecks at ports came in the form of what was dubbed "Wran's armada". The ironic term described the large number of ships queuing at the ports of New South Wales in the early 1980s. The essential problem was that in 1982 NSW coal ports could load around 30 million tonnes of coal per annum. This was less than half of the capacity proposed in 1975 and commitments were running at 40 million tonnes per annum.[58] One economist claimed that infrastructure bottlenecks in the coal industry had lost Australia $1400 million of export earnings and the State $14 million in forgone royalties and $50 million in freight. In contrast with New South Wales, Queensland's railways and ports were better able to cope with the expansion during the resources boom, partly because the Queensland government had insisted on companies financing almost all the infrastructure required. During the resources boom, the Queensland government's receipts from exporting coal became considerable, and between 1979 and 1982 higher freight rates for 10 coal projects were negotiated. In 1980–81, the Queensland Treasury took $35 million in coal royalties and profits from rail freights of between $70 and $110 million.[59]

By 1982 the world recession, which had extended to the coal industry, to some extent alleviated the problem of port congestion. The *Sydney Morning Herald* commented on how the world recession was affecting the coal rush of the resources boom:

> The world-wide recession has halted the frenzied growth mid-stream. Those projects which were planned to come on stream in the next few years now hinge on one critical factor: if their production is

> covered by contracts they have a chance of surviving; if not, then they will almost certainly join the swelling ranks of coal projects which have been set aside for another, better day. Mothballed. Put on ice, as the saying goes.[60]

Thus MIM was able to proceed with new projects at Collinsville, Newlands and Oaky Creek because it had contracts for all of them. BHP, on the other hand, was forced to mothball a joint venture with AMAX at Boggabri and a steaming coal prospect at Bargo in New South Wales. Similarly the NSW Electricity Commission developed problems with a $450-million project at Mount Arthur and a $100-million project at Birds Rock in the western coalfields, while White Industries was forced to sell 30 per cent of its interest in the Ulan mine in the central west of New South Wales, 45 kilometres north-east of Mudgee.

The Aluminium Rush

After a slump in the period from 1972 to 1975, Australia's bauxite and alumina industry expanded in the middle to late 1970s. Then it began a major new foray into smelting aluminium. In the 1960s and 1970s Australia had focused on mining bauxite and converting it into alumina for sale to aluminium smelters overseas. Only to a more limited degree were Australian companies smelting aluminium. By the mid-1970s Alcoa had two alumina refineries in Western Australia; Comalco had organised a consortium, Queensland Alumina Limited (QAL), to produce alumina from Weipa bauxite in Gladstone; and Swiss Aluminium and CSR had built an alumina refinery in the Northern Territory to refine Gove's bauxite. But there were only three aluminium smelters in Australia in the mid-1970s – Comalco's at Bell Bay in Tasmania, Alcoa's at Point Henry in Geelong and Alcan's at Kurri Kurri in New South Wales. Comalco's other aluminium smelter had been built on the south island of New Zealand to take advantage of cheap hydroelectric power. Before the rapid development of the Bowen Basin, Comalco had not thought that Queensland could produce electricity cheaply enough.

Demand for the Australian bauxite and alumina grew at a rapid rate from the mid-1970s.[61] In 1976 the industry's total exports, largely of bauxite and alumina, exceeded $600 million, about 15 per cent of Australia's exports of mineral products.[62] By 1980 Australia mined more than one third of the world's bauxite and produced about one quarter of the world's alumina but was a much smaller provider of aluminium metal, contributing only about three per cent of the world's annual total.[63]

In Western Australia, the bauxite and alumina industry (Alcoa of Australia) moved steadily southwards in the 1970s to take advantage of large reserves of bauxite south of the alumina refinery at Pinjarra. Bunbury, one of Western Australia's oldest ports and an outlet for rural and timber products, was developed to cater for the expanding mining industry. In the 1970s the WA government organised a program of works that deepened the port's shipping channel so that it could also provide an additional outlet for Alcoa's alumina. Once the expanded port was opened in 1976, Alcoa had the option of transporting alumina from Pinjarra on the State rail system either to the newly created port of Kwinana or the established port at Bunbury.[64]

Geoffrey Blainey has observed that the development of Bunbury assisted Alcoa's expansion. It also created opportunities for its competitors in Western Australia. Alcoa had staked claims to the most promising bauxite reserves in the Darling Range but had seen no point in paying the government rent for lower grade bauxite outside its leases. Taking advantage of this opportunity, Rupert Murdoch, head of News Corporation and owner of some Australian newspapers, acquired 5000 square kilometres of bauxite leases east of Alcoa's and attempted to organise an aluminium venture with BHP to develop them in the 1970s. The name of the new venture was Alwest. BHP, which had previously joined forces with Reynolds Metals Company to bid unsuccessfully for bauxite leases at Gove, brought with it that American company, the fourth largest aluminium company in the world, into Alwest. The Alwest concept was to locate a bauxite mine at Mount Saddleback, about 120 kilometres north of Bunbury, and to send each year 700,000 tonnes of Collie coal to Bunbury. There, a power station would generate cheap electricity for an alumina refinery that would be built at Bunbury, or perhaps inland at Worsley, to treat the bauxite.

The Whitlam government held up the Alwest plan mainly because of environmental considerations.[65] But by 1974 Charles Court had replaced John Tonkin as Premier of Western Australia and sought to resurrect the project. Court's idea was that Alwest would not build a separate alumina refinery but rather that Alcoa and Alwest should jointly build and manage a refinery at the rural town of Wagerup, south of Pinjarra. Court's idea was that, like the partners in QAL, Alcoa and the partners in Alwest would each provide finance for the jointly operated refinery and take their revenue and profit from the alumina that would be shipped from Alcoa's wharf at Bunbury. In addition, Court suggested an exchange of leases – Alwest would give Alcoa a large part of its area near Mount Saddleback and in return Alcoa would provide Alwest some of its leases near the new refinery. With Court's support, Alcoa and Reynolds Metals Company signed a memorandum of agreement at the end of 1976 to contribute capital costs of $666 million to build a new alumina refinery at Wagerup with a total yearly output reaching as high as perhaps four million tonnes of alumina, making it the biggest alumina refinery in the world.[66] But in 1977 the American partners abandoned the joint venture because of fears that it might violate US anti-trust laws. After that, they went their separate ways. Alcoa in 1979 decided to proceed alone with building its third alumina refinery at Wagerup, while BHP and its main partner, Reynolds Metals started building their own alumina refinery at Worsley in the following year. Alumina production at Worsley would start in 1984, increasing to a production rate of more than one million tonnes of alumina per annum by the end of the decade.[67]

By the early 1980s, the energy crisis was transforming the world aluminium industry. Higher oil prices made Australia's coal-fired electricity plants attractive as a source for providing the power necessary to smelt aluminium metal from alumina. North America and Europe were running out of new major sites for hydroelectric power, the cheapest type, and nuclear power was being looked at more cautiously because of a nuclear accident at Three Mile Island in the United States in 1979. Prices of both oil and natural gas had increased during the energy crisis and were gravely affecting the aluminium industry in North America and Japan; the Aluminum Co. of America (Alcoa) would close down a smelter in Texas powered by natural gas in 1982. In Japan the aluminium industry began to be driven out of smelting because of its heavy dependence on oil or gas, making its aluminium twice as costly to produce as metal imported from the United States. By 1981 Japan would shut down 45 per cent of its smelting capacity.[68]

At the same time that oil companies were beginning the rush into the Australian coal industry, international aluminium companies and Australian firms began to see that new aluminium smelters would not be built in Japan, the United States or Europe but in energy-rich areas, including Australia.[69] Australia became a favoured destination for aluminium smelters in the late 1970s, resulting in a rush into the aluminium industry that was a major dimension of the resources boom.

In 1978 Comalco announced plans that at last fulfilled hopes that Queensland politicians had harboured since the 1950s by beginning to construct an aluminium smelter at Gladstone with an initial capacity of 206,000 tonnes per year, rising eventually to 412,000 tonnes.[70] Meanwhile, throughout 1978, Wran had been actively seeking to entice aluminium companies, such as the Californian company Alumax and the French company Pechiney, into New South Wales. He appeared to have succeeded in the first instance with Alumax.[71] In 1979 it announced its intention to build a $500-million aluminium smelter near Lochinvar, 11 kilometres west of Maitland in the Hunter Valley. It was joined at the end of the year by BHP, which decided to take a 35 per cent interest in the planned smelter.[72]

By 1978 power stations at Munmorah, Vales Point, Wangi and Liddell used 10.7 million tonnes of Hunter region black coal to generate 84 per cent of all the electricity generated in the State.[73] To provide for the needs of aluminium smelters, the State Electricity Commission developed an accelerated capital works plan to increase NSW electricity-generating capacity by two-thirds. It had been forecast that the three proposed aluminium smelters would consume two-thirds of the electricity produced by one of the two larger power stations being located in the Hunter Valley. Given the long lead-times in constructing power stations, NSW Energy Minister, Pat Hills, devised a program which he described as matched "nowhere in the world except perhaps in France which has more than ten times the population of NSW".[74] Taking advantage of reforms to the procedures of the Loan Council in 1978, the State borrowed enormous sums to construct three large coal-fired electricity stations at Eraring, Bayswater and Mount Piper at a cost of $3 billion to increase the State's generating capacity to 6600 megawatts. But by giving away too much electricity too early, the Wran government lost the opportunity to auction its electricity to competing aluminium companies, a problem that was compounded when Alumax abandoned its proposed venture.[75]

By 1980 there were many companies talking to Australian State governments about the possibility of building aluminium smelters. Nabalco (Swiss Aluminium and CSR) was contemplating a smelter in New South Wales that was blocked after the government earmarked all the available electricity for Alumax. Alcan was talking to the Queensland government about another possible smelter in that State. BHP and Reynolds Metals Company were looking at possibilities in Western Australia; Shell and Billiton were discussing Victoria; and Pechiney was investigating a smelter in Queensland.[76] BHP underlined the importance of aluminium to its operations by establishing a new aluminium division, along with steel, minerals, and oil and gas, in July 1980.[77]

Only some of the planned aluminium developments were completed during the resources boom. Alcan dropped the idea of a new aluminium smelter in Australia but in July 1980 clinched a $600-million long-term contract with Nippon Light Metal that enabled it to expand its existing smelter at Kurri Kurri.[78] Although Nabalco did not succeed in building an aluminium smelter, one of the partners in Nabalco was Gove Alumina, majority owned by the Australian industrial and mining conglomerate CSR. In 1980 Gove Alumina and the Australian Mutual Provident Society (AMP) took a 50 per cent stake in a planned new aluminium smelter at Tomago in the Hunter Valley and persuaded Pechiney to take a 35 per cent minority interest. On 23 January 1980 CSR ensured that it would be able to raise its $210-million share of the financing of the smelter by securing a deal worth $600 million to supply Sumitomo Light Metal Industries with about 400,000 tonnes of aluminium metal over 12 years from 1983.[79] The deal confirmed CSR and Gove Alumina as new entrants into the world aluminium industry.

In 1978 Alcoa of Australia announced the expansion of its Point Henry smelter at Geelong from 104,000 tonnes of ingot a year to 161,000 tonnes at a cost of $110 million.[80] The expansion of Point Henry would be completed in 1981, but in the meantime Alcoa had decided on building another much larger aluminium smelter at Portland in southwest Victoria. The oldest port in Victoria, Portland remained a relatively undeveloped town of a few thousand people until modernised in 1961 by breakwaters that could protect 70,000 tonne ships from the westerly winds of the southern Indian Ocean. In a speech to the Victorian Parliament on 25 March 1980 Liberal Premier Dick Hamer announced the benefits of what he expected to be the largest industrial venture Victoria had seen.[81] It would take three years to construct the first stage of a smelter that would have a capacity of 130,000 tonnes per year and, by 1988, after four separate stages of expansion, it would become one of the world's largest aluminium smelters, with a capability of 530,000 tonnes each year. The Victorian government seized on the Portland development as its prize in the scramble by the Australian States for aluminium smelters during difficult economic times for Victoria.

Like New South Wales, Victoria was in economic decline in the late

1970s. The international recession of 1974 had ended a long period of economic growth as its manufacturing industry faced stronger competition from abroad and capital expenditure on Victorian industry declined for the first time in decades. The numbers employed in manufacturing were in long-term decline, having fallen from 32 per cent in 1966 to 24 per cent in 1976.[82] Victoria's economic performance in the late 1970s, measured by manufacturing turnover and employment, was in fact poorer than that of New South Wales as a consequence of the high concentration in Victoria of the depressed textile, clothing and footwear industry and the automobile industry.[83] In this context, the prospect of a major industrial development based on the resources industry was immensely attractive. The Liberal–National Party government was able to agree to Alcoa's Portland project because, like New South Wales, the Victorian State Electricity Commission had embarked on a multi-billion dollar expansion of brown coal-fired power stations in the 1970s. The government had approved in 1976 the giant Loy Yang station, scheduled for completion in stages between 1983 and 1992. As an inducement for Alcoa to proceed with its aluminium project, the Hamer government agreed in 1979 to offer Alcoa a concessional rate of power until 1989. In turn, Alcoa agreed to contribute about one-fifth of the capital cost of the transmission lines to Portland and for its concessional electricity tariff to rise in line with other charges.[84]

The Alcoa-Portland project, though it would eventually succeed in the later 1980s, became a casualty of the coming to an end of the resources boom. In 1982 the Victorian SEC was facing capital commitments that had blown out to $820 million. It was consequently forced to raise its electricity charges by 20 per cent for ordinary users and 25 per cent for Alcoa. At the same time, with the onset of international recession, Alcoa was facing a deterioration of demand for its products along with the unexpected rise in electricity prices. Sir Arvi Parbo, now chairman of Alcoa, criticised both the SEC and the Victorian government for reneging on its agreement by imposing a discriminatory tariff. Though reluctantly agreeing to the construction of the first stage of the Portland project, Alcoa suspended the project in 1982 because of low aluminium prices.[85]

Victoria's problems with Alcoa were matched in New South Wales when a series of plant failures brought that State to the verge of a breakdown of its electricity-generating system. The New South Wales SEC's problems were undoubtedly exacerbated by its breakneck expansion to accommodate the expanding aluminium industry. The Wran government sought to solve its problems by selling off part of the electricity-generating industry to the private sector, announcing in 1981 that a consortium of private companies would finance the Eraring Power station on the western shore of Lake Macquarie.[86]

Uranium

In 1976, as the coal and bauxite/aluminium industries were preparing to boom again, the uranium industry was still in a position of uncertainty. The Whitlam government's *Environmental Protection (Impact of Proposals) Act* 1974 required impact statements for major development projects and public hearings where Commonwealth decision-making was involved. The first public inquiry under the act was into sand mining on Fraser Island, 200 kilometres north of Brisbane, as a result of which the federal government banned mining on the island in a landmark environmental decision.[87] In July 1975 Whitlam set up a higher-level inquiry into the proposal by the Australian Atomic Energy Commission and private mining companies to mine and mill uranium at the Ranger site in the Northern Territory. The Ranger Uranium Environmental Inquiry was charged with inquiring broadly into the environmental aspects of the development. It was headed by Justice Russell Fox of the Australian Capital Territory Supreme Court and included as members Charles Kerr, Professor of Preventive and Social Medicine at Sydney University, and Graeme Kelleher, a civil engineer.

The Fox Commission began hearings on 9 September 1975 and

continued after Sir John Kerr's dismissal of the Whitlam government in November 1975. On coming to office, the Fraser government was keen to promote the development of mineral resources, including uranium, and may have been tempted to try to close down the Fox Enquiry, as it had ended a Royal Commission on Petroleum, by cutting off its funds. But militating against this option was that the Ranger Commission had been set up by statute and was not subject to the discretion of ministers. The Fraser government therefore stayed its hand in announcing a policy on uranium until the Fox Commission had handed down its findings.

In October 1976 the Fox Commission issued its first report, recommending that with proper regulation and controls it would be safe to mine and sell Australian uranium.[88] While the Fraser government took this first report as support for its wish to go ahead with the industry, the Fox commissioners were at pains to make clear that they were actually recommending proceeding with caution after due public discussion and debate. Their main continuing reservations with establishing a uranium export industry were the need for adequate safeguards to prevent the diversion of uranium to nuclear weapons and the perceived weaknesses of safeguards regimes established under the Nuclear Non-Proliferation Treaty (NPT).

The Fraser government moved to address these concerns in May 1977 by announcing that Australia would export uranium only to countries that accepted the safeguards and inspections of the International Atomic Energy Agency. It also announced that countries importing Australian uranium would have to agree neither to enrich Australian uranium beyond 20 per cent nor to reprocess spent fuel without the Australian government's prior consent. In the meantime, a large popular movement against uranium mining had started to mobilise. Within the Australian Labor Party opposition to uranium mining also grew, with South Australian Premier, Don Dunstan, joining those opposed to development in 1977. At the 1977 ALP conference, Dunstan and the Leader of the Victorian Labor Party, Clyde Holding, successfully moved that a future federal

Labor government would prohibit the mining and exporting of uranium until the party was satisfied that the problems of disposal of waste and proliferation of nuclear weapons had been satisfactorily addressed. It was in these circumstances of growing public concern that Prime Minister Malcolm Fraser announced on 25 August 1977 the government's approval for uranium exports. Fraser presented the decision as a sober response to Australia's obligations as a good international citizen:

> The government has made this decision with a deep sense of international responsibility. Were it not for that sense of responsibility, were it not for our obligation to provide energy to an energy deficient world, we would not have decided to export uranium. Commercial considerations were not the dominant motive in our decision. In themselves they would not have been sufficient.[89]

Fraser's Minister for Trade and Resources, Doug Anthony, on the other hand, stressed the economic importance of uranium mining, predicting earnings in excess of $2 billion by the year 2000 and thousands of jobs for Australians.

By the time that the Fraser government had announced its uranium decision, it had steered through parliament another Whitlam initiative, the *Northern Territory Land Rights Act* 1976, which empowered Northern Territory Aboriginal people, acting through newly established Land Councils, to negotiate the terms and conditions of mining on Aboriginal land. The legislation was the result of the movement for land rights for Aborigines that had developed in the early 1960s as a result of local Aboriginal opposition to bauxite mining at Gove. The act did not give traditional owners of land a veto over mining operations already in existence, like Ranger and Narbalek, but only the power to negotiate on the terms and conditions of mining. The Northern Land Council (NLC) negotiated an agreement on Ranger with the joint venturers, Peko-Wallsend, Electrolytic Zinc and the Australian Atomic Energy Commission representing the Commonwealth government. As a result of these negotiations, the NLC secured 4.25 per cent royalties and a $2 million upfront payment. Monies from mining were distributed according to a formula specified in the Land Rights Act. The three major Aboriginal Land Councils, the Northern, Central and Tiwi Councils, received 40 per cent of royalties; Aboriginal peoples living in areas affected by mining received 30 per cent; and the remaining 30 per cent was disbursed to other Aboriginal people living in the Territory.

Despite claims that Aboriginal peoples would obtain great largesse from uranium mining, the sums they obtained were modest.[90] The Gagudju Association, for example, devoted royalties from Ranger to such projects as the Cooinda Motel in Kakadu National Park and only $1000 per annum to individuals.[91] Ratification of the Ranger agreement was complicated first by NLC chairman, Galarrwuy Yunipingu, describing it as "rotten" and then by two other NLC members taking legal action to prevent its ratification.[92] Only after the intervention of the Minister for Aboriginal affairs, Ian Viner, was the "Ranger Agreement" adopted in 1978. Similar Aboriginal concerns about royalty payments along with environmental concerns delayed an agreement with Queensland Mines on Narbalek, but in 1979 the Fraser government was finally in a position to issue mining leases for both Ranger and Narbalek. In the 1980s, when advocates for Indigenous peoples pressed for the legal regime in the Northern Territory to be extended to the States, the mining industry would conduct one of its most successful political campaigns against a system of national land rights.

The Fraser government in the meantime sold off the AAEC's interest in Ranger on the grounds that only if overseas interests had direct equity in the venture would they be prepared to enter into long-term contracts. As a result of the sale, a new company, Energy Resources of Australia (ERA), was formed to mine Ranger. Peko-Wallsend and Electrolytic Zinc each held 30.5 per cent of ERA, a Japanese consortium 10 per cent, a German and Danish company 15 per cent, while 14 per cent of the company was floated on the Australian stock exchange.[93] When it came time for the new company to negotiate uranium contracts, the overseas equity in the project did not assist the company to secure a good outcome. The price of uranium had already started to drop from its peak in 1976 and Ranger negotiated its contracts after the Three Mile Island accident. Of the other Northern Territory uranium projects, Noranda sold its interest in Koongarra to Denison Australia, subsidiary of a Canadian company, and the project was never developed. Pancontinental's Jabiluka project was held up because of its potential to cause severe environmental damage until the company switched to an underground mine, which received Aboriginal approval in 1982.[94] Although there had been high hopes in the early 1970s that uranium exports might vie with iron ore, export receipts from uranium reached only $415 million or 2.2 per cent of total exports in 1981-82.[95] The failure of the uranium industry to reach its potential was partly due to the inability of Australia to enrich uranium but also to the Indigenous rights issue and the increasingly influential environmental campaigns that culminated in the Hawke Labor government's "Three Mines" policy. This policy allowed uranium mining only at Ranger, Narbalek and Olympic Dam in South Australia (to be discussed in Chapter Eight).[96]

Because mining was largely the responsibility of the States, it was State governments that enacted the first environmental legislation and that empowered bodies like the Joint Coal Board in New South Wales to have regard to the impact of coal mining on the environment. In 1974, however, the Whitlam government's federal Environmental Protection (Impact of Proposals) Act became law. The legislation introduced the environment as a formal factor in Australian government decisions by requiring environmental impact statements for certain projects and, in some cases, public inquiries. The States, which had the power to issue production leases, also had their own power to seek environmental studies and hold public inquiries.[97] The Fraser government resisted pressure from the States and private enterprise to return environmental powers to the States.

In the 1960s and 1970s there were major public environmental campaigns against sand mining in Myall Lakes and against coal-loading facilities on the New South Wales South Coast (Clutha and Botany

Bay). In the mid-1970s the strength of the public campaign induced the Fraser government to halt sand mining on Fraser Island using its export controls powers. From then onward, major mining projects such as the uranium enterprises in the Northern Territory, coal projects in the Bowen Basin, iron ore proposals in the Pilbara and the natural gas project for the North West Shelf were required to make environmental impact statements.[98] The Australian Mining Industry Council accommodated itself to the new environmental regime, although it was very critical of the inadequacy of compensation offered by the Fraser government to the mining company involved at Fraser Island.

Iron Ore

During the 1970s resources boom, the iron ore industry developed at a pace dictated by the slower rate of growth of the Japanese steel industry, the main customer. Between 1950 and 1970 world production of steel increased at an average annual rate of nearly six per cent. But in the years from 1971 to 1977, the annual growth rate fell to an average of 2.4 per cent.[99] The Australian iron ore industry also had to compete during the resources boom with the Brazilians, who received a significant boost during the period. From 1971 the Japanese steel mills and their allied trading houses started sponsoring new iron ore projects in Brazil by making long-term contracts with, and minority investments in, three local firms – Mineracoes Brasileiras, Nibrasco and Capenama. As a result 13 new projects had developed in the Brazilian iron ore industry by the mid-1980s.[100] But despite slower growth between 1976 and 1983, Australian iron ore exports were still earning $1482 million or 7.2 per cent of merchandise exports in 1982-83.[101]

Indeed, coal and iron ore together were the main drivers of growth in Australian exports in the middle to late 1970s. Coal export receipts rose threefold, from $A352 million in 1973-74 to $A1282 million in 1977, while iron ore receipts almost doubled, from $A473 million to $A901 million, over the same period.[102] But Australian iron ore mines ran at 20 to 30 per cent below their capacity of 120 million tonnes per year. This reflected both the sluggishness of the Japanese steel industry and also unusually severe industrial troubles. The WA Industrial Commission, iron ore companies and governments were all concerned about the state of industrial relations in the Pilbara in the first half of the 1970s. In 1976 adverse comment started to come from Japan. The journalist Max Suich commented in the *National Times* in November 1976 that "[t]here is a general feeling of insecurity in Japan about the stability of Australian supplies of raw materials, founded in the problems of strikes in the iron ore and coal mines and their ports and railways".[103] When Saburo Tanabe criticised industrial relations in the Pilbara in September 1976, Court interpreted the remarks as an attack on the trade union movement, which he made clear with an attack of his own. Reacting to Court, in January 1977 the WA Trades and Labor Council sent a delegation to Japan led by the President of the Australian Council of Trade Unions (ACTU), Bob Hawke. Hawke promised Tanabe to try to introduce industrial relations arrangements in the Pilbara that would head off disputes before they reached the strike stage.[104] His efforts, however, unsuccessful. Man-hours lost to industrial action in the Pilbara were 65,538 in 1975; in 1979 they had risen to over one million.[105]

Hamersley Iron, with its two mines at Mount Tom Price and Paraburdoo, accounted for about 46 million tonnes of capacity and Mt Newman a similar figure. Robe River expanded its capacity from about 15 million tonnes per year to nearly 20 million tonnes per year by the end of the 1970s.[106] By the early 1980s, despite its slowing rate of growth, Australia achieved the position of the world's biggest exporter of iron ore, although it would be overtaken by Brazil later in that decade.[107] Australian iron ore mines had acquired close to half of the Japanese iron ore market and 20 per cent of the world iron ore trade, which was not far short of Saudi Arabia's share of the world oil trade.[108] Moreover, despite the downturn in the steel industry in the late 1970s, large Australian companies such as Hamersley

Iron actually became more profitable in the second half of the 1970s than they had been earlier in the decade. Hamersley increased its net profit from $42 million in 1976 to nearly $58 million in 1977.[109]

In the mid-1970s, the iron ore industry had nurtured high hopes of expansion. Port Hedland became one of the world's busiest ports. In 1975 the Hamersley railway carried more freight than the NSW and Victorian railway systems combined.[110] The established Pilbara mining companies and new players seeking to enter the industry were negotiating with the Japanese mills to take up to an additional 70 million tonnes of ore per year in the 1980s on top of the roughly 80 million tonnes that Australian producers were supplying in the late 1970s. The object of establishing a new mine was most urgent for the Goldsworthy Associates, who were facing the prospect of running out of iron ore from their existing mines at some time in the 1980s. Consequently, they made vigorous efforts in the second half of the 1970s to persuade the Japanese mills to back them in opening a new mine known as "Area C", 92 kilometres west-north-west of Mount Newman. This area was estimated to have proven reserves of 340 million tonnes of the type known as "Marra Mamba".[111]

In 1977 the reshuffled Goldsworthy partners wanted Japanese backing to invest $300 million in a new mine at Area C that would produce six million tonnes in 1979 rising to 18 million tonnes in 1981.[112] In order to entice the mills to sign contracts, they were willing to give the Japanese between 20 and 30 per cent ownership of the new mine.[113] But Lang Hancock and Peter Wright had an even more ambitious new project. Hanwright had 50 per cent ownership with American mining company Texas Gulf Sulphur of a project to mine 239 million tonnes of proven deposits at Marandoo, 45 kilometres east of Mount Tom Price. The Marandoo partners wanted to raise $A679 million to develop a mine capable of producing 10 million tonnes of ore per year. Like the Goldsworthy partners, Hanwright and Texas Gulf Sulphur were willing to offer the Japanese one-third ownership of the project. Meanwhile, BHP, which was a partner in the Mount Newman Mining Company, had large reserves of limonite ore of its own at Deepdale that it hoped to develop. "The Big Australian" believed that it could be producing 15 million tonnes of iron ore per year in the third year of operation of a Deepdale mine.[114] In September 1975 BHP launched its Deepdale plan by exercising an option to purchase, for $40 million, a 50 per cent interest in the adjacent Robe River project's railway and its stockpiling and ship-loading facilities at Cape Lambert.[115]

What was needed in the case of each new mine was for the Japanese mills to sign a long-term contract and perhaps also invest money. Opening a new mine capable of producing 20 million tonnes of ore per year now required investment of about $1 billion. The other, cheaper option which the Japanese mills had before them was to support the expansion of the existing capacity of established mines by a few million tonnes per year, which could be achieved at one tenth of the cost of building a new mine. While Australian companies were urging them to take additional tonnages of 70 million, the Japanese expected to contract for only about 25 million extra tonnes from existing and potential partners in the 1980s.[116] By the late 1970s it became evident that iron ore was in oversupply, even without the establishment of new Australian mines.[117] Brazil's expansion was in large part the cause.

Despite Australian urging, the Japanese steel mills declined to take a position on backing a new Australian iron ore project in the late 1970s and early 1980s. Nonetheless, three of the established Pilbara miners, Hamersley Iron, the Mt Newman and the Robe River Associates, all expanded their operations. Hamersley Iron was the first, initiating the construction of a beneficiation plant at Mount Tom Price in 1976 at a cost of $375 million.[118] This was able to upgrade or mix low-grade with high-grade ore to a point where it was marketable. Mount Newman soon followed suit, in developing a beneficiation plant capable of producing an extra five million tonnes per year.[119] Both Hamersley Iron and Mt Newman, as large mining companies by world standards, were influenced more by the long-term view and less by the short-term vagaries in the markets to which they were tied. Constructing the beneficiation plants meant that, even though both companies were facing a lower-than-expected Japanese demand, there was nonetheless increasing activity in the Pilbara in the late 1970s.[120]

The progress of the Robe River project had been complicated by the collapse in the early 1970s of Mineral Securities, one of the largest shareholders in Robe River Limited, the Australian group in the Robe River venturers. The WA entrepreneur, Alan Bond, subsequently bought a large parcel of Robe River shares, which he elected to sell in 1976. Hancock made a bid for Bond's shares by offering to sell Robe River Limited one-third of his royalties from Hamersley Iron's exports and one quarter of the iron ore venture in Wittenoom, including his share of the not-yet-developed Marandoo deposits.[121] He was still chafing at the Tonkin Labor government's decision to take away his temporary reserves in the Angelas and later to give the Robe River project one of them. By purchasing a stake in Robe River Limited, Hancock hoped to fulfil an ambition for the joint development of Marandoo and the West Angelas. Harking back to the Daniel Ludwig plan for the Pilbara in the early 1960s, Hancock's plan was to build a unified rail system to a port at Ronsard, 80 kilometres west of Port Hedland, that would service all the new iron ore mines in the Hamersley Ranges.[122] In the end, however, Bond decided not to accept Hancock's offer but instead to sell his shares to the old firm Burns Philp and Company.[123]

In the meantime, the Robe River Associates approached the Japanese steel mills seeking sufficient contracts to justify a $65-million expansion of their operations. Essentially, the company wanted to increase its capacity to sell iron ore fines to the Japanese from 3.5 million tonnes per year to 13 million tonnes per year while keeping its sales of pellets to about five million tonnes per year.[124] This proved important for the long-term future of the Robe River project, which had been launched as a pelletising operation in the early 1970s. But the OPEC oil price rises in the 1970s made pellets increasingly uneconomical because oil was an essential component of the process. One of the great achievements of Robe

River's early management was its success in persuading the Japanese mills to convert pellet contracts to fines contracts.[125] The large Japanese equity in the operation, via Mitsui and Co., helped the Robe River Associates to convince the mills to make these vital changes.

In June 1977 Mitsui, along with Nippon Steel Corporation and Sumitomo Metal Co., increased Japanese equity further by paying $21 million for the five per cent stake in Robe River, owned by Garrick Agnew's company, Mount Enid Iron. This represented the biggest new investment that the Japanese steel industry had made in the Australian iron ore industry and fell into a pattern of Japanese customers attempting to obtain an equity holding in Australian resource projects as they were doing in Brazil. In 1976 the Robe River joint venture also received from BHP about $40 million for the sale of a half share in the Robe River infrastructure. On top of that, in 1977 the Robe River Associates were awarded contracts from the European steel industry and the Japanese mills that allowed them to proceed with a $65-million expansion program designed to lift the project's capacity from 15 million tonnes of iron ore per year to about 20 million tonnes. The first phase of the project involved spending $25 million on accommodation for extra workers while other spending was on additional rolling stock, mining equipment and maintenance facilities.[126]

Thus the three biggest established Pilbara mining companies managed to weather the downturn in the steel industry in the second half the 1970s and to expand their operations, although the Goldsworthy Associates were unable to initiate their Area C project. But in mid-1978 the Japanese steel industry gave notice of significant changes in the iron ore trade. Pointing out that the Japanese steel mills had helped the Australian iron ore industry out of its revaluation problems in 1973 and 1974, they argued that they were now in a difficult economic position having made losses in 1977 of $700 million. Consequently, they wished to recoup these losses by reducing the cost of iron ore procured from Australia under long-term contracts.[127] The original contracts between Australia and Japan had been based on fixed quantities and prices. But in the period of inflation and currency instability after 1972, yearly reviews of prices had been incorporated. The Japanese gave notice of their intention in 1978 to cut back quantities of iron ore purchased and to limit price increases.[128]

The pressure of the Japanese mills on Australian iron ore companies in 1978 led to a bitter dispute between the federal Liberal-National coalition government in Canberra and the Court government in Perth. The federal Minister for Trade and Resources and Country Party leader, Doug Anthony, retained the essentials of the system of mineral export controls that Rex Connor had introduced in 1973. This system involved discussion and broad agreement between the government and mining companies on contract prices before individual iron ore companies negotiated with the Japanese steel mills. After negotiations, Anthony could either approve or disapprove the contracts.[129] In 1978 the Japanese mills successfully divided three companies negotiating contracts – Hamersley Iron, Mt Newman and the Mount Goldsworthy Mining Associates (MGMA). The mills convinced Mt Newman and MGMA to accept prices below levels that the Brazilian iron industry had negotiated with the Europeans earlier in the year. It did so by giving them what the *Australian Financial Review* described as a "kickback" in the form of an unscheduled upward revision of Mt Newman's largest iron ore contract.[130] Anthony held off on approving the contracts in the hope that Hamersley Iron, which had not been offered any sweetener, could negotiate a better deal. Anthony's decision irritated Court, who was in Japan during the negotiations. The WA Premier telexed Prime Minister Malcolm Fraser on 4 August 1978 urging him to allow the deal to proceed. Court explained to Fraser that "as a result of my discussions with the Japanese steel mills I reached certain basic understandings for the long-term continuity of our market share and paved the way for companies quickly to conclude negotiations".[131]

While Court was pressing for a resolution, Anthony was deeply dissatisfied with the single-buyer approach that now affected almost all of Australia's iron ore exports. He noted to Fraser that West Germany purchased iron ore on behalf of the Western European steel industry and settled the price with the Brazilians in what became a "benchmark" in the world market.[132] The Japanese, he explained, maintained close communication with the European buyers and used those negotiations as the basis for their own reduced offer to the Australians. Prices to other markets in the region – China, South Korea and Hong Kong – were related to the Japanese prices. Anthony concluded:

> This year's "negotiations" with the [Japanese steel mills (JSM)] who take nearly 80% of our iron ore exports, in effect represented a scramble for tonnage already pre-determined by the JSM. Price reductions have not resulted in an additional ton of iron ore being sold; indeed, as noted above, cuts below minimum contractual tonnages have been the order of the day.[133]

Confronted with an unpalatable dilemma, Anthony considered disapproving the contracts and trying to bluff the Japanese mills into giving more. In the end, he decided to approve them, while also registering with the Japanese government his deep dissatisfaction with the way that the negotiations had been conducted.[134] The result of the deal was that the landed cost of Australian iron ore into Japan was significantly below the price negotiated by Brazil early in 1978. Anthony complained to the Japanese government that "if I accept this – and it is very unsatisfactory – we will have to do some heavy thinking on future Australian marketing policy".[135] At the same time Anthony informed Australian iron ore producers that he was looking to make comprehensive changes in how Australian iron ore was sold.

Anthony's initiative elicited an outraged complaint from Court who remonstrated:

> I believe what is suggested is not only disastrous from a trading point of view and our credibility as a trading nation, but is contrary to the basic private enterprise philosophies which the Commonwealth and Western governments profess to espouse.[136]

Court had fought against Commonwealth guideline prices for iron ore in 1966 and 1967 largely with the aim of getting started the last of the foundation Pilbara projects, Robe River. In 1978 he wanted to limit federal intervention in iron ore negotiations in the hope of maintaining Australian market share in the world iron ore trade and of obtaining Japanese help to initiate the Goldsworthy Area C project. With the objective of settling the serious dispute that had developed over iron ore pricing, Court suggested a meeting in the Pilbara between Commonwealth and State governments to discuss the issue.[137] But Anthony, incensed by the Japanese steel mills' negotiating tactics, introduced tougher policy guidelines for Commonwealth control of mineral exports without cabinet approval in October 1978.[138] Court reacted by expressing his "dismay and bitter disappointment" about the Fraser government's failure to consult him.[139] In Court's rather exaggerated view, Anthony's action amounted to "nationalisation of our mineral exports". He warned Fraser that Anthony's policy prescriptions would stifle large-scale investment in long-term iron ore projects and would end up in the federal government having to allocate, or at least influence, iron ore tonnages between competing private companies. The Japanese steel mills also retaliated against Anthony's plans by cutting Australia's share of Japanese iron ore imports by five per cent in 1979.[140] The combined actions of Court and the Japanese steel mills won the day, with Fraser agreeing in 1979 to the Commonwealth allowing mining companies more latitude in negotiations with the Japanese.

Despite the easing of Commonwealth controls over iron ore exports, the Japanese did not make a decision to support a large new project in the Pilbara. By 1980 two contenders were vying for Japanese support after BHP had decided not to pursue for the time being its Deepdale option and CRA had purchased the Marandoo deposits from Texas Gulf Sulphur. MGMA was still pursuing the Area C project. At the same time, the Robe River Associates, who also faced the prospect of running out of iron ore in the 1990s, wished to develop the West Angelas. While the WA government had persistently pressed for the development of Area C, the Japanese mills were more inclined to support Robe River's West Angelas project.[141] But neither new project was developed during the resources boom. The onset of worldwide recession in 1982, and the shift from steel to the greater use of aluminium, ended hopes of a new mine in the Pilbara until the end of the 1980s.[142]

The North West Shelf Natural Gas Project

In the 1960s the WA government had given a small Melbourne-based company, Woodside (Lakes Entrance) Oil NL, exploration leases over 160,000 square kilometres off the north-west coast of Western Australia. In 1964 international oil companies the Royal Dutch Shell Group and Burmah Oil joined Woodside Oil to form the North West Shelf Venture participants. A little later, Chevron Asiatic Limited, a subsidiary of Chevron Corporation, at that time Standard Oil of California, acquired half of the Shell interest in the North West Shelf. British Petroleum (BP) took a half interest in Burmah's stake.[143] In the later 1960s Woodside merged with Burmah to form Woodside–Burmah Oil and to acquire a 50 per cent stake in the venture, while Shell, British Petroleum and California Asiatic Oil each retained 16.65 per cent shares.

Towards the end of 1971 the North West Shelf joint venturers discovered major reserves of natural gas at North Rankin, 130 kilometres off the north-west coast of Western Australia and 3500 metres below the seabed. The following year, the consortium based its planning on selling liquefied natural gas (LNG) overseas because the WA market was not sufficient to support such a large development project.[144] The North West Shelf project was put on ice, however, after the election of the Whitlam government, which legislated in the *Seas and Submerged Lands Act* 1973 to give the Commonwealth sovereignty over Australian territorial seas and resources to the extent of the continental shelf. Whitlam's Minerals and Energy Minister, Rex Connor, had the idea that, using the Commonwealth's powers under this legislation, the Australian pipeline authority could buy North West Shelf gas at the wellhead and transport it across Australia for industrial use, such as providing motor spirit for vehicles and powering alumina plants and iron ore pellet plants in Western Australia. Moreover, Connor believed that when exploration permits for the North West Shelf expired they would revert to the Crown in right of the Commonwealth, and that some of these areas would then be available to the planned Petroleum and Mineral Authority.[145]

Discouraged by Connor's intentions, the joint venturers mothballed the North West Shelf project in 1973, but in 1974 were spurred back into action after gas prices doubled in the wake of the OPEC oil price increase. After the Burmah Oil Company made large losses on its fleet of tankers, BHP and Shell joined together in August 1976 to form North West Shelf Development Pty Ltd, a company that would buy Burmah's share in the North West Shelf project, while Woodside–Burmah became Woodside Petroleum.[146] In August 1976 the reshaped joint venturers received assurances from the Fraser government that it would allow them to export at least half of the proven and probable gas reserves to ensure the viability of the project. For its part the WA government agreed that the other 50 per cent should be reserved for Western Australia and that the State government would be ultimately responsible for building the pipeline that would transport the natural gas from Dampier to Perth. On 24 August 1977 the Commonwealth and WA governments both gave their formal approval to the North West Shelf natural gas project, which at that time was predicted to cost between $2.5 and $3 billion. In doing so, the Commonwealth gave its approval for the joint venturers to export LNG. The ALP federal Opposition supported the Fraser government in this decision in part because of the influence of Paul Keating, shadow minister for natural resources. Formerly a protégé of Connor, Keating saw the necessity for the project to export a large proportion of natural gas.[147]

The joint venturers considered markets for the sale of the natural gas in the United States, South Korea and Japan but in the end opted for Japan after five major electric utilities and three gas utilities expressed a firm interest in purchasing the entire output of LNG from the project.

The State Energy Commission of Western Australia (SECWA) agreed to take 10.9 million cubic metres a day of gas for 20 years commencing in 1984, with some to be used in the Pilbara iron ore province and the rest for domestic and industrial use by customers in Perth and the south-west of Western Australia. The contracts were "take or pay", meaning that SECWA would have to pay for a contracted annual quota regardless of actual domestic demand.[148] Among the biggest potential users of the gas was a planned new aluminium smelter in Western Australia, which Alcoa of Australia had been discussing with Court. SECWA also undertook the responsibility to finance and construct a 1500-kilometre onshore pipeline from a treatment plant to be built on the Burrup Peninsula to consumers in Perth and the south-west of Western Australia.

In the early 1980s the joint venturers constructed an offshore production platform on the North Rankin field 235 kilometres from Dampier that would begin production in 1984 at the same time as SECWA began supplying gas to domestic customers. The joint venturers then planned to build a second platform which would operate from 1986 to lift total gas production to the level required to meet export contracts to Japan. A third platform would be established on the Goodwyn field in the 1990s. A pipeline would carry gas and liquids from the platforms to a gas treatment plant at Withnell Bay on the Burrup Peninsula. The joint venturers were responsible for providing accommodation and amenities and arranging for a fleet of specialised tankers to transport the LNG to terminals in Japan.[149]

By July 1981, when the Japanese utilities signed memoranda of intent for the sale of North West Shelf natural gas, Japan had begun to scale back its demand for energy. In response, the joint venturers rescheduled the construction program and spread the risk in the North West Shelf project. In 1982 Woodside Petroleum divested part of its interest by inviting Japanese trading houses Mitsubishi Corporation and Mitsui and Co. to come together and take a 16.75 per cent interest in the project, while Shell and BHP increased their direct participation by acquiring another one-twelfth interest from Woodside Petroleum. The project thus restructured moved to complete Australia's largest ever single commercial undertaking – what had become a $12-billion venture to provide natural gas to Western Australia and export markets in Japan. The first domestic gas was delivered in August 1984 from what was the "biggest natural resource project in the world and the largest engineering project in Australian history".[150] The North West Shelf project was bigger than the iconic Snowy Mountains Hydro-Electric Scheme, which was estimated to have cost $5 billion in mid-1980s dollars. The second phase of the North West Shelf Project from mid-1985 to 2000 cost $9.5 billion in mid-1980s dollars.[151]

The Rise and Fall of the Resources Boom

By 1982-83, the major mineral industries that had developed in the 1960s – coal, bauxite and iron ore – were earning over one quarter of Australia's export income. Between 1980-81 and 1984-85 the share of the minerals sector in Australia's merchandise exports increased from 30 to 44 per cent.[152] The coal, iron ore and bauxite/alumina industries, the largest Australian mining industries, had all expanded during the resources boom and were supplemented by a uranium export industry and a world-ranking natural gas project in the North West Shelf. One of the corollaries of the resources boom was a significant change in the arrangements for government borrowings through the Loan Council, the vehicle coordinating Commonwealth and State borrowings. This took place in 1978.

In 1976 the State premiers had expressed concern to Fraser that the requirements of many large development projects associated with the resources boom could be impeded unless a system of special additions to normal borrowing could be approved. After two years of study, the Loan Council adopted arrangements for financing of projects which could not reasonably be accommodated from resources available to the government and which had special significance for development. On 6 November 1978, Fraser announced that the Loan Council had approved an entirely new borrowing program. The projects approved by the Loan Council covered all six States and provided for total borrowing of $1767 million over eight years commencing with $158 million in 1978-79.

The first group of projects approved by Council was designed to promote further substantial private investment in Australia's natural resources. They included $89 million for improvements to coal-loading facilities at Port Kembla and Balmain and $75 million for improvement to those facilities at Hay Point. A further $416 million was approved for the construction of the Dampier–Perth pipeline, and $41 million was allocated to provide railway and water supply facilities to the Worsley Alumina project. The second group of projects was to provide public investment to increase Australia's electricity generation capacity, which would help in meeting the cost of the new aluminium smelters being constructed in Queensland, New South Wales and Victoria: $343 million was provided for the Loy Yang Power Station located near Tralalgon in the La Trobe Valley, powered by brown coal; $75 million for the Tasmanian Hydro-Electric Power Development; $200 million for a new power station at Eraring on Lake Macquarie; $130 million assisting completion of the Gladstone Power Station; and $111 million for power generating in the Pilbara region.[153] A hallmark of the resources boom was thus the mobilisation of public borrowings into projects that assisted the resources industries.

One newspaper, *Business Week*, commented in 1980 that:

> [w]hile most of the industrial world slides into recession, Australia is riding an upsurge of investments in basic resources that should keep its economy expanding rapidly through the decade to come. Unlike previous Australian boom-and-bust cycles – based successively on agricultural products and on minerals such as iron ore and bauxite – the bonanza that is now gathering momentum is likely to prove relatively invulnerable to the ups and downs of the world economy and the roller-coaster swings of international commodity markets.[154]

Business Week's prediction proved to be disastrously wrong in the short term. By 1982 structural changes in the Japanese economy presaged the end of the resources boom. A world recession and falling oil prices dashed the hopes of those who had optimistically predicted that the boom would extend into the 1980s. The Japanese steel industry did not recover from the period of slower growth in the 1970s and reduced its anticipated orders of iron ore and coking coal from Australia. The mills, moreover, attempted to diversify their sources of supply by turning for iron ore to Brazil and, to a lesser degree, India. From 1983 they ruthlessly forced down prices and quantities of iron ore and coking coal. With oil prices dropping during the global recession, demand for steaming coal grew but not at the rate predicted in the late 1970s, while the demand for other energy products, natural gas, flattened out.[155]

In his election speech in 1980, Fraser had indicated prospective investment of a massive $29 billion in the resources industries, six times larger than the sum that Connor wanted to invest in the resource industries during the first OPEC crisis. This degree of investment did not materialise, as plans for aluminium smelters and new coal projects were wound back. In the meantime, inflation and unemployment increased and wages spiralled upwards. Average wages jumped by 13 per cent per year in 1980-81, 1981-82 and 1982-83 and 229,000 jobs were lost between January 1982 and April 1983. Unemployment peaked at 10.4 per cent in September 1983 – the first time that the unemployment rate had exceeded 10 per cent since the Great Depression. At the same time Australia's rate of inflation was 11 per cent. One of the reasons for the increase in wages, according to the Treasury:

> … was an apparent development of exaggerated expectations about the increases in national income that would be generated by the resource investment "boom" and, perhaps even more importantly, the speed with which this additional income would accrue. As a result, wage claims were formulated in anticipation of benefits which had not, in fact, yet began to flow.[156]

The Fraser government was the major casualty of the end of the resources boom. It lost office in March 1983, partly because of this.

Drilling for natural gas, North West Shelf, 1979. NAA: A6135, K12/6/79/7

Aerial view of Portland aluminium smelter, 1987. NAA: A6135, K26/2/87/43

8

Between Booms – Mining in the 1980s and 1990s

After the recession of 1982 and 1983, the euphoria of the resources boom dissipated. In its place were cautious and often pessimistic views about Australia's resources industries and their ability to maintain Australia's prosperity. From about 1983 to 2000 worldwide production of resources exceeded demand for them, minerals earned low prices and mining companies often recorded poor results. In the period from 1973 to 1999 the real rate of return for the mining industry globally, apart from the high-performing oil and gas sector, was a modest five per cent.[1] Some began to argue that the intensive development of Australia's mineral resource industries, particularly during the resources boom, had been unwise and that generation and exploitation of knowledge would be increasingly important in the creation of wealth.[2] Others called attention to what they felt should be Australia's real mission – to establish a world competitive manufacturing industry, particularly in the high technology area.

Despite this pessimistic view, the mining industry played an important role between the resources boom that ended in 1982 and the start of a much bigger boom – the China boom from 2002 to 2012. Though the 1980s and 1990s were a time of downturn in commodity prices, the mining sector remained one of Australia's strongest sources of export income. In the 1980s, for example, exports from the resources industries were broadly equivalent to, and sometimes exceeded, the rural sector's share of Australia's exports and consistently outperformed manufacturing and services.[3] By the early 1990s Australia's mines were providing

40 per cent of the nation's export income.[4] The iron ore industry generally encountered more difficult times in the 1980s and 1990s than in the 1960s and 1970s, but coal continued as Australia's most significant single export. The aluminium industry also made something of a recovery after the recession of the early 1980s. And gold surged in the mid-1980s to become one of Australia's top five commodity exports in the late 1980s and early 1990s. Reflecting its importance, the mining industry as a whole came to play a more activist role in politics in the 1980s and 1990s, offering its views in a more systematic and powerful way on matters such as economic and industrial reform, Aboriginal affairs and environmental matters.

King Coal

By the beginning of the 1980s, rising levels of international demand had made coal Australia's chief export commodity, ahead of wool, Australia's staple for most of the twentieth century. But the world recession of 1982 and 1983 that marked the end of the resources boom ushered in harder times for the industry, with lower prices and tonnages. The Australian coal industry attributed its slower-than-expected growth to Japan's having deliberately overstated its long-term need for coal. When Japanese crude steel output slid to about 100 million tonnes per year in the early 1980s, Australian coal producers realised that Tokyo had encouraged an excess of supply in Australia and in Canada, where five new metallurgical coal mines were brought into production in competition with Australian coal mines between 1981 and 1984.[5] The Japanese mills had been able to achieve a state of world oversupply because only they had full information on exactly how many contracts they had written around the world.[6]

A harbinger of the harder times for the Australian coal industry was the coal negotiations in Tokyo in March 1983. These resulted in reductions by 20 per cent in the price of Newcastle soft coking coal and by 18 per cent in hard coking coal. This trend of price-cutting continued relentlessly until the end of the decade.[7] The reason for the cutbacks was that the

Japanese steel industry, though the most efficient in the world, was by then operating at less than 70 per cent of its capacity.[8] In such conditions the Japanese mills insisted that they had to import coal and iron ore in reduced quantities and at lower prices. The result of coal price negotiations in 1983 was that returns to Australia from the sale of coking coal fell by $180 million in that year. This meant almost certain unprofitability for underground mines in both New South Wales and Queensland and only marginal gains for the large open cut miners.[9]

The Labor government led by Bob Hawke was elected in the very month coal prices were being negotiated. Hawke's senior ministers included Lionel Bowen, Deputy Prime Minister and Minister for Trade, Paul Keating, the Treasurer, and Senator Peter Walsh as Minister for Minerals and Energy. As ALP spokesman on Minerals and Energy in the 1970s, Keating had gained an intimate understanding of the mining industry, its openness and lack of dependence on tariffs. His experience of the mining industry influenced his later efforts to reduce Australian protection and make the Australian economy more open to international influences.[10] With Prime Minister Bob Hawke, Keating introduced major changes to economic management in response to Australia's deteriorating manufacturing industry, chronic balance of payments problems and falling commodity prices. The most important of these was the December 1983 decision to float the Australian dollar. From the time that it was floated, market forces determined the value of the Australian dollar, which exhibited volatility in relation to currencies of other countries. Because of its association with world prices for minerals and energy and farm commodities, the Australian dollar would fall against the US dollar in the 1980s and 1990s.

The fundamental changes in Australian economic policies of the 1980s were not due to a sudden realisation that Australia had taken a wrong turn in the early twentieth century when Australian governments had adopted policies to stimulate manufacturing through a protectionist tariff. Rather, the rise of Japanese demand for Australia's resources in the 1960s and 1970s presented Australia with new options that it did not have in the less favourable economic climate of the first half of the twentieth century. With these options came a realisation that adopting more outward-looking policies was more attractive and feasible than clinging to the more insular policies of previous decades.[11]

Because in 1983 coal was now a buyer's market, Australia risked losing market share to other countries. "The Japanese steel mills", declared Trade Minister Lionel Bowen, "appear to be saving a staggering $US480 million from lower coal prices from all supplies, of which about $US175 million comes from Australian coal supplies".[12] Bowen admitted that greater care would have to be taken in examining Japanese assessments of their coal requirements in the future.[13] Despite the government's best efforts, however, contracts for coking coal in the 1980s changed, evolving away from a base price plus escalation for various cost increases to regular reviews of both prices and quantities. Many of the coal contracts in the 1980s contained provisions that meant that, if buyer and seller could not agree on a price, the buyer was relieved of any obligation to purchase any coal from the producer. As Stephen Bunker and Paul Ciccantell have argued, the "shift to annual price negotiations opened the door for the Japanese steel mills to take full advantage of the excess capacity that their long-term contracts and equity investments" had created.[14]

The poorer outcome for the Australian coal industry in the negotiations in 1983 and 1984 affected State governments as well as the industry. In 1984 the Queensland government found itself at loggerheads with the coal industry because of the producers' pressure for lower freight charges. For more than a decade the National-Liberal Party coalition led by Joh Bjelke-Petersen had been relying on freight from coal as an expanding and predictable source of its revenue. As well as increasing the prosperity of the State, coal had underpinned the political fortunes of the coalition government. Coal freights by themselves were not far below State income from all other taxes, including mineral royalties, in the Queensland state budget. But the negotiations of 1983 and 1984 meant that coal companies found it difficult to bear the cost of escalating freight charges. For example, CRA complained that rail freight was by far the largest item in the operating cost of the Blair Athol mine. Along with other Queensland coal producers, it pressed for relief but the Bjelke-Petersen government resisted because its budget strategy, and indeed its whole political philosophy, was structured around the ever-growing source of revenue from coal freights. Bjelke-Petersen's boast that Queensland was the "low tax" state in Australia had arisen largely because of the income generated by coal freights.[15]

Australian coal companies suffered lower prices and cutbacks in contracted quantities throughout the 1980s, but they still managed to capture new export markets. During the first half of the 1980s, they achieved a 19 per cent per year volume growth rate, higher than the coal industry in any competing country. They also increased market share and achieved a growth in aggregate export earnings consistently higher than the world average and greater than Australia's average performance during that time.[16] The Australian black coal industry was rationalised and changes in work practices were introduced. Notable among these were new shift arrangements with the emergence of a "drive in, drive out" culture (DIDO) in which workers lived in coastal areas and drove to work. These developments and the completion of investments begun earlier saw the output of raw black coal leap by 50 per cent in four years to reach 182 million tonnes in 1986-87.[17] During a period when all Australia's mineral and rural commodity exports were suffering, Australia's second minerals rush continued in the coal industry. By 1987 Australia was exporting 102 million tonnes or nearly one third of world coal exports.[18] Over the period from 1980 to 1997 coal exports from Australia increased almost fourfold from 43 million tonnes to 157 million tonnes.[19]

Despite the harder times in parts of the Australian coal industry, the Australian government set its face against subsidising the industry. Hawke's Resources Minister, the Western Australian Peter Walsh, told an Australian Coal Conference in Surfers Paradise in 1984 that the Australian government compared favourably the highly efficient Australian coal industry with its British counterpart, which was receiving astronomical subsidies in an effort to prop up employment. The challenge for Australian coal, Walsh argued, was that of an internationally competitive industry adjusting itself to slower world growth.[20] Australia's gains were achieved largely at the expense of the United States, which had been dominant in the world coal trade in the 1950s and 1960s. By 1984 Australia was the world's largest seaborne exporter of coal and in 1985 it became the largest coal exporter in the world.[21] According to statistics compiled in 1984 by the Chase Manhattan Bank, Australia's rise to leadership of world coal exports in 1984 was secured on the basis of virtually static US tonnages.[22]

These gains led to the US government exerting political pressure on Japan to take more American coal. In November 1983 the Prime Minister of Japan, Yasuhiro Nakasone, met with US President Ronald Reagan and the two leaders agreed to encourage industry to substitute coal for oil and examine ways for the Japanese to take more American coal. Following the summit meeting, a Joint Japan/US Energy Group was established. Later a "Standing Technical Committee", consisting of representatives of private industry from both countries with government observers, was set up.[23] Signs of the American pressure being exerted on the Japanese government became apparent in 1984 when the Japanese Ambassador in Canberra cautioned Hawke's advisers that, historically, the United States had once been the sole supplier of coal to Japan and that it might choose to deploy this fact in support of its case for a larger share of the market.[24] Bowen reported to Hawke in September of that year that:

> [i]n the past, MITI have always evaded the issues of resources trade by claiming that they are commercial matters. However, yesterday we stressed our alarm at the apparent linking of Government and commercial US committees and our concern that some non-arms length arrangements might be put in place in regard to the coal trade between USA and Japan.[25]

The Japanese mills believed that American coal could not compete in terms of quality and price with the Australian product. But the US government had hopes that American coal exports to Japan could be increased by long-term contracts. In 1985 the PC Corporation in the United States reached an agreement to supply one million tonnes of steaming coal at prices above then existing world prices.[26] Other contracts of a similar nature followed. The Australian government saw them as a threat, prompting Hawke to assert on a visit to Japan in February 1984 that Australia had a "legitimate expectation that our market share will not be reduced in the name of diversification, at a time when so many Australian mines, developed for the Japanese market, are operating below capacity".[27]

In November 1984, Utah Development considered itself fortunate to have negotiated an agreement for the sale of 19 million tonnes of coal to the Japanese steel mills from its fields at Goonyella, Saraji and Peak Downs. This sale represented the accumulated shortfall of long-term contracts signed earlier. The harder times for Utah in the early 1980s had one particularly significant ramification: the decision by that company's American owners to sell out to BHP, thus significantly lifting Australian ownership of the coal industry. In 1981, Jack Welch, the chief of the American conglomerate General Electric Company (GE) that owned Utah, decided on a five-year program of acquisition and divestment. Divestments included Utah's coal assets in Queensland and its iron ore mine in Western Australia. GE had purchased Utah for US$2.3 billion in 1976, the largest acquisition seen in the United States to that time. But Welch, when he took over General Electric, was uncomfortable about the company having a minerals division that was prone to cyclical fluctuations. Matters came to a head in 1981 when the Japanese steel mills refused to accept contracted shipments of Utah's coal and a strike had disrupted the company's production. After this, Welch deputed to his executives the task of finding a buyer for Utah.

John Burlingame, Welch's deputy, first raised the prospect of a sale in a conversation with BHP chairman, James McNeill, on 23 August 1982. Burlingame valued all of Utah's assets at $US3.5 billion, $US500 million more than BHP's assets were then valued at. Despite this daunting challenge, on 10 December 1982 BHP's board made a decision that historians of the company describe as the one "that marked the turning point in The Big Australian's modern history".[28] It was a turning point as significant in the company's history as its discovery of oil in Bass Strait. BHP's directors decided that if they did not purchase Utah they risked remaining as a company with "faltering steelworks and an ageing oilfield in Bass Strait, a company that was primarily dependent on one country – Japan – for most of its revenue".[29]

BHP's board seized the opportunity to purchase Utah. By June 1985 GE had sold all its interests in Utah to Australian parties so that Australian ownership in the Central Queensland Coal Associates (CQCA) mines and facilities increased to 88 per cent and in its Gregory Mines to 100 per cent.[30] Just as BHP had diversified into aluminium during the resources boom, so it would consolidate its position as the biggest miner of coal in Australia in the 1980s. The decision to buy Utah marked a further step in BHP's transition from steel making back to mining. This shift had been graphically illustrated in 1982 when BHP announced that its Port Waratah Steelworks would have to be restructured if they were to survive. The steelworks had been a backbone of the Newcastle area's economy since 1915. Luckily for the economy of that city, the void created by BHP's faltering steelworks had been partly filled since 1960 by the Hunter Valley's rapidly increasing coal exports.[31]

As BHP was digesting Utah's Queensland mines in 1985, the Australian government had become seriously concerned about the trend to lower coal prices. Consequently, it intervened in negotiations between Japanese purchasers and four companies: Coalex, R.W. Miller, Buchanan Borehole Collieries and Warkworth. In this instance the government insisted that no coal purchases be made at less than the Australian equivalent of $US36 per tonne.[32] It established parameters in which the companies were to negotiate and exerted strong pressure on any company that looked like breaking ranks. Critics thought that this government intervention was extremely dangerous, with Colombian coalmines offering alternatives to

Australian coal. But the strategy worked. The four companies were able to maintain their coal tonnages and also won a 12.25 per cent rise for their products. Excluding political considerations, Australia remained the most attractive source of coal for Japan in the 1980s.[33] This was evidenced in 1989 by Australia's selling 30 million tonnes of coking coal to Japan, as against America's 10.1 million tonnes and Canada's 17.6 million tonnes. In 1986 the Australian coal industry was still the biggest weapon in Australia's export armoury, earning $5300 million in revenue in 1985-86 up from $4600 million in 1984-85. Indeed, coal remained the growth sector in exports at a time when Australia's current account deficit – the difference between exports and imports plus invisibles such as freight and interest charges – approached a disastrous $1400 million. This situation prompted Treasurer Paul Keating to warn of the danger of Australia's becoming a "banana republic" in 1986. The problem, according to Keating, was due to the parlous state of Australian manufacturing and low commodity prices.[34] But with the assistance of the devalued dollar, Australia's coal exports increased from 88 million tonnes in 1985 to 99 million tonnes in 1989. Australia was prevented from achieving even larger coal sales in the late 1980s because of adverse global developments in terms of pressure on prices and tonnages. Japanese Prime Minister Nakasone told the National Press Club on a visit to Australia in 1985 that he did not expect the position of Australia as a supplier of primary commodities to Japan to decline. But in 1987 competition from China and particularly South Africa was increasing the pressure on the Australian coal industry. By the late 1980s, South Africa was exporting nearly 50 million tonnes of coal annually through a terminal completed at Richard Bay in 1981.[35] Japanese power utilities took advantage of this strong competition to try to force cuts on Australian steaming coal suppliers of between 10 and 20 per cent below the prices negotiated in 1986-87. After two months of acrimonious negotiations, the Chugoku Electric Power Company persuaded its Australian suppliers to accept an eight per cent price cut, thus lowering the base price of steaming coal to $US41.16 per tonne. Other steaming coal suppliers were forced to accept prices one fifth less than those of the previous year. The ACA predicted $400 million less in Australian receipts from steaming coal in 1987-88.[36]

Coking coal exporters fared just as badly. Pleading a decline in steel output, the Japanese mills convinced Australian mines in January 1987 to accept a 10 per cent price cut and reductions in tonnages of up to 20 per cent. In 1988 Nippon Steel Corporation told Australian producers that Japanese steel production would fall from 97 million tonnes to about 90 million tonnes by 1990.[37] Faced with these projections, Australian producers settled for prices seven per cent lower in real terms because of changes in the yen-dollar exchange rate. Although the Australian government branded the Japanese trading policies as objectionable and at first refused to grant export licences, it was compelled to back down when BHP-Utah was forced to accept the same price cuts as the smaller suppliers.[38] By contrast with Connor's position in 1973, the Hawke government felt it could not afford to gamble with an implicit threat by the Japanese steel mills to shift their coking coal business to Canada.[39]

After 1988 the Hawke government abandoned interventions in price negotiations on coal with the Japanese steel mills. In a stronger position, the mills were able to force the price of both Australian and Canadian coal downwards, with adverse effects on the coal mining industries in both countries in the late 1980s and early 1990s. In Canada the Balmer, Quintette, Line Creek and Greenhills mines were forced into bankruptcy in the early 1990s and 20 mines were forced to close down in New South Wales in 1987 and 1988.[40] Between 1987 and 1996 producers of Australian black coal experienced a 16 per cent decline in the average real price for their coal.[41] But these price reductions were not catastrophic for an industry as resilient as Australian coal had become. Between 1987 and 1991 the Australian coal industry as a whole continued to increase its output of saleable coal by taking advantage of the spare capacity created by investments made in earlier years in transport infrastructure and mines.[42] Coal remained Australia's leading export commodity throughout most of the 1980s, making up 16 per cent of total merchandise exports in 1986-87.

Although Japan remained Australia's largest export market overall in the 1980s and 1990s, its share of Australia's total exports declined from 25 per cent in 1980 to 20 per cent in 1997.[43] By the 1990s the mighty Japanese economy was suffering a serious economic reversal with significant consequences for Australian resource exports. The end of the long post-war economic boom in Japan is commonly dated to the collapse of real estate and share bubbles in 1991 beginning a long period of deflation.[44] The economic historian Robert C. Allen attributes the fundamental cause of the end of the era of high-speed growth in Japan to the elimination of conditions that had allowed such growth from the 1950s to the 1990s. He argues:

> Japan grew rapidly by closing three gaps with the West – in capital per worker, education per worker and productivity. This was done by 1990, and Japan was then like any other advanced country: it could grow only as fast as the world's technology frontier expanded – a per cent or two each year. The post-1990 growth slowdown was inevitable.[45]

Over the whole period from 1980 to 1997, Japan's dominance of the Australian coal market declined along with its share of Australian exports. But the Australian coal industry was able to compensate for losses in its trade with Japan by gains with European countries and Asian countries other than Japan. From the mid-1970s onward both South Korea and Taiwan established new state-owned steelmakers, the Pohang Iron and Steel Corporation (POSCO) and the China Steel Corporation (CSC). Because both Korea and Taiwan lacked raw materials for their steel industries, they obtained their iron ore and coking coal from regional suppliers like Australia. In 1984 the People's Republic of China launched a modernisation and expansion program for its own state-owned steel industry. While the Chinese steel industry was self-sufficient in metallurgical coal, it was by then having a quarter of its iron ore needs met by regional suppliers.[46] Thus, while Japan's share of Australian coking coal exports declined from 76 per cent to 41 per cent between 1980 and 1997, the shares of Europe and the rest of Asia in Australia's coking coal exports increased from 13 and 10 per cent respectively in 1980 to 18 and

35 per cent in 1997. Australia was able to increase its supply of good quality coking coal to Europe because high-cost European mines were having trouble meeting the demand.[47]

By 1997 Asia had become the destination for 91 per cent of Australia's steaming (thermal) coal exports compared with 53 per cent in 1990. Japan, South Korea and Taiwan were the biggest consumers of Australian thermal coal. By contrast, Europe's share had fallen from 36 per cent in 1980 to six per cent in 1997.[48] Thus compensating for Japan's decline in the 1990s was the rise of other Northeast Asian economies, South Korea, Hong Kong, Taiwan and China. In 1989 Australian economist, diplomat and policy adviser Ross Garnaut called the Hawke government's attention to what he called the "Northeast Asian ascendancy" in Australian trade. Because of this Northeast Asian ascendancy, coal remained one of Australia's biggest exports to Asia at the turn of the century. But while coal still remained Australia's main single export-earning commodity in the 1990s, it had declined as a percentage of Australian merchandise exports from 16 per cent in 1985-86 to 10.1 per cent in 1996-97.[49] Then in 1997, the Asian Financial Crisis further dampened the regional demand until about 2003. By the end of the 1990s, the economic slump in Japan and the East Asian Economic Crisis in 1997-98 had induced in many a sense of pessimism about the prospects of coal into the 2000s.

The coal industry had become more concentrated than earlier. Four companies produced half of Australia's saleable coal by 1997. These were BHP, which through its acquisition of Utah had gained almost total control of Queensland's metallurgical coal; Rio Tinto, which had taken control of Coal and Allied Industries' mining operations and by 1992 controlled three-quarters of NSW metallurgical coal exports; the multinational oil and gas company Shell; and Oakbridge, a publicly listed Australian company based at Singleton in New South Wales producing semi-soft coking and thermal coal.[50]

As shipments of coal for export continued to rise in the early 1980s, the NSW State Rail Authority was moving more than 30 million tonnes per year through Newcastle, Balmain and Port Kembla. By 1996 Port Waratah Coal Services announced a capacity of 56 million tonnes per year and the ability to increase its capacity to 80 million tonnes if required.[51] Appreciating the enormous potential of the coal industry, NSW Rail spent $20 million on 100 new 100-tonne coal wagons in 1985-86 and committed $90 million to upgrading rail lines in the Hunter Valley.[52] During the 1990s New South Wales moved first to corporatise, and then privatise, its coal railway network. In January 1989 the freight operations of State Rail were transferred to Freight Rail, which was made responsible for all freight services. By 1990 35 million tonnes, nearly two-thirds of all NSW rail freight, moved in and out of the rail network based at Newcastle. Coal made up the bulk of this traffic as tonnages increased from 23 million tonnes in 1986 to 43 million tonnes in 1996. As they moved from mines to port, coal trains comprised as many as 84 gross wagons hauled by up to four diesel-electric locomotives. In 1991 16 of these trains per day reached the Port Waratah Coal Services Terminal and 12 more reached the Kooragang Island Terminal. They formed part of a coordinated system of the Hunter Valley Coal Chain directed by computerised monitoring and management controls.[53]

Freight Rail was corporatised in 1996 as FreightCorp, an entity whose principal coal services carried coal for 13 customers from the Hunter Valley to the Port of Newcastle and from mines in the Southern and Western fields centred on Lithgow to Port Kembla. FreightCorp would subsequently be sold to a private joint venture between Patrick Corporation and Toll Holdings in 2002. In the 1980s and 1990s Queensland Rail continued to operate the State's coal freight via four rail supply chains: the Newlands, Goonyella, Moura and Western systems, stretching over more than 2000 kilometres. In the first half of the 1980s Queensland Rail rebuilt, connected and extended the central Queensland network to support heavy coal haulage. As part of this project over 1600 kilometres of track was electrified in one of the nation's largest engineering projects of that time.

Changing Fortunes in the Iron Ore Industry, 1983-2000

The journalist and mining historian John McIlwraith described the 1980s as the period when the iron ore industry lost its innocence. The industry learnt in that decade that the world's steel industry did not owe it a living.[54] Australian iron ore exports had risen continuously from 1967 to 1974, when they peaked at 82 million tonnes, but they were only 71 million tonnes in 1981.[55] In the years of international recession in the early 1980s exports were about 75 million tonnes annually, around 7 per cent of Australia's merchandise exports.[56] By 1984-85 they had got back to 90 million tonnes, but in the following year exports slipped back again to 83 million tonnes. Prices too were adversely affected. From 1982 to 1988 the benchmark price of iron ore fell every year and ended up nearly one third below its starting point.[57] In real terms, prices fell by almost one third between 1983 and 1986 and would fall by a further 16 per cent by 1988.[58] Australia's share of the world iron ore export market fell from 22 per cent in 1977 to 19.4 per cent in 1981, after which Australia came second to Brazil as a world iron ore exporter. Exports to Australia's principal market, Japan, declined from 55 million tonnes in 1984 to 42 million tonnes in 1987. By 1987-88 iron ore exports to all markets were again 90 million tonnes but were only worth 4.6 per cent of total merchandise exports, half their relative significance in the 1970s.[59]

In the United States, the steel industry worked at half its capacity in 1982, and Europe's steel industry functioned at between 50 and 55 per cent.[60] The Japanese steel industry also faced difficulties with production falling below 100 million tonnes per year, although it was still stronger than its counterparts in Europe and the United States. The global downturn in the steel industry, beginning in the 1970s and continuing in the 1980s, put paid to the Western Australian government's aspiration for new steel industries developing in Australia from the Pilbara iron ore discoveries. In the early 1970s Court had hoped that Australia would be producing 50 million tons of steel per annum from Pilbara iron ore. This did not happen and, although BHP would use Mount Newman ore in its own manufacture of steel, BHP's steel division declined in the 1970s and 1980s, hastening the company's return to its historical roots as a miner. The general assumption in the 1960s and 1970s that the Pilbara could be host to value-added processing was hampered by economic considerations. The Pilbara was an expensive region where the cost differential relative to Perth was typically double in the 1960s and it did not improve much in succeeding decades. The mining operations remained very competitive as labour costs per tonne of iron ore mined were reduced over time because of technological improvements and the benefits of economies of scale. But the 1960s policy of building accommodation for employees became too expensive and gave way over time to the cheaper "fly in, fly out" (FIFO) mode of operations. Rather than building new towns, as mining companies did until the mid-1980s, they increasingly began to fly workers in from major cities and regional towns in parallel with the "drive in, drive out" (DIDO) practice in the coal industry.[61]

FIFO and DIDO applied particularly to the coal and gas fields of Queensland and the iron ore and gas region of north-west Western Australia. The practices benefited mining companies that had struggled for decades to attract workers to new mining townships. They also boosted established regional townships such as Cairns, Townsville and Mackay, which became "virtual mining towns" as they accommodated swelling numbers of FIFO and DIDO workers.[62] The cost of the new practices, however, was the detrimental effects on mining workers, their families and mining towns like Tom Price and Paraburdoo. Workers who flew in for 15-hour shifts in towns like Paraburdoo in the Pilbara or Moranbah in Queensland had little stake in the towns built to serve the mines. These towns and their communities consequently suffered social decay. Moreover, the actual locations of the FIFO and DIDO workers were "hollowed out by the absence of adults and the attenuated dynamics of single-parent families".[63] Writing in 2012 historian Erik Eklund argued:

> No longer will capital and the state enable a place, a town, a company project to develop *in situ*. There were few corporate towns built after Kambalda in 1966, and the era of the large private town such as

Broken Hill or Mount Morgan has long faded.[64]

In 1983 the Australian iron ore industry was working at only two-thirds capacity and both the Mount Newman Mining Company and Hamersley Iron were forced to accept a fall in prices of 12.7 per cent, an unprecedented reversal. Mt Newman struggled to find resources to invest in the industry in the early 1980s and consequently was not in a position to respond to rising demand when the world economy started recovering in the mid-1980s. Some estimated that because of production problems at Mount Newman the company would miss out on $100 million in sales in 1985.[65] Indeed, BHP was even forced temporarily to buy iron ore from its competitor, Hamersley Iron. But when rapid depreciation of the Australian dollar provided much needed relief for exporters in 1986, Mt Newman invested approximately $330 million in mine re-equipment and in further dredging of Port Hedland.[66]

The Pilbara returned to greater profitability in the late 1980s and early 1990s as the Japanese steel industry temporarily recovered. Moreover, losses in the Japanese market were recompensed to some extent by increased exports to other Asian countries. In 1987 China was taking almost 10 per cent of Australian iron ore exports, South Korea 7.8 per cent and Taiwan 4.6 per cent.[67] Iron ore exports to Europe also increased as European iron ore mining companies struggled to stay afloat in times of depressed prices. Consequently, the European share of Australian iron ore exports increased from 17.5 per cent in 1983 to 21 per cent in 1987.[68] Iron ore exports as a proportion of Australian total merchandise exports rose from 4 per cent in 1988-89 to 5.1 per cent in 1991-92. But the later 1990s saw the Japanese steel industry again go into decline, with the Japanese economy growing annually at less than one per cent and moving away from heavy industry to knowledge management.[69]

The Goldsworthy joint venturers were always restricted by the smaller size of their deposits as compared with those of the other Pilbara companies. In 1973 they developed the Shay Gap/Sunrise Hill and Nimingarra operations east of Mount Goldsworthy, requiring an extension of the Goldsworthy railway line. As the production of these sites declined and eventually ceased, they developed deposits still further east at Yarrie in 1993. By this time, there had been significant changes in the ownership of the Goldsworthy joint venture. BHP had acquired a 39 per cent interest in 1984. Six years later, the Big Australian acquired 100 per cent ownership of the Goldsworthy project and subsequently sold a 15 per cent interest in it to two Japanese companies: CI Minerals (Itochu) and Mitsui.[70] In 1986, BHP purchased the holdings of both CSR and AMAX in the Mount Newman Mining Company, raising its interest in the company to 85 per cent. The acquisition, costing $880 million, placed the overwhelming majority of the Pilbara iron ore industry in the hands of two of Australia's largest companies, BHP and CRA, which by 1984 wholly owned Hamersley Iron.[71]

Another problem that began in the early 1970s and reached a crescendo in the 1980s was industrial relations in the Pilbara. By the 1980s industrial disputes were the most significant factor causing loss of iron ore production. From 1976 to 1981 the four major Pilbara miners had suffered 3905 strikes and lost 5,755,126 man-hours.[72] In 1986 labour problems climaxed when the Robe River Associates took steps to challenge what were beginning to be called "restrictive work practices" in the industry.[73]

In the first half of the 1980s Robe River's profitability had held up well as a result of marketing advantages conferred by equity partner Mitsui and favourable spot prices. After 1985, however, the higher Japanese yen imposed new constraints on the company's profitability. In 1986, Charles Copeman, the chief executive of Peko-Wallsend, now the majority owner of the company, took decisive action to challenge prevailing work practices.[74] He sacked some of his senior management, declared all existing agreements null and void and sought to transfer the workforce from a State to a federal award, a move opposed by the State Labor government. Copeman was a Queensland-born Rhodes scholar and member of a new organisation, named the H.R. Nicholls Society, which had been formed in 1986 with mining executives such as Ray Evans and Hugh Morgan of

Western Mining Corporation prominent among its members. The society supported the deregulation of the Australian industrial relations system, the abolition of the award system, the widespread use of individual contracts and the lowering of minimum wages. It was part of a wider movement that developed in the mid-1980s, named the "New Right", which opposed the consensus politics of the Hawke government and called attention to the damage wrought by government intervention, tariff protection and, most of all, trade unions and the arbitration system. New Right thinking had an attraction for elements of the mining industry, which was vulnerable to environmental regulation, strike action and Aboriginal land claims. Moreover, as an exporter dependent on imported equipment, the mining industry was sympathetic to lower tariffs.

Copeman, a forceful advocate of New Right thinking, was convinced that the Robe River project would not survive without sweeping reform of its industrial relations. His move against "restrictive work practices" triggered a bitter industrial dispute that was hard fought by workers and management. The company imposed a 400 per cent increase in rent for workers who went on strike. After the WA Industrial Relations Commission (WAIRC) ordered a moratorium on industrial action and an investigation of work practices, Peko sacked 64 employees who had contested changes to their working arrangements.[75] When the WAIRC ordered their reinstatement, the company sacked its entire workforce of 1160 people.[76]

There was no quick end to the dispute and pressure mounted on Copeman to settle it. The WA Labor government thought that Peko was jeopardising the State's reputation as a reliable supplier of iron ore and Peko's Japanese partners were also becoming worried. Bob Hawke, moreover, branded Copeman and other militant employers as "political troglodytes and economic lunatics".[77] The Robe River dispute was eventually settled in a peace accord brokered with the assistance of the Australian Council of Trade Unions. For some, the initiative of Peko-Wallsend in leading the attack on Pilbara work practices had been a brave action taken out of economic necessity. The effect of Peko's actions could be measured, as the Productivity Commission later did, by substantially improved labour productivity at Robe River after 1986.[78] For others, including Peko's Japanese partners and federal and State Labor governments, Peko-Wallsend had won a pyrrhic victory by overreaching and in so doing endangering the reputation of the Pilbara as a reliable supplier.[79]

In the longer term, the Robe River dispute proved to be a watershed in the industrial relations history of the Pilbara. After the dispute, Robe River's management moved its workforce from State awards to common law individual contracts in the late 1980s. Later, when Hamersley Iron's workforce struck over the employment of a non-unionist in 1992, Hamersley Iron also moved its workforce to individual contracts, taking advantage of new State laws in Western Australia that introduced statutory individual contracts. Union power was reduced across the Pilbara, once a centre of union strength in the 1970s and 1980s. As individual contracts became the norm in the iron ore industry, BHP followed suit in 1999.[80] Mining was always something of a frontier in Australia's industrial relations system and had particularities that "set it apart from the mainstream industrial landscape".[81] Historian Malcolm Knox argues that "[e]ven when strikes were tearing these communities apart, there was a fundamental common ground between workers and bosses, built on geographic isolation and knowledge that both sides were equally helpless in their dependence on foreign investment and global metal prices".[82] What made the mining industry particularly amenable to individual contracts was its need to offer higher wages and conditions so as attract workers to remote locations and to persuade them to stay there. Individual contracts proved to be a framework acceptable to workers to provide for these higher wages and conditions. By March 2007 nearly five per cent of the Australian workforce would be covered by Australian Workplace Agreements (formalised individual agreements) and 32 per cent of mineworkers would be covered by them.

Another important development in the 1970s was the end of the era of cheap oil. The OPEC oil price rises of the 1970s took the price of using oil in the energy-intensive process of making iron ore pellets from about $10 per tonne to about $100 per tonne.[83] This made the already marginal pellet plants of Hamersley Iron and Robe River uneconomic. Their consequential closure was more significant for the Robe River Associates than Hamersley Iron because the former had been conceived as a pelletising operation. But the Robe River Associates surmounted the problems in a number of ways. First, they had concluded an agreement with BHP in 1970 to bolster the limited iron ore reserves of 150 million tonnes that they had at Robe River with 157 million tonnes from BHP's East Deepdale area. In return, BHP acquired an option to call on 50 per cent of the Robe River railroad and as much as 100 per cent control of Cape Lambert. BHP would exercise this option in 1975. Armed with 300 million tonnes of reserves through the deal with BHP, the Robe River Associates had access to a much larger source of iron ore fines. Robe River's management, as we have seen, successfully persuaded the Japanese mills to convert the pellet contracts to fines contracts.[84]

BHP further assisted Robe River in 1987. This was three years after American mining company Cleveland Cliffs, adversely affected by the decline in the US steel industry, sold its stake in the Robe River mine to Australian mining company Peko-Wallsend Ltd for $US54 million.[85] Threatened by the possibility of a takeover by entrepreneur Robert Holmes à Court, BHP resolved to rationalise its iron ore operations. Deciding to concentrate on Mount Newman, BHP sold to the Robe River Associates its main Deepdale deposits, including Mesa J, which became the cornerstone of the Robe River project from the 1990s. The Robe River Associates obtained the additional reserves and also bought out BHP's rights to use its railway and port for a mere $42 million.[86] Later, the Robe River Associates expanded their operations to develop the West Angelas deposit, one of the areas of which Lang Hancock had been divested by the Tonkin government. This deposit became operational in 2002.

In the 1970s and 1980s, as we have seen, Chairman of CRA, Sir Roderick Carnegie, spearheaded an ambitious plan to "Australianise" CRA. He achieved such success that by 1986 RTZ's shareholding in its Australian subsidiary had dipped below 50 per cent. In the 1970s and 1980s RTZ had operated more as a mining finance house than an operator of mines. But in the 1980s and 1990s, CRA lost some of its prestige when its copper mine in Bougainville, once its most prized asset, was forced out of production by a long and bitter civil war.[87] Throughout those years RTZ was also changing. The multinational company acquired British Petroleum in 1989 and set up a joint venture with Freeport in a copper mining project in West Papua in 1995. By that time RTZ was in a position where it directly controlled many mining operations and one of them in what had previously been regarded as CRA's sphere of influence. In the view of RTZ chief executive Bob Wilson:

> The operation was up there in West Papua, traditionally CRA's doorstep – they'd never even had a look-in at a deal there. It became clear from both sides that not only were we heading for a potential conflict of interest but it was also going to be a problem for CRA's aspirations, certainly outside Australia, if we couldn't find a constructive way to resolve this.[88]

Wilson devised a solution. He and CRA chief executive Leon Davis persuaded the Australian government to agree to a merger between RTZ and CRA in 1995. A dual listed company under single management, RTZ-CRA, was created, although with shares maintained in different entities. The Rio Tinto Group became commonly known thereafter simply as Rio Tinto. The 1990s was a period of extraordinary growth for the company. In 1989 Anglo American Corporation, a multinational mining company headquartered in London and with major operations in South Africa, had been the largest global mining company.[89] By 1997, through mergers and takeovers, including the 1995 merger with CRA, Rio Tinto had taken Anglo American's place as the world's largest mining company, with assets of $US13 billion, a turnover of $US9.2 billion and profits of $US1.2 billion.[90]

By 2000 Rio Tinto had also made the transition from the third-largest iron ore miner in the world to the second-largest iron ore company, behind only Brazil's Vale. This was mainly a consequence of Rio Tinto's successful takeover of North Limited (previously Broken Hill North) in 2000. North Limited had earlier merged with Peko-Wallsend to take up the majority position in Robe River and in 1999 had launched a bid to have the Hamersley Railway declared "open access", which was always possible under its agreement with Western Australia. In response, Rio Tinto announced in June 2000 a cash offer for all of North Limited's shares, provoking strong resistance from Robe's minority shareholder, Mitsui, and the Japanese steel industry. But Rio Tinto overcame the opposition, successfully took over North Limited and thereby became a partner with Mitsui Iron Ore Development, Nippon Steel Australia and Sumitomo Metal Australia in Australia's third-largest iron ore mining firm, Robe River Iron Ore Associates. As a consequence of the merger, Rio Tinto almost doubled its iron ore production and became owner of an integrated Hamersley-Robe River railway system and the second-largest iron ore producer in the world.[91]

Meanwhile, against the tide in Australia's iron ore industry in the 1970s and 1980s, Hamersley Iron pioneered the first new iron ore mine in the Pilbara since Paraburdoo in 1971-72. This was a joint venture with the People's Republic of China's China Metallurgical Import and Export Corporation (CMIEC) and Hamersley Holdings to develop 200 million tonnes of high-grade hematite ore over approximately 20 years at Channar. CMIEC was an import/export corporation operating within the Chinese Ministry of Metallurgical Industry (MMI) and was responsible for importing ore and associated steel-making raw materials in China. First proposed in 1983, the project's construction began in 1988 to mine Channar ore by open cut methods and convey the crushed ore 20 kilometres overland to Paraburdoo where it would be blended with Paraburdoo materials into lump and fines. Like the foundation Pilbara projects, foreign capital became essential to get the project started, especially after the Australian

Taxation Office ruled in 1988 that Australian banks, which had initially contemplated financing the project, would not be eligible for the income tax benefits. The joint venturers formed a jointly owned finance company to enter into a 12-year $US170-million debt funding arrangement with a group of 13 international banks, which lent the funds on to the joint venturers. In turn, the joint venturers lent those funds on, in Australian dollars, to the partnership.[92] The establishment of Channar resulted from successful collaboration between the Australian and Chinese governments with political support from the highest levels. The Channar venture laid the basis for the much larger Chinese purchasing of iron ore that took place in the 2000s during one of the biggest booms in Australian history.

Aluminium, Gold and Olympic Dam

The decision not to proceed with the Portland aluminium smelter in 1982 had been one of the harbingers of the end of the resources boom. But only a year later, in 1983, Alcoa of Australia sought to revive the initiative. It proposed that the Victorian government, now led by the ALP's John Cain, should take a 45 per cent shareholding in the Portland aluminium smelter with Alcoa remaining as the managing partner. The resurrection of the Portland project was aided by rising prices for alumina and ingot that permitted Alcoa to take its new Wagerup alumina refinery out of mothballs. On 31 July 1984, two years after the closure of Portland, Cain and Alcoa of Australia signed an agreement based on electricity prices that would vary with the fluctuation of world aluminium prices but never fall to a position where the Victorian State Electricity Commission was selling electricity below cost. As workers returned to build the smelter, Alcoa and the Victorian government found other partners in the venture. With Alcoa holding 45 per cent and the Victorian government now 35 per cent, the First National Resource Trust, a subsidiary of the National Australia Bank, took up 10 per cent equity. The People's Republic of China, through its China International Trust and Investment Corporation, took up the other

10 per cent. China agreed to take its profits in the form of 10 per cent of the ingot produced by the smelter. By November 1986, 26 potlines – the long buildings containing "pots" or electrolytic cells in which aluminium is made – were working and, on 9 February 1987, Cain formally opened the Portland smelter.[93]

Completion of the Portland plant coincided with a rise in the world price of aluminium. After the dismal years of global recession in the early 1980s, the Australian aluminium industry as a whole staged a remarkable recovery later in the decade. Between 1983 and 1986 Western Australian production of bauxite had increased by 50 per cent so that by 1986 that State accounted for 60 per cent of Australia's bauxite production. Geoffrey Blainey noted that in 1989 Alcoa alone was producing almost 50 per cent of the alumina and more than 40 per cent of the aluminium made in Australia. Indeed, Alcoa itself had become by then Australia's second-biggest earner of income, earning more than five per cent of Australia's export receipts.[94] In 1989 Alcoa's net contribution to Australia's balance of payments was $4.5 million per day. In addition, Alcoa and Worsley contributed $825 million to the Western Australian economy each year.[95]

The most astonishing improver in the mineral industry in the second half of the 1980s and early 1990s was gold. Gold mining had been important on two earlier occasions in Australian history: in the nineteenth century gold rush and in the Depression years of the 1930s in Western Australia.[96] The third occasion that gold became significant was as part of the second minerals rush during the late twentieth century and specifically in the 1980s and early 1990s. In the 1960s the rise of inflation and balance of payments problems in the United States had the effect of reducing America's gold reserves so that they fell to $US10 billion in 1971, half of their level in 1960. In 1971 President Richard Nixon had little choice but to abandon the Bretton Woods system of fixed exchange rates and its associated policy of direct convertibility of the US dollar into gold. America's abandonment of the gold standard ended the era of a fixed gold price, which had previously been at $US35 per ounce.[97] Henceforth a range of factors would determine the price of gold. These included fluctuations in supply and demand, changes in interest and exchange rates and variations in levels of international confidence. In the uncertain times of the 1970s and 1980s, gold became the ultimate store of value.[98] After recovering an average price of $US110 per ounce in August 1976, gold increased to $US206 per ounce in August 1978 before more than trebling to an average of $US673 per ounce in 1980. In 1975 South Africa was responsible for 60 per cent of all gold mined throughout the world. Its mines, however, were deep and difficult to work, and the system of apartheid meant that many investors were reluctant to invest there. Among other countries well positioned to take advantage of the increase in the price of gold were the United States, Canada and Australia.[99] These three countries accounted for 60 per cent of global exploration activity in 1991, with most of this exploration directed towards gold.[100]

Depreciation of the Australian dollar in the mid-1980s meant that the price of gold for Australian producers leapt from $372 per ounce in 1982 to $638 in 1987.[101] Because of this startling price increase, investment in Australia's gold mining industry increased, enabling it to produce 119.5 million tonnes in 1987-88, exceeding output during the peak years of the nineteenth-century gold rush.[102] Western Mining Corporation, which had begun as a gold miner before diversifying into bauxite and nickel, became Australia's largest gold producer. Because of the extraordinary profitability of gold, WMC earned profits of $83.8 million in 1987.[103] Gold mining took place around previously worked higher grade deposits, whose surrounding areas had become more economical, in Western Australia, Queensland and the Northern Territory. The industry also took advantage of an increasing number of new gold discoveries.[104]

Australian producers had the luxury of being able to sell as much of their extra gold production as they liked without worrying because these exports did not have the same effect on the world price that commodities like iron or coking coal had. As Arvi Parbo told John Horgan:

> … we could not possibly double or treble the production of

> iron ore or coking coal or nickel or aluminium unless the current consumption increased accordingly. Gold is obviously in a different situation.[105]

The Australian gold industry was also aided by the fact that earnings from gold continued not to be subject to the corporate tax rate until 1991. In mining gold, Australian companies were literally digging up foreign exchange from the ground. By 1987 Australia's annual production of gold was earning $1.8 billion. This made gold Australia's fourth largest commodity export after coal, wool and wheat, having moved up from seventh place in 1980.[106] In the late 1980s the government announced that from 1991 income earned from gold mining would no longer be exempt from taxation. This prompted a rush of investment to take advantage of the period before the new tax provisions were implemented. The wave of investment made the Australian gold mining industry very efficient by world standards, with production costs per ounce around $300 in 1987-88.[107] It was only by the latter part of the 1990s that the price of gold began to weaken as many central banks decided not to hold such a large portion of gold in their reserves.[108]

At the same time as the gold boom of the 1980s, another big development in Australian mining was occurring with the development of the Olympic Dam mine at Roxby Downs in South Australia. In the 1970s geologists from Western Mining Corporation (WMC) had been able to discover the world's seventh-largest known copper deposit and its largest deposit of high-grade uranium ore because of the extraordinary resources that WMC devoted to exploration. In the 1960s and 1970s, WMC was spending on average 56 per cent of its annual profits on exploration. These amounts equalled the exploration budgets of CRA and BHP, much bigger companies, in absolute terms.[109] WMC built on the work of geologist Douglas Haynes, who had posited that common types of continental basalts became potent sources of copper when altered by heated water in the earth's crust. Adopting this theory, the company set out to search for copper, not in narrow and easily missed veins, but in sedimentary rocks in which copper had been deposited. From five possible targets containing altered basalt in the Kimberley, the northwest Northern Territory, northwest Queensland, southern New South Wales, and several areas in South Australia, WMC's geological team concentrated its search around a pastoral property in South Australia called Roxby Downs, 550 kilometres north-west of Adelaide.[110]

The company was aided in its exploration work by the fact that South Australia was well ahead of other Australian States in systematically conducting geophysical surveys. After two years of exploration WMC geologists found copper interlocked with uranium, gold, silver and other minerals within a very complex ore body. In January 1979 Norman Shierlaw, now a journalist writing for the *Advertiser*, published an estimate that mining at Roxby Downs could produce minerals worth $54 billion (over $200 billion in 2015 dollars):

> If the whole project is developed to its logical conclusion, the benefit to Adelaide and all South Australians will be equal to if not greater than the potential advantage to Perth and Western Australians of the three billion dollar expenditure on the North West Shelf natural gas program. We know what the development of the iron ore province has already done for Western Australia and coal for Queensland.[111]

Shierlaw hoped that the Olympic Dam mine would have the same transformative effect on South Australia as mining had had in Queensland and Western Australia.

Since developing Olympic Dam would ultimately require the expenditure of billions of dollars, even before copper could actually be exported, WMC set up a team to conduct a tender for a joint venture partner in mid-1978. There were some, however, who thought that only the South Australian government had the financial resources to develop such a huge mining operation. One was Sir Ben Dickinson, the influential retired head of South Australia's Department of Mines. In late 1978 Dickinson wrote to South Australian Premier Don Dunstan urging him to nationalise the mine on the grounds that WMC did not have the resources to develop it. Another was Norm Shierlaw, who also pressed Dunstan to take a stake in Olympic Dam by following the precedent the

State government had earlier set when it took an 18 per cent interest in gas in the Cooper Basin through the South Australian Oil & Gas Corporation. Shierlaw estimated that comparable State government equity in Olympic Dam, in addition to royalties from the entire project, would amount to about one quarter of the state budget.[112]

WMC, however, proved that it was able to develop Olympic Dam in collaboration with an international partner from private enterprise. Sounding out interests from BHP, BP, Shell, Mobil, Texaco, Exxon and Anaconda, WMC eventually decided on BP as its venture partner. The explanation for BP's successful bid was that BP agreed to act as banker for Western Mining's share of the development costs. Moreover, BP would borrow against the project – without recourse to other assets of Western Mining – and agreed to be repaid only when the company began to generate profits. BP undertook to pay $5 million at the commencement of the joint venture; contribute $50 million to the estimated cost of exploration, metallurgical testing and other parts of the feasibility study for the project; finance Western Mining's share of developing the mine; and spend $10 million on exploring Western Mining's tenements across the Stuart Shelf in South Australia.[113] BP's funding arrangements were crucial in providing bridging finance to compensate for the reluctance of uranium buyers to sign contracts until mines were actually built.

In 1979 the project was given further impetus when a Liberal Party government led by David Tonkin replaced the South Australian Labor government. The Labor government had responded to the growing tide of opinion against allowing further uranium mining and opted against allowing the development of Olympic Dam. The Liberal government won the election in large part because of its favourable stance to the development of the mine. Within weeks it started negotiating what was called an Indenture Agreement, setting out the rights and obligations of the joint venturers and the State government in the development of Olympic Dam. But with Labor and Liberals evenly divided in the Legislative Council, South Australia's Upper House, it looked as though the South Australian Parliament would reject the Indenture Agreement. Only by the defection of a Labor member of the Legislative Council would the parliament eventually approve the agreement. The approval of the agreement was followed by years of exploratory work at the end of which, on 8 December 1985, the joint venturers were able to advise the South Australian government that they would develop Olympic Dam to produce 55,000 tonnes of copper, 2000 tonnes of uranium oxide and about 90,000 ounces of gold per annum.

Commencing construction in March 1986 they aimed for the first production in mid-1988. In 1993, five years after Olympic Dam had been commissioned, BP would sell to Western Mining its share of Olympic Dam for $US240 million in addition to $190 million in order to repay the balance of BP's loans to Western Mining. David Upton, the historian of Olympic Dam, writing in 2010, noted:

> In current dollar terms, the total cost for BP's stake was about $US650 million. Today this seems like a remarkable bargain for half of the world's greatest mineral deposit, but the price was struck when Olympic Dam was only a fraction of its current size and in an era of depressed commodity prices. The China-led resources boom was still more than a decade away.[114]

After the mine was opened in 1988, poor prices for copper and uranium prevented it from reaping the full reward for the huge investment made in its exploration and development.[115] The 1980s and 1990s were a time of depressed uranium prices, which only started to recover with the China boom in 2003.[116] The history of Olympic Dam is emblematic of the period between booms in the 1980s and 1990s. On the one hand it is the story of persistence and ingenuity of an Australian mining company in finding and developing the world's largest mineral deposit. On the other, it is a story of how that company failed to recoup the full rewards of its efforts because of depressed prices for the output of the mine. Olympic Dam was developed during a decade when there was a significant public campaign against nuclear weapons and nuclear testing and an associated campaign against uranium mining. In 1983 about 140,000 people marched in Palm Sunday anti-nuclear rallies in Australia's largest cities. The Australian Labor Party was divided on the issue of uranium mining and when the Hawke government decided to allow Olympic Dam to go ahead on 31 October 1983, a member of the cabinet, Stewart West, tendered his resignation in protest. When the matter came before caucus, Gerry Hand, on behalf of the Left, proposed a commitment to phasing out Australia's participation in the uranium industry. The motion was defeated by 55 votes to 46, reflecting the division in the Australian community about uranium mining.[117]

The Mining Industry, Politics, the Aboriginal Question and the Environment

In the period between booms in the 1980s and 1990s, the mining industry became more active in the realm of politics. During the 1960s, when the mining industry was experiencing boom times, the industry had not seen any need to adopt an activist stance in politics. Things began to change for the mining industry in the 1970s with the election of the Whitlam government, the growing prominence of issues such as the environment and Aboriginal affairs, and particularly the publication in 1974 of Tom Fitzgerald's assessment of the mining industry and taxation. At that time the chief executive officer of the mining industry's peak body, the Australian Mining Industry Council (AMIC), was former Department of Trade official G. Paul Phillips. Phillips found AMIC ill-equipped to respond to Fitzgerald's devastating critique. Another problem with AMIC was that Australian companies like Western Mining Corporation felt that the peak mining body was too dominated by foreign interests.

By the late 1970s, however, mining companies, both Australian- and foreign-owned, had come to feel that they could no longer adopt a position of complacency on political affairs and their relationship with powerful interest groups. As Arvi Parbo, one of the leaders of the mining industry in the 1980s and 1990s, put it:

> The realisation that what happens in the public policy arena is today often more important in determining the success or otherwise of a mineral enterprise than what happens in the mine or the mill has brought about fundamental and permanent changes in the industry. The most senior executives now devote the majority of their time to non-technical issues through industry associations, or directly in the name of the company, or else in activities outside the industry. An increasing number of those executives have non-technical backgrounds.[118]

Parbo proved his point by combining the chairmanship of Western Mining Corporation, Alcoa of Australia and BHP with getting the Business Council of Australia (BCA) established in the 1980s. Likewise, Hugh Morgan, the son of Bill Morgan and Managing Director of WMC, devoted much of his time to conservative think tanks the Institute of Public Affairs and the Centre for Independent Studies. In April 1981 Morgan took over the presidency of AMIC and encouraged it to think seriously about the long-term issues facing the mining industry. Under Morgan's leadership, AMIC took three important steps: it initiated annual mining statements prepared by Coopers & Lybrand to gain quantitative knowledge about the industry; it formed a public relations committee to conduct surveys of public opinion; and it launched its first national public relations campaign in 1982. This campaign, "backbone of the nation", was designed to improve the image and public credibility of the mining industry.[119]

With the downturn in the 1980s in the demand for Australia's commodity exports, both farm and mineral, some advisers to government saw Australia's dependence on commodities as a weakness and urged that Australia should deliberately foster the export of manufactured goods. A constant refrain of mining industry spokesmen in response to such criticisms was that the minerals industries remained vital for Australia and that any new manufacturing industries that were developed would take time. For example, Parbo told Haileybury College in Melbourne on 21 July 1986 that, as 80 per cent of Australia's exports in 1986 were rural and mineral:

> The mineral industry is clearly here to stay: it is such an important part of the economy that we cannot afford to neglect it. Development of new industries is desirable, but these will be complementary and not a substitute to the mineral industry.[120]

The reliance that the mineral industry placed on floating exchange rates after 1983 was dramatically illustrated by a letter from Parbo to an academic from the University of New England in 1986. Asked about the effect of the 1986 devaluation on the mineral industry, Parbo commented:

> [t]he short answer to your question is that the main effect in the last two years has been to enable us [Western Mining Corporation] to stay in business. Put in another way, had there been no devaluation, we would now be out of business, as I suspect would be just about everyone else in the mineral industry.[121]

In 1983 the Australian Labor Party committed itself in its election platform to establish uniform national land rights laws and to employ the Commonwealth's constitutional powers against States that did not enact suitable legislation. Consistent with their federalist philosophy, the Liberal and National (formerly Country) parties maintained that Aboriginal land rights should be left to the States. By 1986 the number of Aborigines and Torres Strait Islanders in Australia was 227,638, about 1.42 per cent of a population of 15.9 million Australians.[122] Aboriginal people, like other Australian citizens, did not have ownership rights of minerals; all mineral ownership rights were vested in the Crown and State, territorial and federal governments received royalties from mining activities in their jurisdictions. In some cases, such as Queensland and Western Australia, these royalties constituted more than 20 per cent of total State revenue. States such as these were loath to give up such a major source of revenue by placing them under Aboriginal control.

The ownership of minerals in Australia contrasted with the position in the United States where individuals possessed rights to minerals below the surface as well as to those on surface land. Also in the United States, the federal government and the American States recognised the tribal governments of American Indians and their customary laws. American Indians possessed not only an absolute right to consent to mining on their land but also the power to tax mineral development in various ways. This was not the case in Australia where the legal fiction of *terra nullius* applied from the time of colonial settlement until the early 1990s. According to this doctrine, Australia was regarded as unoccupied or belonging to no one: Aboriginal claims to sovereignty were rejected and their customary laws were not accepted by Australian courts until the High Court's 1992 Mabo decision, which recognised native title to land. Before Mabo, since Aboriginal people did not have statutory power to extract economic benefit from mineral development, they sought to benefit by deploying the power to deny mining companies access to their lands as a way of gaining bargaining power for commercial negotiation.

The main reason why the Australian mining industry opposed national land rights legislation in the 1970s and 1980s was to prevent Aboriginal peoples from being able to deny mining companies the right to access their land for the purpose of exploration and mining. The mining industry held the position that, because the Crown owned minerals in Australia, taxes levied on mining should be paid to governments and not Aboriginal people. If governments wished to improve the welfare of Indigenous people, argued the mining industry, this should come from the general budget and not primarily from the mining industry, which had no specific obligation to provide for their welfare. Aboriginal representatives countered that obtaining rent from mining activity on Aboriginal land was one of the few ways they could raise capital since most Aboriginal land was inalienable (i.e. non-transferable). Regarding the *Aboriginal Land Rights (Northern Territory) Act* 1976 as a disaster for mining in the Northern Territory, the Australian Mining Industry Council was determined to prevent the extension of land rights legislation to Western Australia, whose newly elected Labor government under Brian Burke commissioned an inquiry by Paul Seaman QC into Aboriginal land rights in 1984.

After G. Paul Phillips retired as Executive Director of AMIC in 1984, Hugh Morgan acted for 16 months as both President of AMIC and its executive director. In February 1983 Morgan brought in the lawyer James Strong as Executive Director. Strong had had experience with both bauxite mining at Gove and Aboriginal issues in the Northern Territory. Morgan and Strong together adopted what historian Ronald Libby described as a strategy of mixing with politicians "eyeball to eyeball".[123] A key to the success of Morgan and Strong was their successful collaboration with the Chamber of Mines of Western Australia.[124]

With AMIC conducting an overall national campaign against land rights, the Western Australian Chamber of Mines launched a specific campaign in Western Australia in 1984 that became a landmark in public relations for the mining industry. Costing as much as a million dollars, it was, in Libby's description, "one of the first large-scale public advocacy campaigns carried out by an industry in Australia and it relied heavily upon television advertising".[125] The campaign used such arguments as:

> Beware Our Future Prosperity Hangs in the Balance. The mining industry does not disagree with the principle of Aborigines owning land. We simply believe that all rights in land should be equal. This means that regardless of who owns the land, Aborigines or non-Aborigines, the minerals below the ground are owned by the Crown on behalf of all people.[126]

The pressure from the mining industry and the Liberal Party Opposition led the Western Australian Labor government to oppose the Hawke Labor government's proposed national land rights legislation. Opposition by both the Burke government and the mining industry saw an eventual compromise worked out in the ALP in 1986. In return for Burke agreeing not to attempt to water down the Aboriginal land rights plank in the federal ALP platform, the left-wing of the ALP agreed to withdraw its insistence on over-riding national land rights legislation. To the consternation of AMIC, however, Hawke also agreed to allow an Aboriginal mining veto over new mining operations to remain in the Northern Territory Land Rights Act.

By 1989 the promise of uniform land rights had been abandoned in large part because of the mining industry's campaign.[127] But a few years later on 3 June 1992 the High Court of Australia asserted in its Mabo decision that the common law recognised native title where there was continuous association with the land and where that title had not been extinguished by a valid act. The judgment upheld the legal argument first developed in defence of Aboriginal interests protesting the impact of mining at Gove in 1968.

Mining companies, along with pastoralists, feared that Aboriginal groups would use the judgment to make claims over much of inland Australia. The mining industry was particularly hostile to the Mabo decision. Hugh Morgan argued implausibly that the judgement had "plunged property law into chaos and given substance to the ambitions of Australian communists and the Bolshevik Left for a separate Aboriginal state".[128] The possible impact of the decision on mining was soon evident when the Wik people on the Cape York Peninsula made a native title claim on land that included Comalco's bauxite mine. Queensland Labor Premier Wayne Goss feared that this claim put at risk the financing of expansion of Comalco's Boyne Island aluminium smelter at Gladstone. He therefore called for Comalco's bauxite lease to be validated and the Wik people to be compensated for the loss of title to their land.[129]

Despite strong pressure from elements of the mining industry, pastoralists and some State governments, the Australian government, now led by Paul Keating, held firm to the position of establishing a legislative framework to manage native title claims. By mid-October 1993 the government had reached agreement with representatives of the pastoralists, the mining industry and the Aboriginal people to legislate for such a framework. The *Native Title Act* 1993 provided that Aboriginal people could apply to the Federal Court for recognition of pre-existing native title rights and interests in areas of land. Aboriginal people did not get a right of veto over mining development on native title land, but they would have a right to negotiate with mining companies or governments and to have matters settled by a native title tribunal.[130] Though decisions of the tribunal could still be overruled by the government, the earlier passage of the Whitlam government's *Racial Discrimination Act* 1975 meant that compensation would have to be paid to Aboriginal people for leases that had been allowed on Aboriginal land between 1975 and the bringing down of the Mabo judgment. The native title legislation thus provided a mechanism for Indigenous Australians to assert their rights over their traditional lands in negotiations with mining companies.

Another milestone came a few years later in 1995. In that year, the Chairman of CRA, Leon Davis, threw his influence behind the acceptance of native title and respect for traditional owners. In a speech to his peers in the mining industry in 1995, Davis proclaimed:

> Let me say bluntly, CRA is satisfied with the central tenet of the Native Title Act. In CRA we believe that there are major opportunities for growth in outback Australia which will only be realised with the full co-operation of all interested parties. This government initiative has laid the basis for better exploration access and has increased the probability that the next decade will see a series of CRA operations developed in active partnership with Aboriginal people.[131]

Davis's remarks proved to herald a turning point in the mining industry's stance to Aboriginal peoples. Davis was prescient in recognising that parties to native title claims were not anti-development. As historian Paul Kauffman explains:

> With the incentive of returns from commercial activity on land that might have native title over it determined, Indigenous people have often been pro-development. Indigenous interests are increasingly accepting that the expectations of profits are a pre-requisite for investment. Through their Native Title Representative Bodies they are seeking a fair share of the profits … The "right to negotiate" provisions of the current *Native Title Act* are in fact less confrontationist than the more powerful de facto property rights contained in the "right of veto" or "right of consent" available in the *Aboriginal Land Rights (Northern Territory Act)*.[132]

In 1997 a landmark agreement, the Yandi Land Use Agreement, was signed by Hamersley Iron and Gumula Corporation representing the Yinhawangka, Banyjima and Nyiyaparli people. This was the first major land use agreement in the Pilbara after the Mabo decision. The agreement, a forerunner of many others in the Pilbara and elsewhere in Australia, provided economic and development opportunities for Aboriginal peoples through a package of benefits including employment and training, education, cultural heritage management and business development.[133] Following Mabo, mining companies and the general community came to accept that acceptable mining agreements should provide for educational assistance, training and employment of Indigenous peoples.[134]

In the late 1980s the environment had become such a potent political issue and the environmental vote so important that it was critical in the Hawke Labor government's winning a fourth term in 1990.[135] On the eve of the election Hawke had announced in a televised conference his "absolute commitment on preserving the environment". Not long after the election, issues of the environment and Aboriginal affairs combined when a proposal was submitted to develop the mineral deposits of Coronation Hill, where uranium had been discovered in 1953 and later gold, in what was to become Kakadu National Park. Environmental investigations concluded that a mine at Coronation Hill would neither diminish the grandeur of its cliffs and gorges nor disrupt the movement of birds. But when the local Aboriginal people, the Jawoyn, objected that a mine at Coronation Hill would violate a sacred site, a bitter dispute divided the Hawke government. In June 1991, the final decision on whether or not to mine was made by Hawke who, though an agnostic, decided to ban the mine at Coronation Hill on the grounds of its sacredness to local Aboriginal people.[136]

Conclusion

While the periods between booms from 1974 to 1976 and from 1983 to 2001 were not mining booms, they were still years in which the resources sector remained crucially important for Australia. Just as historian Ian McLean argues that the discovery of gold contributed to a half-century of economic growth from 1850 to 1900, *The Second Rush* argues that a second great minerals rush substantially influenced the half-century of relative prosperity in Australia from 1960 to 2012 across periods of resource booms and interludes between them.

The second period between booms in Australia's second minerals rush began with an international recession in 1982-83. Aggregate growth rates following this recession were a little better in the 1980s than they were in the 1970s (3.5 per cent compared to 3 per cent). But there was no return to the level of prosperity associated with the 1960s. Nonetheless, the Australian economy would not have been able to grow at the rate that it did in the 1980s and 1990s had it not been for the growth of Australia's mining industries and their ability to generate export income for Australia.[137]

The period between booms was punctuated by a severe stock-market downturn in 1987, another recession in 1990-91 and a financial crisis in East Asia in 1997-98. The mining industry endured a period of indifferent commodity prices and of cutbacks by its major customer, Japan, on long-term coal and iron ore contracts. Nonetheless, the period between booms was also a time when coal remained for the most part Australia's chief single commodity export earner, when the industries based on bauxite staged a revival in the mid- to late 1980s, and when gold became an important export earner for the first time since the Depression of the 1930s.

Because of such factors as industrial relations, the emergence of Brazil as an iron ore super-power and economic vicissitudes in Japan, the iron ore industry faced much tougher times in the 1980s and 1990s than in the 1960s and 1970s. Even so, it remained one of Australia's most important export industries. The mining industries began the period between booms as the leading Australian export sector and would remain one of its leading export sectors in the 1980s and 1990s. The period between booms was also a time when the Australian Mining Industry Council adopted a much more activist stance on issues like economic management, industrial relations, Aboriginal affairs and the environment and emerged as one of the most powerful lobby groups in the country.

9

The China Boom

China's economic growth, urbanisation and industrialisation were the mainstay of the third and most dramatic of the resource booms during Australia's second great minerals rush. The China boom spanned the decade from about 2002 until 2012, when prices of Australia's mineral exports began to fall. It followed the Japanese-driven minerals boom from the early 1960s to 1973 and the resources boom from 1977 to 1982. Of the three booms during the second minerals rush, the China boom was the most significant, largely because of the breadth and intensity of China's demand for Australia's mineral resources in the 2000s: the boom almost doubled the value of Australian mining exports and mining's share of nominal GDP in Australia.[1] Mining amounted to nine per cent of GDP in 2003–04 and 18 per cent in 2011–12 in terms of dollar values; and in real terms the China boom added three per cent to Australia's GDP in those years.[2]

The China boom that began in the early part of the 2000s was greatly intensified by the Global Financial Crisis (GFC) of 2007–08. China responded to the worldwide economic crisis that developed in 2007 and 2008 with a massive stimulus package that transformed an upswing in Australia's mining sector into what some believed to be a "commodity super cycle" that would last for decades.[3] The heights that the Australian mining sector reached during the China boom can be appreciated by the fact that mineral and fuel products accounted for 68 per cent of merchandise exports in 2009. Not since the gold rush of the nineteenth century had minerals figured so prominently in Australia's export profile. The leading mineral exports were iron ore, metallurgical black coal, refined gold, crude oil and related products, thermal black coal, and liquefied natural gas (LNG). Measured as a percentage of total Australian exports of goods and services, in 2009 minerals and fuels accounted for 55 per cent, services for 19 per cent, manufacturing for 11 per cent and rural products for a mere 10 per cent.[4]

A phase of economic growth and industrialisation in China in the 2000s drove the Australian resources boom in the same way as Japan's "economic miracle" underpinned the minerals boom and the industrialisation of Northeast Asian economies such as Taiwan and South Korea supplemented a faltering Japanese economy in the 1980s and 1990s. A common attribute of Taiwan, South Korea and Japan was that their economies were complementary to Australia's. They were all abundant in labour and poor in resources, while Australia had a rich endowment of minerals necessary for the industrialisation of East Asia. China was different because it had its own reserves of minerals and particularly coal and iron ore, two of the main commodities in Australia's second minerals rush. Indeed, until 2000 China was better known to the rest of the world as an exporter rather than an importer of mineral commodities.[5] In the 2000s, however, after its accession to the World Trade Organization (WTO), China increasingly sourced imports of iron ore and coal and other mineral commodities from Australia.

This process propelled China into becoming Australia's largest trading partner, and iron ore into the position of Australia's top single export earner. During the China boom iron ore also became a barometer of Australia's economic success in the same way that wool had been in the first half of the twentieth century. From the mid-2000s, Australia's economic wellbeing would become more closely tied with China than it had been with Japan from the 1960s to the 1980s. In the period before the China boom, some had seen Australia's dependence on minerals exports as a serious weakness. During the China boom many looked at minerals in a different light – as an endowment that was helping to make Australian living standards higher than those of Americans, Canadians, the British and most Europeans. The last time that Australians had managed such a feat was during the nineteenth-century gold rush.[6]

In the boom conditions of the first decade of the twenty-first century, mining companies considered that they would be in a favourable position for a generation. With this frame of mind they sought to obtain the best possible prices for iron ore and hence to move trade away from stable but lower-priced long-term contracts to the higher prices available in the spot market. On the buyers' side, Chinese steel companies, worried by increasingly high resource prices, pushed strongly to take controlling interests in mining assets abroad. In the process they stirred nationalist sentiments in Australia and elsewhere. The actions of buyers and sellers militated against a sense of partnership between Australia and China in the development of Australia's iron ore, a partnership that the Hawke government had established with the opening of the Channar mine in the 1980s.

Once the China boom ended in 2012, with the waning of this sense of partnership, Chinese steel companies drove a hard bargain on iron ore prices. During the China boom, Australian mining companies and governments hoped that the good times would last for years and that Chinese steel production would exceed one billion tonnes per year – just as in the early 1970s mining companies had hoped that Japanese steel production would exceed 150 million tonnes per year. It did not happen with Japan in the 1980s and 1990s and it would not happen with China after 2012. The "commodities super cycle" turned out to be an aberration created by the Chinese fiscal stimulus during the GFC feeding directly into state-owned and other heavy industries. From about 2012 China reverted to a long-term strategy with seven per cent growth as the desirable goal, a balancing of exports and domestic consumption, and a rebalancing of heavy, light and services industries. This change in the pattern of Chinese economic growth after 2012 saw an end to the Australian resources boom and harder times for key Australian mining sectors, particularly iron ore and coal. The iron ore industry faced ruthless price-cutting and the coal industry, a high emitter of carbon, was particularly exposed to the world's response to the challenge of climate change.

China's Economic Transformation and Australia

China's modern economic transformation began in 1978 with the emergence of Deng Xiaoping as paramount leader. After establishing diplomatic relations with the United States in 1979, Deng began opening the Chinese economy to the world. He liberalised the agricultural sector, began to import technology, encouraged foreign direct investment and foreign trade and established special economic zones in some coastal provinces.[7] After joining the International Monetary Fund and the World Bank in 1980, China became the largest recipient of aid from both global institutions. By the late 1980s Deng's reforms had led to the expansion of light industry, the development of stock markets and the diversification of the banking sector. Deng also decentralised state-owned trading corporations and manufacturing corporations so that they were able to manage their own export and import policies within the context of the government's goals.

In the first decade of Deng's reforms between 1979 and 1989, China's trade increased tenfold in absolute terms and more than doubled as a percentage of GNP from 11.4 per cent to 26.3 per cent. China exported agricultural products, textiles and light industrial goods in those years. Its top three markets were Hong Kong (still at that time a British Crown Colony), Japan and the United States.[8] Between 1978 and 1999 China's GDP grew at an annual rate of 9.5 per cent.[9] By 2000 the market economy had won out over the socialist economy in China. Historian Odd Arne Westad argues that "[f[[or China's population it was clear that they were living in a new society in which market forces were dominant".[10] China's post-Mao transformation into a capitalist economy was as significant a development for Australia as had been the Chinese Communist defeat of the Nationalists in 1949. The 1949 Communist victory encouraged the Japanese to look to countries like Australia rather than China for their raw materials, in so doing driving the twentieth-century stage of Australia's second minerals rush. China's late twentieth-century embrace of capitalism would take Australia's second mineral rush to its pinnacle in the first decade of the twenty-first century.

Despite China's economic development in the last two decades of the twentieth century, few were predicting that Chinese demand would spark another resources boom in Australia at the beginning of the twenty-first century. When the Reserve Bank of Australia held its annual conference in 2000, there was not a single mention of China while the United States, which was then Australia's top two-way trading partner, was mentioned 93 times. China at that time ranked only sixth of Australia's export markets, behind Japan, the United States, New Zealand, South Korea, and Taiwan. Australian exports to China, at around $4 billion, were less than one quarter of exports to Japan and half of exports to the United States.[11] In 2000 Japan was importing 125 million tonnes of iron ore compared with China's 70 million tonnes.[12] Having lost the mantle of Australia's top two-way trading partner in 1996-97, Japan would regain it from the United States between 2003–04 and 2006-07.[13]

At the turn of the new century, the Australian economy was recovering from the East Asian economic crisis of 1997-98, which had particularly affected Indonesia and South Korea, and from the worldwide downturn precipitated by the bursting of the US technology (dot.com) bubble. At that time, Japan was still the biggest purchaser of coal and iron ore and was in the middle of its third technical recession in a decade. In previous downturns, such as at the end of the resources boom in the early 1980s, the effect on Australia was that the prices of its rural and mineral commodities plunged while those of its imported manufactured goods held firm. The exact reverse occurred in the first decade of the twenty-first century. This was largely due to China's entry into a new phase of economic development. It was to be one of the most extraordinary phases of economic growth in world history. When Britain started to industrialise in 1780, it took more than half a century to double British GDP per head. Such a doubling took Japan 34 years and South Korea 11 years, starting in 1966. By comparison, China doubled its GDP per head every eight years in the 1980s and 1990s and every seven years in the 2000s.[14]

In December 2001, China finally succeeded in gaining entry into the WTO after 15 years of trying. To achieve this goal China had made substantial concessions to the United States on tariffs, agricultural subsidies, intellectual property rights and reform of the banking sector.[15] China's WTO membership was advantageous to a policy it had launched in 2000 called "going global", the goal of which was to transform Chinese businesses into transnational enterprises. Registering some of China's largest firms on the Hong Kong and New York stock exchanges raised billions of dollars for Chinese companies to acquire or merge with foreign holdings.[16]

China soon surpassed Japan as the third-largest trading economy after the European Union (EU) and the United States. Then, in the years between 2004 and 2007, China doubled the size of its exports and overtook the United States as the world's largest exporter. In doing so, China became the workshop of the world, producing electronic goods, domestic appliances, automobile components and textiles.[17] By the first decade of the twenty-first century China had developed the world's largest reserves of foreign exchange and was one of the biggest recipients of foreign direct investment. China's energy consumption surpassed that of the European Union in 2004 and that of the United States in 2006.[18] In 2007 China was responsible for more than 10 per cent of the growth of the world economy.[19]

China's rapid economic growth generated, in turn, a deep thirst for resources. Between 2000 and 2010 China's use of copper trebled, steel output increased fourfold and production of aluminium increased even more rapidly.[20] Chinese usage of electricity, powered by both hydro-electricity and coal, also surged. Although China had its own mineral resources, they were insufficient to feed this rapid industrialisation and export growth. A Chinese White Paper on mineral resources noted in 2003 that there was "a fairly large gap between the supply and demand in oil, high-grade iron, high-grade copper, fine quality bauxite … We shall open still wider to the outside world".[21] In 1998 China was a modest consumer of minerals and metals, accounting for 10 per cent of global consumption of minerals and metals. By 2008 it was consuming between one quarter and one third of minerals and metals and nearly one half of industrial metals.[22] Indeed, from 1998 to 2013 China accounted for all the global growth in commodities such as copper, nickel, seaborne iron ore and coal. As economist David Humphreys has argued, this "represented a degree of dominance in minerals use unmatched since the early 1950s when the United States similarly dominated world markets".[23]

When Deng Xiaoping became leader of China, its urban population had stood at around 18 per cent of the total population. After China accelerated its industrial production and exports in the 2000s, increasing numbers of Chinese moved from the countryside to the cities so that by 2013 urban Chinese represented half of the total population of more than one billion people.[24] China's increasing focus on exports spurred industrialisation in the east and south of the country, particularly in Guangdong, and along the Yangtze River Delta in Zhejiang, Jiangsu and Shanghai. Constructing urban housing and infrastructure such as roads, bridges, railways, office buildings and sewerage systems created a huge demand for steel. To encroach as little as possible on agricultural land, Chinese buildings became increasingly taller, greatly increasing their use of steel.[25] Until 2000 light manufacturing underpinned China's growth. Up to the turn of the century, steel production had risen at the modest rate of 7.5 per cent per year, adding between six and nine million tonnes to world steel production annually. But in 2001, as China's cities and urban infrastructure developed, steel production increased by 23 million tonnes. In 2002, 30 million tonnes were added and in 2005 75 million extra tonnes of steel were produced.[26]

China's own reserves of iron ore were of lower grade and located far away from the steel-producing centres on its eastern seaboard. Consequently, China's steel industry began to turn its attention to Australia and other suppliers for iron ore and coal. China had evolved from importing only three per cent of its iron ore consumption in 1981 to more than half by 2003.[27] The increase in Chinese imports of iron ore after 2003 was even more dramatic for Australia. In 2000 China imported 70 million tonnes of iron ore. This represented 15 per cent of the global market for seaborne iron ore.[28] But over the course of the next decade China totally transformed the seaborne iron ore market. By 2013 China was importing more than 800 million tonnes of iron ore, 65 per cent of global trade in that commodity.[29] Australia's iron ore exports to China were valued at $1.2 billion in 2000-01. In 2005 China overtook Japan as the top buyer of Australian iron ore, and by 2013 Australian iron ore exports to China were estimated at $57 billion out of total Australian iron ore exports amounting to $75 billion.[30] Iron ore became Australia's main single export during the China boom just as coal had been Australia's main single export for most of the 1980s and 1990s. In 2003, when the China boom was beginning, iron ore exports were worth half of coal exports. By 2012, exports of iron ore were worth one fifth more than coal exports.[31] From 2006-07 onwards China was Australia's top two-way trading partner.[32]

A similar phenomenon occurred with bauxite and alumina. Despite lacking alumina, China became the world's largest producer of aluminium metal in the first decade of the twenty-first century. By using low-cost hydro-electricity, China was producing 22 million tonnes of aluminium metal, or 46 per cent of global output, in 2013. From importing less than two million tonnes of alumina in 2000, China was sourcing seven million tonnes, or half its requirements of alumina, from abroad in 2005. Though China had substantial reserves of bauxite, it was not the kind suitable for producing alumina. From that year, China also began to import bauxite for processing into alumina in China. Starting at two million tonnes in 2005, China was importing 70 million tonnes of bauxite in 2013.[33] While Chinese purchases of Australian bauxite amounted to nearly nothing in

2004, by 2007 China was Australia's top customer for that mineral. By that time Australia was the world's leading producer of bauxite and alumina and the world's fifth-largest aluminium producer. In 2008 Australia mined about 31 per cent of the world's bauxite and produced 33 per cent of its alumina and eight per cent of its aluminium metal.

China had been largely self-sufficient in coal, was ranked as the world's largest coal producer and was an exporter of coal into the 2000s. But by 2009 China had become a significant net importer of thermal and coking coal. The Australian coal industry benefited from this shift since Australian black coal had the advantage that it was low in sulphur. China's imports of coal from Australia rose from a modest four million tonnes a year in 2007 to 47 million tonnes in 2009. This was nearly one fifth of Australia's coal exports and made China second only to Japan as an importer of Australian coal.[34] China was, however, a lesser destination than Japan for Australian exports of liquefied natural gas (LNG). These exports increased from $2.2 billion in 2003-04 to $16.3 billion in 2013–14 when Japan continued to take the majority of Australian LNG exports by volume and value.[35] The North West Shelf project was an important part of Australia's second great minerals rush examined in this book. LNG projects that are being developed at the time of writing in 2016 have the potential to make Australia the largest world exporter of LNG. Starting with Woodside's $15 billion Pluto project in 2007, nine new LNG projects were approved across northern Australia, including Chevron's Gorgon project on Western Australia's Barrow Island. These investments give Australia the potential to become the world's largest exporter of LNG. But these developments are beyond the scope of *The Second Rush* and the focus of the concluding chapter is the China boom from 2002 to 2012.

In the year 2000, Australian trade with China overall increased by 50 per cent, followed by another 25 per cent leap in 2001. With exports almost trebling in the five years up to 2004, China became Australia's second-ranked market.[36] After the end of the China boom in 2013, Australia was the largest producer of iron ore in the world with 32 per cent of global production as opposed to Brazil's 19 per cent; the largest producer of bauxite with 29 per cent of global production to Indonesia's 19 per cent; the second-largest producer of gold with 9 per cent of global production to China's 14 per cent; and the fifth-largest producer of coal with five per cent of global production. Australia was also the fifth-largest producer of copper and the fourth-largest producer of nickel.[37]

Towards the end of the China boom, Paul Cleary wrote:

> The mining boom that got underway in the middle of the last decade has already made us the richest citizens in Australian history. The extra increase amounts to about 15 per cent of our economy, or about $190 billion each year. It is as though we inherited the entire economy of New Zealand in one fell swoop.[38]

The boom vastly increased Australians' buying power, measured as the ratio of export prices to import prices, the terms of trade.[39] While real GDP per capita increased by 12 per cent between 2002 and 2010, real gross domestic income per capita increased by 22 per cent. This was because the China boom significantly added to Australian incomes even though this was not fully reflected in conventional measures of GDP.[40] The Governor of the Reserve Bank of Australia, Glenn Stevens, illustrated the point graphically in a speech delivered in 2010. Stevens noted that:

> … five years ago, a ship load of iron ore was worth about the same as about 2,200 flat screen television sets. Today it is worth about 22,000 flat-screen TV sets – partly due to TV prices falling but more due to the price of iron ore rising by a factor of six.

At the end of the China boom, the World Bank estimated that average incomes in Australia were one third higher than American incomes. The last period in which average incomes in Australia were higher than in America was the golden age of the second half of the nineteenth century.[41]

Federal and State governments reaped significantly higher revenue from the boom. In 2003-04 the mining industry paid net company tax of $2.8 billion; in 2008-09 it paid $13.4 billion.[42] The consequent increase in Australia's budget surpluses up to 2007-08 was in the view of John Edwards, "not very far short of the rise in company tax from mining in the period 2003–04 to 2007–08".[43] The gift of additional revenue from the mining boom between 2003 and 2007 provided the means for the Howard government to lower personal income tax rates, extend superannuation concessions and increase spending on social transfers.[44] But as the Australian Treasury got used to the revenues from the mining boom, it then began to over-estimate mining revenues from the Global Financial Crisis onward, thereby compounding Australia's budgetary problems as it fought to ward off a recession.[45] The high prices engendered by the Chinese fiscal stimulus disguised Australia's budgetary problem in the period up to 2012. After the end of the boom, as the budget moved into deficit, a major challenge for Australian governments was to restore the federal budget to balance following the tax cuts made in the early years of the commodity price upswing.[46]

The China boom commenced at about the same time as the merger of BHP with multinational mining company, Billiton, in 2001. By the 1990s BHP was already a global mining company, one third of whose assets were held outside Australia. Before the boom started, BHP had an interest in Escondida, one of the biggest copper mines in Chile, and another at Ok Tedi in Papua New Guinea, oil prospecting concessions in Vietnam and Indonesia, coal mines in Indonesia and the United States, a gold mine in West Africa and exploration ventures in diverse countries.[47] BHP's merger with Billiton was the culmination of a process of consolidation of major mining companies that had been underway in the world mining industry throughout the 1990s and into the 2000s. In 2000, for example, Alcoa acquired Reynolds Metals and in 2004 Alcan completed a $5-billion takeover of the French company Pechiney. Alcan would itself soon be merged into a new mega-company, Rio Tinto Alcan.[48]

There were several factors at work in this consolidation. One was the generally poor performance of companies whose operations were based on a single commodity. A second was that larger and more diversified mining companies were better able to cope with the increasing cost and complexity of large-scale resource projects. A third factor was the emergence of mining managers who were motivated to achieve shareholder value rather than remaining attached to particular assets. The rationale for the merger of BHP with Billiton was BHP's fear of becoming a marginal player in the world mining industry.[49] Another of BHP's motivations was to take advantage of China's economic transformation. Billiton was a leading producer of aluminium as well as the world's biggest producer of chrome and manganese alloys, with operations in Australasia, Africa and South America.[50]

The new company, BHP Billiton, was a diversified mining company, listed on both the London and Australian stock exchanges, with a market capitalisation of $60 billion. BHP's shareholders controlled about 58 per cent of the equity and Billiton's 42 per cent. The merged company instantly became the world's second-biggest mining and metals group after the American company Alcoa, and the largest mining company within a few years. The consolidation that was at work in the mining industry was exemplified by the fact that in 1990 the top five mining and metals companies made up less than one quarter of the world mining industry's total equity value of $US150 billion. BHP, the largest of those companies, then had a market capitalisation of around $US9 billion. But in 2009 the top five companies accounted for almost half of the resource equity of $US250 billion. BHP Billiton remained the largest company but was four times as large as BHP had been in 1990.[51]

A few years later, BHP Billiton made another important acquisition that would help equip it to respond to the China boom. WMC Resources (formerly Western Mining Corporation) was by this time the world's third-largest nickel producer, supplying eight per cent of global demand, and was the owner of the prized asset Olympic Dam.[52] In 2004 Olympic Dam was producing 225,000 tonnes of copper, 400 tonnes of uranium and almost 90,000 ounces of gold. WMC Resources itself was earning tidy profits of $515 million in the first half of 2002. Nonetheless, the chief executive of WMC, Andrew Michelmore, was well aware that his company was susceptible to a takeover. It was a price-taker (or uninfluential in affecting prices), heavily weighted to the London Metal Exchange and subject to considerable volatility. In 2004 newly formed Swiss-based multinational Xstrata made an offer to purchase WMC at $6.45 per share – thus valuing the company at $7.4 billion. WMC felt that Xstrata's offer undervalued the company. On 24 November Michelmore reinforced the point when he gave notice that the mineral reserves at Olympic Dam were nearly 30 per cent per cent higher than previously thought. By this new reckoning, Olympic Dam had the world's fourth-largest remaining copper and gold resources. Although Xstrata would raise its bid to $8.4 billion, BHP Billiton trumped its rival on 8 March 2005 by announcing a bid of $9.2 billion for the Australian company. The purchase of WMC Resources brought to BHP Billiton the world's largest mineral deposit, Olympic Dam, just as one of the world's biggest resources booms was beginning.[53]

The Expansion of the Iron Ore Industry in the Pilbara

Three companies, Vale, BHP and Rio Tinto, accounted for more than two-thirds of iron ore exports to Northeast Asia at the beginning of the twenty-first century. On the consumers' side, five companies dominated Japan's steel industry and one dominated South Korea's. The Chinese steel industry was less concentrated. At the turn of the twenty-first century it was divided into three tiers. The bottom tier consisted of several thousand small-scale enterprises using antiquated technology and accounting for about 15 per cent of China's steel production. A second tier consisted of around 70 medium sized "key enterprises". This second tier contributed about half of Chinese output, but its members were technologically deficient when compared with leading steel companies in other countries. Then there was a top tier of ten "national champion" firms that had achieved the technological standards of other top world steel-makers but accounted for only one third of national production.[54] Even the "national champion" steel companies were comparatively small by international standards, at only one third the size of world-leading steel companies.[55]

The position of the Chinese steel industry in 2000 was rather similar to that of the Japanese steel industry at the beginning of the 1950s.

Until 2000 the lack of international competitiveness of the Chinese steel industry had not been regarded as a problem because the Chinese government was able to protect it from international competition. But after China's accession to the WTO in 2001, the government had to scale back protection for, and subsidies of, its steel industry. During the 1980s the Chinese steel industry had been self-sufficient in steelmaking raw materials with its own coal and iron ore. But, while large in volume, China's iron ore deposits were low in grade. The result was that the cost in raw materials of Chinese steel was high by world standards in the 1990s. By the turn of the century, an iron ore crisis had developed in China. To feed the Chinese steel boom, mining of Chinese iron ore increased. But because of its high cost, Chinese steel companies increasingly shifted their purchases to overseas iron ore suppliers, including such countries as Australia to which the Chinese had begun directing foreign investment. Chinese imports of iron ore rose from 70 million tonnes per annum in 2000 to 628 million tonnes per annum in 2009.[56]

The Chinese position in the world iron ore market differed from Japan's in another way. Japanese companies had formed many joint ventures with Australian and South American mining companies from the early 1970s. China, by contrast, had relatively fewer at the start of the twenty-first century, notwithstanding its pioneering partnership with Hamersley Iron at Channar. Nor did the Chinese mills have the same capacity as the Japanese steel companies to function as an organised group in annual price negotiations. Despite the Chinese steel industry's growing size, the Japanese steel mills maintained leadership of annual iron ore "benchmark" negotiations on long-term iron ore contracts for much of the first decade of the twenty-first century.

At first the major Pilbara producers were slow to respond to the onset of the China boom. In 2000 and 2001 BHP was consummating its merger with Billiton and also dealing with the adverse effect on the world economy of the September 2001 terrorist attacks in the United States. At the end of that year, however, the company established its biggest network office in Shanghai.[57] In 2002 BHP Billiton's sales to China were $US742 million. They rose to $US1.2 billion in 2003 and then to $US1.1 billion in the first half of 2004 alone. In March of 2004, BHP Billiton Chief Executive, Chip Goodyear, initiated the kind of deal with Chinese steel mills that BHP had been pursuing with Japanese companies in the last three decades of the twentieth century. Goodyear locked four Chinese steel mills – Wuhan Iron and Steel, Maanshan Iron and Steel, Jiangsu Shagang group and Tangshan Iron and Steel – into a long-term iron ore contract worth $US9 billion over 25 years in return for Chinese equity in a new mining operation. The four mills each took 10 per cent of the Jimblebar mine (formerly McCamey's Monster), east of Mount Newman.[58]

BHP Billiton's Board also took the decision to invest $US213 million to expand the production of its iron ore mines and an additional $US351 million to enhance its port and rail facilities. By the end of 2004, BHP Billiton's total sales to China were $US2.4 billion. China was making up one tenth of BHP Billiton's total sales and accounting for more than one fifth of its sales growth.[59] For five consecutive years up to 2006, BHP Billiton broke the records for the highest six-monthly net profits for an Australian company.[60]

The acceleration in Chinese demand for iron ore in the first half of the 2000s gave iron ore producers unprecedented market power in negotiations as global supply struggled to keep pace with demand. Iron ore mining companies had a very strong hand in annual price negotiations and, feeling that this strong position would last for a generation, sought to reap the maximum profits from Chinese steel companies.[61] Following on a nine per cent rise in 2003, prices rose by almost 17 per cent in 2004 before an unprecedented increase of more than 70 per cent agreed in 2005 between Vale and Nippon Steel.[62] China considered this last price increase catastrophic. The Chinese steel industry criticised the deal as one settled by Japanese firms whose declining market share meant that they were now incapable of negotiating for the North East Asian industry as a whole. In response to a sharp fall in the profitability of Chinese steel triggered by higher iron ore prices, the Chinese government issued a new steel policy in 2005. Its first component was to develop new iron ore suppliers, principally by encouraging Chinese steel firms to invest in iron ore projects overseas with the assistance of the state-owned banking system. Australia's relative proximity to China made it a natural target for Chinese investment. By investing in Australian iron ore, China hoped to achieve direct control over a substantial proportion of its imported iron ore. In 2011 the China Iron and Steel Association (CISA) stated that it hoped to increase the share of imports from China-invested operations in Australia from 10 per cent to between 40 and 50 per cent over a period of five to ten years.[63] The second component was to foster better coordination between Chinese steel companies with the goal of improving the Chinese position in annual price negotiations. So between 2006 and 2009 the Chinese government took action to limit the number of Chinese companies allowed to import iron ore and to place restrictions on each company's allowed imports.[64] The objective of these measures was to try to lower iron ore prices on long-term contracts by reducing the competition among Chinese iron ore importers.

But the Chinese government's efforts were hampered by market forces and the actions of the buyers and sellers. While the top 10 Chinese steel mills continued to purchase iron ore under long-term contracts from Australian and Brazilian mining companies, the smaller Chinese companies bred a vibrant spot market in which the short-term price rose as much as 40 percent above the long-term contract price level. By 2005 almost half of Chinese iron ore imports were being made through spot purchases.[65] The Chinese government and the steel industry did not welcome this development. The increasing tendency of the big mining companies in Australia to take advantage of higher spot prices led to complaints from Chinese industry and from the Chinese Premier, Wen Jiabao, who called in April 2006 for iron ore contracts to be placed in the hands of governments.

Early in 2006 the Chinese government announced that in future leading Chinese steel firm Baosteel would settle iron ore prices on behalf of Asian steelmakers.[66] In the 2007 annual contract negotiations, Vale had seized the position of leading iron ore negotiations from Rio Tinto. The Brazilian miner settled on a price rise of only 9.5 per cent of the price achieved in the previous year. This left the Australian miners deeply dissatisfied. When Vale in February 2008 again settled on a price rise of 65 per cent, well below the 140 per cent rise requested by the two Australian firms, Rio Tinto and BHP Billiton refused to follow.[67] Rio Tinto contended that spot prices for iron ore were double those which Vale had agreed and notified its intention to sell a further 15 million tonnes on the spot market.

Rio Tinto's decision greatly upset the Chinese steel industry. The Chinese Iron and Steel Association (CISA) complained that Rio Tinto was deliberately under-filling long-term contracts by claiming *force majeure* and then supplying additional tonnages on the spot market.[68] Indeed CISA interpreted Rio Tinto's use of the spot market to achieve more than the price reached with Vale as a "breach of a long-standing rule of the market, according to which everyone agreed to the price of the first settlement".[69] Consequently, CISA upped the ante by threatening that Chinese steel enterprises would boycott any mining companies that switched from long-term contracts to spot prices. Baosteel's negotiator, Ding Shouhu, indicated to Rio Tinto executives that China was prepared to cut its steel production by 50 million tonnes a year to enforce a boycott of Rio Tinto and BHP Billiton ships.[70] In the end, however, these threats were not carried out and the Chinese steel companies reached a compromise with the iron ore miners. They agreed at the end of 2008 on a record-breaking increase for the iron ore producers of 96.5 per cent in lump ore and 79 per cent for iron ore fines. This worked out at a weighted average rise of 85 per cent for Rio Tinto and a little less for BHP Billiton.[71]

CISA tried to put a brave face on the outcome by commenting that the deal with Rio Tinto at least preserved the system of long-term contracts and annual price negotiations. This was not the view of new

BHP Billiton chief executive, Marius Kloppers. Kloppers wanted to help create a new system in which iron ore producers negotiated long-term contracts that fixed quantities exported but that also allowed prices to be adjusted quarterly on the basis of the spot price prevailing in the previous quarter.[72] A breakthrough for Kloppers came in 2009 when Chinese steel companies failed to reach an agreement with the major iron ore producers. Consequently the iron ore companies concluded an agreement with Japanese, Korean and Taiwanese steel companies on prices for their 2009–10 contracts, much to the consternation of Chinese steel companies, which were holding out for lower prices. A sign of the high stakes in these negotiations was that China arrested four Shanghai-based Rio Tinto executives, charging them with stealing secret Chinese information valuable for the iron ore negotiations.[73] Although the ensuing negotiations with Chinese mills were inconclusive, large Chinese steel companies increasingly took advantage of spot market prices, which had begun to fall as a result of the global financial crisis. By 2009 spot market pricing accounted for 60 per cent of the global iron ore trade.[74]

In these circumstances, the Chinese steel industry was forced to accept quarterly index pricing along the lines advocated by Marius Kloppers, thus ending a system of "benchmark" pricing that had been in place since 1973. Under the new pricing regime, the regional price of iron ore continued to rise, reaching $US170 per tonne in mid-2011. The China boom had thus transformed the global iron ore trading system, a system that Australian producers dominated during that period.[75] Initially the Chinese had hoped to use the old system of long-term contracts and annual price negotiations to exercise market power in favour of lower prices. This did not happen. The emergence of large numbers of smaller Chinese steel producers and the actions of the large mining companies pushed the system in precisely the opposite direction.[76] The new system of index pricing reduced the power of the Chinese steelmakers and was more advantageous to Australian mining firms during boom conditions.[77] At the end of the boom, however, the loss of a sense of partnership between buyers and sellers of iron ore saw the Chinese steel companies have no compunction in driving down the price of iron ore to levels that threatened to put some Australian companies out of business. Kloppers' win in 2009 turned out to be a pyrrhic victory. In the long term the iron ore companies had forfeited the protection offered by long-term contracts and stable prices. At the end of the boom, the big mining companies, Rio Tinto and BHP Billiton, saw their salvation in expanding production to maintain their market share against higher-cost producers both in Australia and elsewhere.[78] State and federal governments also faced a mammoth task in adjusting to the falling away of revenue from mining generated during the boom.

The Rise of Twiggy Forrest and Fortescue Metals

The China boom saw the entry of new players in the Pilbara to supplement the production of the iron ore duopoly, BHP Billiton and Rio Tinto. They included Hancock Prospecting, owned by Gina Rinehart, the daughter of Lang Hancock who had died in 1992, FerrAus, Aquila Resources, BC Iron, Brockman Iron and Atlas Iron. The most remarkable of the new iron ore companies was Andrew "Twiggy" Forrest's Fortescue Metals, a company formed in 2003. Forrest's great-great uncle was Western Australia's first premier, Sir John Forrest, who had presided over the State's first gold mining boom in the late-nineteenth and early-twentieth centuries.[79]

Like his famous forbear, Andrew Forrest pursued grand dreams and was prepared to take huge risks to realise them. As a young boy he overcame the affliction of stuttering through the mentoring of Ian Black, the son of "Scotty" Black, an Aboriginal. Forrest, like Hugh Morgan, held strong Christian beliefs that led him in later life to support the cause of ending slavery. He was a generous donor to charitable causes and to Australia's leading university, the University of Western Australia. Like Rio Tinto's Leon Davis, Forrest was a staunch advocate of employing Aboriginal people in the mining industry. He developed into mining from stockbroking in the 1990s when he set out to develop large nickel fields in Western Australia but was forced to back out when costs became prohibitive.[80] Undeterred, Forrest decided in 2003 to move into the iron ore business by establishing Fortescue Metals. The obstacles in Forrest's way were formidable. The barriers to entry in the iron ore trade were so high that it was estimated to cost in excess of $3 billion to finance the infrastructure needed to become a viable producer.[81] In 2004 Fortescue's market value was $60 million compared with the more than $100 billion combined worth of BHP Billiton and Rio Tinto.[82] But the China boom and Forrest's imagination and tenacity enabled Fortescue Metals to overcome these huge obstacles.

By the 2000s the large Pilbara iron ore deposits that could be identified relatively easily, such as Mount Tom Price, Mount Newman and Jimblebar, had already been found. The conventional wisdom until the China boom was that potential iron ore bodies in the Pilbara ended close to any surface outcrop.[83] Consequently, the rate of iron ore discoveries had declined since the 1960s and 1970s. For several decades the iron ore majors had held leases over areas that remained undeveloped in the Chichester Ranges. After years exploring these areas, BHP and Rio Tinto both reached the conclusion that the Chichester Ranges had insufficient ore to warrant the infrastructure required for mining.[84] Forrest, however, had the confidence that he would be able to find ore where others had failed. Spurred on by his foresight about the strength of Chinese demand for Australian iron ore, he acquired leases that the majors had discarded in 2004 and 2005.[85] Rio Tinto acquiesced in letting Forrest try to develop the leases, perhaps relying on its ability to buy them back if serious reserves were proved and on the knowledge that Fortescue Metals would have to build a port and rail system to develop them.[86]

Forrest's geologists counted on the fact that major deposits of iron ore, though not visible to the eye, might yet lie below the surface. They used the "Tiger's Tail" model to look for indications where a small part of the ore – the tail – might be above ground while the bulk of the deposits, the tiger, lay below the surface.[87] Such was the reasoning that led the Fortescue geologists to discover massive quantities of ore in the Chichester Ranges. Whereas deposits such as Mount Whaleback and Mount Tom Price were like icebergs with large sections above the ground, the Chichester deposits were found to lie in vast, flat blankets stretching for tens of kilometres and yielding reserves of billions of tonnes of ore.[88] Despite this breakthrough, however, the ore was found to be low-grade and so riddled with shale and other impurities as to be worthless.

At this point, luckily for Forrest, John Clout, a geologist regarded as one of Australia's foremost experts on iron ore, achieved a crucial breakthrough for Fortescue Metals. He identified small amounts of microplaty hematite – the same variety of ore as at Mount Tom Price and Mount Newman – in the iron ore in the Chichester Range. Clout argued to Forrest that the challenge was to find enough premium ore to blend with Fortescue's low-grade ore and justify a mine.[89] This was the same kind of breakthrough that had underpinned Western Mining Corporation's unconventional thinking in the 1960s about Jarrahdale bauxite. Forrest agreed with Clout and ordered his geologists to peg and drill every inch of his leases, which by the end of 2004 covered almost 40,000 square kilometres. Forrest's biographer found that four years later, when Fortescue Metals began exporting iron ore, it had leases over 52,000 square kilometres, making it a bigger holder of iron ore tenements than either BHP Billiton or Rio Tinto.[90]

Predicting early the magnitude of the iron ore trade between Australia and China, Forrest negotiated initial sales agreements with Chinese steel mills and made arrangements with Chinese engineering firms to build port and mine infrastructure. He forged alliances in 2005 with investors including the China Harbour Engineering Corporation (CHEC), the China Railway Engineering Corporation (CRE) and the China Metallurgical Construction Group (MCC).[91] This deal fell over in 2006 when MCC adopted an aggressive strategy and sought a majority equity stake in Fortescue Metals. Forrest was forced to negotiate new arrangements.[92] By 2006, with its exploration funds running out, Fortescue Metals required a cash injection of $2 billion. With the aid of the American banking group Citigroup, the company performed a feat analogous to the foundation Pilbara projects of the 1960s by raising this extraordinary sum in America. At that stage there were only two owners of infrastructure in the Pilbara, BHP Billiton and Rio Tinto, and each had been successful in rebuffing attempts by others to use its rail network. Rio Tinto had actually been prompted to take over North Limited when the latter tried to get access to its infrastructure in 2000.

In 2007 Fortescue Metals achieved a breakthrough when it negotiated its first iron ore sales contract with Baosteel. Fortified by this deal and having received permission from the WA government to build a berth at Port Hedland, Forrest abandoned the effort to persuade the majors to let Fortescue share rail facilities. Instead he built his own 260-kilometre railway system to Port Hedland. In May 2008 Forrest made his first commercial shipment of iron ore to China on the *Heng Shan*. Between then and May 2013 Fortescue Metals shipped more than 200 million tonnes of iron ore and developed the infrastructure to produce more than 155 million tonnes per year. Forrest had achieved the astonishing feat of building the fourth-largest iron ore company in the world during the China boom.

The Global Financial Crisis and the Mining Boom

The latter stages of the China boom coincided with the GFC and the attempt by BHP Billiton to effect one of the greatest mergers in mining history, that of BHP Billiton with Rio Tinto. This initiative provoked Chinese conglomerate Chinalco to pre-empt the merger by taking an equity stake in Rio Tinto. These dramatic events occurred at the same time as the Australian government was reacting to one of the greatest economic challenges in Australia's history: preventing the global financial crisis that had begun in the United States in 2007 from precipitating a depression in Australia.[93]

Two weeks after the election of a Labor government under Kevin Rudd, in November 2007, BHP Billiton made an informal offer to take over Rio Tinto at the prevailing share prices that valued the company at $US140 billion. BHP Billiton had contemplated such a merger in the early 2000s before the acceleration of the China boom.[94] Its rationale was that BHP Billiton had the better of the iron ore bodies in the Pilbara but that Rio Tinto, now strengthened by its absorption of Robe River, had superior infrastructure. But by 2002 what might have been the biggest merger in the history of the world mining industry had come to naught. Not deterred by the earlier failure, BHP Billiton chief Marius Kloppers revived the idea in 2007 to achieve greater economies of scale and larger operations by combining the two largest Pilbara iron ore operations, which between them were now supplying just under 40 per cent of China's iron ore needs.[95] The merger would have created a giant company worth US$350 billion in market capitalisation. Kloppers' case for the merger was that it would deliver US$3.7 billion in annual cost savings, most of which would come from combining the operations of Hamersley Iron with BHP Billiton's seven mining operations in the Pilbara and sharing rail and port facilities.

When Kloppers revived the plan, the Pilbara was in the midst of the greatest boom in its history and the Chinese government and industry were becoming seriously concerned about high resource prices. Liu Yikang, the China Mining Association's deputy secretary general, was aghast at the prospect of a behemoth in the Pilbara, commenting: "It's terrible. If completed, the deal between Rio and BHP would create a mining giant enjoying a monopoly in many mineral resources and a further lessening of competition, and would thus lead to hikes in iron ore prices".[96] The British *Financial Times* quoted Chinese Premier Wen Jiabao saying: "I don't know a lot of things but what I do know is that this deal is not good for me".[97] The high stakes involved in the merger of these two giant mining companies were highlighted by business analyst Ian Verrender, who commented: "This is not just another merger – this is a huge battle for control of vital and scarce resources, the sort of thing that has brought countries to war in the past, and it has coincided with an uprising of Chinese nationalism".[98] The Chinese feared that a merger between BHP Billiton and Rio Tinto would lead to higher iron ore prices, which would affect Chinese steel and manufactured exports and thus endanger China's ongoing economic growth.

BHP Billiton's manoeuver also provoked strong resistance from its long-time competitor, Rio Tinto, whose American chief executive, Tom Albanese, rejected BHP Billiton's offer of three shares for every one of Rio's shares. Albanese had just consummated a $40-billion takeover of the aluminium producer Alcan, calculating that taking over the huge debt that had resulted would deter BHP Billiton from proceeding with the Rio Tinto merger.[99]

In the meantime, BHP Billiton attempted to reassure Chinese industry that a merger would not have adverse effects on China and signed up with Baosteel to export an additional 94 million tonnes of iron ore from the Pilbara. Then, on 30 January 2008, the state-owned Chinese company Chinalco joined forces with American company Alcoa in what was described as the "biggest ever dawn raid" on the London stock market.[100] Chinalco, the world's second-largest producer of alumina and fifth-largest smelter of aluminium metal, was a company that wanted to diversify into other areas of mining. It had bought Peru Copper in August 2008 and was assessing the prospects of buying other mining companies around the world. Joining forces with the American aluminium giant, it purchased a nine per cent stake in Rio Tinto for $US14.05 billion as a means of pre-empting the BHP Billiton offer. Alcoa's motive in assisting Chinalco to block BHP Billiton was to give itself advantages in the Chinese market. The exercise was the largest offshore investment yet made by a Chinese company.[101]

In reaction to the Chinese move, BHP Billiton made its takeover of Rio Tinto hostile, raising the bid to 3.4 shares to one of Rio Tinto's.[102] The chair of BHP Billiton, Don Argus, justified the proposed merger to Rudd on the grounds that companies like BHP Billiton in small countries like Australia had to grow bigger in order to compete in a globalising world. Argus argued:

> The plight of every small economy is that it has too small a home to grow champions of scale. With increased globalization and consolidation, this inevitably results in local companies being absorbed or becoming subsidiaries of multinationals. After being purchased, they quickly get hollowed out and become branch offices … Investment choice in the local market dries up.[103]

In the following year, in February 2009, Chinalco tried to invest a further $US19.5 billion in Rio Tinto consisting of $US7.2 billion in purchases of convertible bonds and $US12.3 billion in minority equity stakes in Rio Tinto's bauxite mine in North Queensland, Hamersley Iron in Western Australia and some of Rio Tinto's key copper mines around the world. The proposed deal would have raised Chinalco's stake in Rio Tinto to roughly 18 per cent and entitled its executives to two seats on the board.[104] Chinalco envisaged that its share of Rio Tinto's output would be marketed by a joint venture, which would allow the Chinese company to gain knowledge of the workings of the iron ore industry and the way that producers priced iron ore. This prospect aroused the same kinds of fears that had arisen in the 1970s and 1980s about Japanese companies taking equity positions in Australian mining operations to gain an insight into their pricing.

Continuing with its own campaign for Rio Tinto, BHP Billiton argued strongly to the Rudd government that the proposed Chinalco deal with Rio Tinto "will have the clear effect, and is intended to have the effect, of substantially reducing the prices paid by the Chinese steel mills for iron ore, particularly that produced in Australia". BHP Billiton further advised:

> If the transaction was approved in its current form, it would result in effective control over Rio Tinto passing to Chinalco and, as a State Owned Enterprise, to the Chinese government. Rio Tinto would likely be the foundation of a powerful China Mining Company. This

> would irrevocably change the face of the global natural resources industry – it would see a major competitor operating on an unlevel playing field with their portfolio based on Australian assets.[105]

Peter Drysdale, an economist at the Australian National University, argued the exact opposite case – that the deal, if successful, would have brought about "the first great Anglo-Australian-Chinese mining and metals company, probably headquartered in Australia. This company would have been positioned to play a lead role in the Chinese market".[106]

But the BHP Billiton criticisms of the proposed Chinalco investment in Rio Tinto found a sympathetic response from the Liberal–National coalition Opposition.[107] Trade unionists such as Paul Howes, national secretary of the Australian Workers Union, also opposed the acquisition. By April 2009 it had also become clear that Australia's Foreign Investment Review Board (FIRB), the advisory board established in 1976 under the Foreign Acquisition and Takeovers Act, was taking an active interest in the Chinalco case. The FIRB was required to review all foreign acquisitions of Australian assets worth more than $A100 million, evaluating them for their consistency with the national interest. In April 2009 the FIRB ruled that it would be anti-competitive for Chinalco to buy into Rio Tinto's aluminium operations given that the company was already China's biggest aluminium producer. The FIRB ultimately informed Chinalco that the deal was "too big and too complex".[108]

While Chinalco's further investment in Rio Tinto was thwarted by the FIRB, it was the GFC that stopped BHP Billiton's attempted takeover of Rio Tinto. The crisis had begun in the United States in the second half of 2007. Following the collapse of Lehman Brothers in the following year, bank lending in advanced economies, which had fuelled consumption and helped China's export growth, was curtailed and global GDP shrank for the first time since the end of World War II. The Chinese economy had become dependent on exports in the boom years and these shrank by one third between the second half of 2008 and the first half of 2009.[109] China's response to the world economic crisis was to introduce in November 2008 a four trillion renminbi ($US586 billion) stimulus program. Amounting to 14 per cent of GDP, this made the Chinese stimulus package the largest of any country relative to the size of its economy.[110] The package was directed to public investment in transport, rural infrastructure and public housing. The effects were remarkable with industrial production rebounding in 2009 as public investment filled the void left by the slowdown in exports and manufacturing.

With the resurgence of the Chinese economy came a spurt in its demand for mineral commodities overseas. While some were predicting the end of the China boom in Australia in the wake of the GFC, this did not happen. Rather, Chinese reaction to the GFC prolonged and heightened the boom in Australia. After its recovery in 2009 the Chinese economy grew by 10.3 per cent in 2010 and 9.3 per cent in 2011. The comparison with other G7 countries is striking. While industrial production in these countries was six per cent below its level in 2007, China's industrial production doubled over the same period. In 2009 China surpassed Japan as the largest producer of cars in the world and in 2014 was assessed by the World Bank to have the largest economy in the world.[111]

The huge Chinese fiscal stimulus was the key to what was being called a "commodities super cycle".[112] The thinking behind the concept was that emerging economies such as China, Brazil and India were growing faster than advanced economies and that the world was on the threshold of a number of decades of material intensive growth.[113] Those who thought in these terms, however, were mistaken about long-term Chinese policy and the effects of the GFC on this policy. The Chinese government had always expected that in the normal course of events China would revert in the second half of the 2000s to a long-term sustainable growth rate of around seven per cent per annum and that it would shift to a more domestically driven economy with a strong emphasis on developing western and central China. The Global Financial Crisis of 2007–08 and China's fiscal response to it delayed this measured shift from export-led and investment-led growth to domestically led growth by half a decade and in the process fuelled investment and oversupply by resources and energy companies in Australia. The reckoning for Australia, when the Chinese stimulus ran its course, would be harsh.

At the start of the China boom, royalties from iron ore to the State of Western Australia were less than $300 million. By 2013-14 iron ore royalties exceeded $5.3 billion and accounted for more than one fifth of the Western Australian government's own-source revenue. Federal coffers also received a windfall gain from higher company tax. The mistake made by Australian governments was to misunderstand the nature of the commodity cycle and assume that the higher taxes and royalties would perpetuate a super cycle for a generation. They adjusted their spending in line with these flawed assessments. At the height of the boom the Australian government also made an effort to raise higher federal taxation from mining. In 2010 the Rudd government took the radical initiative of announcing it would impose a Resource Super Profits Tax (RSPT) – a 40 per cent tax on mining "super profits", triggered when a company's profits exceeded $75 million. A vigorous public relations and advertising campaign by the Minerals Council of Australia, the successor of the Australian Mining Industry Council, and mining companies proved to be one of most successful in Australian history. The campaign dominated the airwaves in May and June 2010 and inflicted such huge political damage on the Rudd government that even some mining workers sided with their employers over the Labor government. The smaller miners participated along with the big miners. Forrest and Gina Rinehart, boosted by the China boom to become Australia's richest woman, addressed a boisterous crowd in Perth on 9 June. Rinehart led the crowd in chanting slogans such as "axe the tax". The public campaign played a large role in the ALP party-room defeat of the generally popular first-term Prime Minister, Kevin Rudd, who was ousted as prime minister by a caucus vote on 24 June. Rudd's successor, Julia Gillard, renegotiated the tax as a Mineral Resource Rent Tax. However, because of design flaws and the end of the resources boom, this tax failed to raise significant revenue, and the Abbott government abolished it after the 2013 election.[114]

The China Coal Rush

Although iron ore exports surpassed those of coal during the China boom, coal was either Australia's top export commodity or second-biggest export commodity during that boom. The China boom in Australia centred on the expansion of both coal and iron ore. As with iron ore, the pressure on coal prices in the 1990s had seen the Australian coal industry become more concentrated in ownership. By the early 2000s, four companies were producing about three quarters of saleable coal output in Australia. The hopes of these companies for better times were realised in the China boom when the price of coal began to rise at about the same time as that of iron ore.

The rise in coal prices was almost as spectacular as those of iron ore. Between 2003 and 2012 the price of thermal coal more than quadrupled, from around $US25 per tonne to well over $US100.[115] Over the 10 years of the China boom from 2002 to 2012 the Bureau of Resources and Energy Economics reported that Australia's production of saleable black coal increased by 46 per cent to reach 401.4 million tonnes in 2012-13. Whereas Australia had been exporting 120 million tonnes of coal in 1991, by 2012 coal exports had more than doubled to 316 million tonnes.[116] At the height of the China boom, the coal industry earned around $48 billion in exports, employed around 46,000 people and attracted huge investment.[117] In the Hunter Valley and Newcastle regions alone, coal production increased by 33 per cent to 141 million tonnes, yielding 105 million tonnes of saleable coal.[118] Newcastle was by now the biggest coal port in the world, servicing the Hunter region, which contained most of the 60 coal mines and produced more than 70 per cent of NSW output.[119] The Global Financial Crisis did not slow down the coal boom and indeed this was the period in which China increased its purchases of Australian coal. By 2011 ten new coal projects were in progress and another 14 in the pipeline. Six port and five railway expansions were also underway in an industry whose new spending was $55 billion in 2010-11 and $73.7 billion in 2011-12.[120]

Australia's coal output increased because of rising demand from Asian countries other than China in the first part of the boom and then as a result of increasing demand from China itself. In the latter stages of the China boom, from about 2008-09 to 2012-13, coal accounted, on average, for about 15 per cent of Australia's total exports of goods and services and was in itself equal to or greater than Australia's agricultural exports. The economist Sinclair Davidson estimated that the "coal economy" – that is, coal mining and related services and industries – amounted to 3.1 per cent of Australia's gross domestic product in 2011-12. He further calculated that the "broader coal economy", defined as the output of the coal economy plus the spending of wages earned in the coal economy – represented 4.2 per cent of GDP or almost $60 billion in 2011-12, the same size as iron ore and agriculture (at the farm gate).[121]

During the China boom Australia's coal producers were dominated by two companies that had been mining coal during the minerals boom, BHP Billiton and Rio Tinto, and three multinational companies that had entered or re-entered the Australian coal industry in the 1990s and 2000s, Anglo American Corporation, Peabody, and Xstrata (later GlencoreXstrata and then Glencore). Between them, the big five companies accounted for production of about 250 million tonnes of coal annually at the height of the boom.[122]

In combination with Mitsubishi, BHP Billiton was Australia's largest coal producer, contributing more than one quarter of Australia's annual coal exports and about 28 per cent of the world's seaborne trade in metallurgical coal. The BHP Mitsubishi Alliance was the successor to Central Queensland Coal Associates, the combination between Utah and Mitsubishi that had developed during the minerals and resources booms. During the China boom, the BHP Billiton Mitsubishi Alliance operated seven mines in the Bowen Basin through the Hay Point Terminal – Goonyella Riverside, Broadmeadow, Daunia, Peak Downs, Saraji, Crinum, Blackwater and Caval Ridge. With Mitsui, BHP Billiton operated two other Bowen Basin mines, the South Walker Mine and Poitrel Mine. BHP Billiton also mined thermal coal at Mount Arthur in the largest individual coal production site in the Hunter Valley.[123]

Rio Tinto operated five mines in Australia during the China boom. Hail Creek was an operation that had first been mooted during the minerals boom in the early 1970s but only began operations under Rio Tinto in 2003. Rio Tinto's other Bowen Basin mine was the Kestrel mine that Rio Tinto had purchased from Atlantic Richfield in 1999. Rio Tinto operated an open-cut mine at Bengalla, four kilometres south of Muswellbrook, which commenced operations in 1999; it operated the Hunter Valley operations of Coal and Allied Industries, the management of which Rio Tinto had assumed when it took over Coal and Allied in the late 1980s; and it ran the Warkworth thermal and soft coking coal mines, which had first begun operating under R.W. Miller and Coal and Allied Industries in 1981.

Peabody had first come to Australia in a consortium with Thiess Bros and Mitsui in the 1960s. In 2000, Irl Engelhardt, the chairman and chief executive officer, floated Peabody as a public company to generate the capital to fund new ventures, including an expansion into China and into more coal-fired power generation in America.[124] After its flotation, Peabody Energy became the largest coal mining company in the world.[125] In New South Wales during the China boom, Peabody Energy operated the Metropolitan mine 30 kilometres north of Wollongong, the Wambo mine 30 kilometres west of Singleton in the Hunter Valley and the Wilpinjong mine 40 kilometres north of Mudgee. It also operated six mines in the Bowen Basin in Queensland. Anglo American Corporation, which was the dominant force in coal mining in South Africa, began mining coal in Australia from about 2000.[126] During the China boom, it operated Capcoal in the German Creek formation in a joint venture with Mitsui to supply coal for export to steel markets. The area had first been developed during the resources boom. Anglo American had also acquired the Callide open-cut thermal coal mine, mainly supplying coal for power generation in Queensland, and the Drayton open-cut thermal coal mine in the Hunter Valley, a mine originally developed by CSR during the resources boom.

In 2001 Xstrata, a small company with its headquarters in Switzerland, persuaded Glencore, the Swiss trading and mining company that owned 40 per cent of Xstrata, to allow the coal assets of its subsidiary, Enex, to be passed to Xstrata.[127] Xstrata thereby assumed the management of the Bulga mine in the Hunter Valley, a mine originally started by BHP as Saxonvale during the resources boom. Xstrata successfully listed on the London Stock Exchange in March 2002. From a tiny company of about 17 people working from the basement of J.P. Morgan's offices near Blackfriars Bridge in London, Xstrata would become a $US40 billion company during the China boom.[128] In 2003 it paid almost $3 billion to acquire the coal and other assets of its Australian rival, MIM Holdings, thereby turning itself into the world's largest exporter of thermal coal.[129] But in 2005, as we have seen, Xstrata was beaten by BHP Billiton in its efforts to take over WMC Resources (formerly Western Mining Corporation).[130] In 2007 Xstrata acquired the Anvil Coal Mine (later Mangoola) in the Upper Hunter Valley from Centennial coal.[131] Five years later, at a time when Xstrata was making one third of its sales to China, it opened its first office on mainland China in Shanghai. In July 2012 Xstrata was taken over by Glencore, a company that itself had taken over the operations of petroleum and minerals trader Marc Rich.[132] Xstrata subsequently traded as Glencore.

Besides the big five coal companies there were about a dozen medium-sized companies Centennial Coal, Yancoal, Wesfarmers, Ensham, Whitehaven, New Hope, QCoal, and Jellinbah Resources, which together accounted for about 70 million tonnes of coal. The coal rush of the 2000s produced a number of millionaire coal barons. One of the most colourful was Nathan Tinkler, a coalmining electrician born in Inverell in Northern New South Wales in 1976. In 2007 Tinkler went heavily into debt to purchase a Middlemount coal deposit for $30 million. He later sold it to Macarthur Coal for $275 million in cash and shares and came out as a 10 per cent shareholder in the company. Macarthur's share value doubled before Tinkler sold out of it in 2008 for $441 million. With the proceeds, Tinkler's company, Aston Resources, purchased the Maules Creek deposit from Rio Tinto for $480 million in 2010. When Aston Resources later listed on the stock exchange it was valued at $1.2 billion and Tinkler had joined the ranks of Australia's mining billionaires. In 2011, when Tinkler's personal wealth was valued at over $2 billion, his company merged with a nearby company, Whitehaven, and Tinkler subsequently launched a takeover bid for the merged company. But Tinkler's fortunes rose and fell with the China boom. When coal prices started falling, Tinkler's net worth shrank precipitately.[133]

Climate Change and Mining

In the 2000s, at the same time as Australia's resources industries were booming, some of them were coming under increasing criticism for contributing to the warming of the earth's temperature. The link between industrialisation, which began in Western Europe in the late eighteenth century, and changes to the earth's climate began to be subject to heated debate in the 1960s and 1970s.[134] Scientific research ascertained that, from about 1800, when the world began burning fossil fuels, the level of carbon dioxide in the earth's atmosphere increased from 280 parts per million to 380 parts per million. In the twentieth century, the period of the greatest burning of fossil fuels, the average temperature of the earth's surface rose by approximately 0.6 degrees Celsius.[135] A major source of increased greenhouse gas emissions is the world's increased population, which has trebled since 1950, and its rising incomes. World GDP has also trebled since 1950, and with increased incomes came increased usage of energy and hence emission of greenhouse gases.[136]

In 1988 a group of scientists met in Geneva to inaugurate the Intergovernmental Panel on Climate Change (IPCC).[137] The IPCC drew its legitimacy from two international organisations: the World Meteorological Organization and the United Nations Development Program. Through conferences, papers, and dialogue, the IPCC set out to understand the world's climate.[138] In October 1990, it reported to the United Nations that the earth was definitely warming but that it would take another decade or so to determine whether this warming was essentially man-made or a product of the natural variability of climate.[139] The IPCC later predicted warming in the twenty-first century in the range from 1.4 to 5.8 degrees Celsius by 2100.[140] An increase at the upper end of this range was predicted to have catastrophic effects such as severe droughts and rapidly rising sea levels.

In response to the first report of the IPCC, the United Nations General Assembly called for an international agreement to limit greenhouse gases, primarily carbon dioxide. The making of the convention would be by the Earth Summit: a United Nations Conference on Environment and Development held in Rio de Janeiro in 1992. The conference reached agreement in April of that year to a common goal: "stabilization of greenhouse gas concentrations in the atmosphere at a level that would prevent dangerous anthropogenic interference with the climate system".[141] Five years later, in 1997, the Japanese city Kyoto hosted an international meeting to reach an agreement to cut greenhouse gas emissions. The conference agreed that participating nations would reduce their output of greenhouse gases emissions by between six and eight per cent below 1990 levels by the years from 2008 to 2012.[142] The United States agreed to a seven per cent reduction, the European Union to eight per cent, Japan to six per cent and twenty other nations to five per cent.[143] Developing countries such as China and India were allowed to increase their greenhouse emissions without limit. Australia's delegation at the Kyoto climate conference secured agreement for an eight per cent increase in greenhouse gas emissions over 1990 levels on the grounds that its resource-intensive economy demanded a special concession.[144]

After the Kyoto conference, energy and industry groups mounted such an effective campaign against the agreement that the United States Senate refused to ratify the agreement. One of the principal explanations of the US stance on the Kyoto Agreement was that it did not apply to the developing world. The Howard government in Australia also refused to ratify the agreement in sympathy with the American stance. The Australian government argued that the Australian economy was growing faster than the OECD average and that its population was increasing at a faster rate than other developed countries. Consequently, Australia argued that the task of reducing emissions to a historical benchmark was greater than for comparable developed countries like France or the United Kingdom. The Howard government did, however, agree to establish a Mandatory Renewable Energy Target as a by-product of its Kyoto negotiations. It came into operation in 2000 and required power generators to supply two per cent of their generation from renewable energy sources.[145]

The challenge of reducing greenhouse gases generally was made all the more difficult by the fact that consumption of electricity was growing worldwide.[146] About 75 per cent of Australia's electricity is generated from coal-fired plants.[147] The country now has 23 coal-fired power stations, which together contribute about one-third of Australia's greenhouse gas emissions.[148] Australian coal exports, moreover, constituted about one third of the world's seaborne coal trade, as Australian coal continued to feed the power plants and steel mills of an industrialising Asia. Throughout the world 40 per cent of electricity is still generated from coal.

From just before the China boom to the present, environmental concerns over the link between climate change and carbon dioxide has emerged as a critical question for the future of Australia's resources-based economy and particularly the mining industry, which has been central to Australia's economic growth since the 1960s. Yet hydrocarbons – oil, natural gas and coal – provide more than 80 per cent of the world's energy use.[149] And some predict that between 75 per cent and 80 per cent of world energy will still be carbon-based in two decades time.

The mining industry under most challenge from the global movement to reduce global greenhouse emissions is the coal industry because it is the largest single source of greenhouse gases in the world.[150] In contrast to other fuels, like oil and natural gas, coal has the advantage of being cheap and plentiful worldwide. There are still an estimated one trillion tonnes of recoverable coal, which is by far the largest reserve of fossil fuel left on the planet.[151]

One proposed technological solution to the link between coal usage and climate change was carbon capture and sequestration (or storage), better known by the shorthand CCS. The idea was to keep carbon out of the atmosphere by capturing it and burying it underground. In July 2005 six countries – the United States, China, India, Australia, South Korea and Japan – agreed to the Asia–Pacific Partnership on Clean Development and Climate. It aimed to accelerate the deployment of new, more efficient power-generating plants in China and India.[152] The proposed system for CCS would, however, be expensive and complex. In the view of energy historian Daniel Yergin it would be like:

> [c]reating a parallel universe, a new energy industry, but one that works in reverse. Instead of extracting resources from the ground, transporting and transforming them, the "Big Carbon" industry would nab the spent resource of CO_2 before it gets into the atmosphere, and transform and transport it, and eventually put it back into the ground. This would truly be a round-trip.[153]

In 2007, in response to growing community concerns about climate change, the Australian Prime Minister, John Howard, announced that he was amenable to establishing an emissions trading scheme. He established a Prime Ministerial Task Force on the subject that included representatives from BHP Billiton and Xstrata to advise him. The press release announcing the group referred to the "possession of large reserves of fossil fuels; as one of Australia's major comparative advantages" that had to be preserved.[154] The Howard government was defeated before it could implement the advice to establish a carbon market. After being elected the Rudd Labor government ratified the Kyoto Protocol and in 2009 expanded the Renewable Energy Target to 20 per cent. Then the Gillard government successfully implemented a carbon-pricing scheme in July 2012 at the end of the China boom, although this scheme was repealed two years later under a new conservative government. Since the China boom, significant international efforts have been made to develop a new legal framework to "shape the transition to a new system of energy and the market".[155] Australia's two leading minerals industries in the second rush, iron ore and coal, have both been adversely affected by the slowing down of Chinese economic growth. Adding to the uncertainty in the coal industry is how a new system of energy use established to mitigate greenhouse gas emissions might affect the coal industry and therefore, by implication, also iron ore. This uncertainty and falling prices mean that financial institutions have become increasingly reluctant to invest in large new ventures such as Adani's Carmichael thermal project in the Galilee Basin in northern Queensland. While some have gone as far as predicting the end of the coal era, at least as it applies to thermal coal, others argue for the continued importance of coal in the global energy mix and hence of the coal industry in Australia.[156]

Conclusion

The China boom was the biggest boom in Australia's twentieth and twenty-first century minerals rush, the second rush. The China boom increased the wealth of Australians and filled the coffers of State and federal Treasuries. Some policy-makers and mining industry strategists saw the China boom as a "commodities super cycle" that would last for decades. This was not to be. When Kevin Rudd briefly returned to office as Prime Minister in 2013, he announced that "[t]he China resources boom is over" and that a "huge adjustment was required".[157] In the wake of the boom, economist Bob Gregory predicted that Australians were facing "the largest negative macro shock to Australian living standards since the 1930s depression" as prices returned to their pre-boom levels.[158] From a peak of $191 per tonne in February 2011 the price of Australian iron ore had fallen to $43 per tonne in November 2015.

The surge of Chinese growth from 2005 to 2011 was not the start of a sustainable long-term trend but a unique historical period in which China sought to establish itself as the workshop of the world. When the

Global Financial Crisis threatened the Chinese economy, its government announced a stimulus package in 2008 that intensified the boom and kept it going for another half a decade. But when the stimulus ran its course, not only did the rate of Chinese growth slow but also there was a rebalancing of the Chinese economy away from exports and investment towards domestic demand and consumption that resulted in a "less material-intensive" form of growth.[159] The end of the China boom thus requires a bigger adjustment than those required by the failure of Japanese steel production to reach 150 million tonnes in the 1970s and the recession of 1982-83 that ended the resources boom. This is because the mining industry and mining exports were much bigger comparatively than during earlier phases of the second minerals rush in the 1970s and 1980s. In addition to the end of the China boom, the mining industries, and particularly Australia's coal industry, face a longer term challenge of adjustment as the world addresses the problem of climate change.

Afterword

The second rush began with the development of new export industries, the most prominent being coal, iron ore and bauxite. They were new because none of them had been significant export industries before World War II. Iron ore had been embargoed between 1938 and 1960; coal, crippled by low profitability and poor industrial relations, had turned inwards during the interwar period; and bauxite had hardly been mined in Australia before the 1950s. But by the end of the 1960s internationally competitive industries were exporting black coal, iron ore, bauxite and alumina. Also important in the minerals boom were other new industries such as uranium and nickel and older industries like copper, which was boosted by the discovery of world-class deposits at Bougainville, and manganese, which was augmented by the finding of massive deposits at Groote Eylandt. Petroleum and natural gas were also major discoveries of the second rush.

Politicians and public servants in Canberra and in the State capitals had not anticipated the revival of mining in the 1960s. Described in 1965 by Sir John Crawford as not a major sector of the economy, the mining industry was already changing Australia by the end of the decade. Australia's parlous balance of payments problem was turned around; foreign investment brought with it new technologies and ways of managing resource projects; new methods of financing were developed; and Japan supplanted the United Kingdom as Australia's major trading partner. Mining substantially influenced Australia's increasing focus on the Asia–Pacific in foreign and trade policy after 1960.

Australia's move to greater economic openness to the world began well before the 1980s with the 1957 Australia–Japan Agreement on Commerce, the end of the embargo on the export of iron ore, the Joint Coal Board's reform of the NSW black coal industry, federal subsidisation of coal ports in the eastern States and the wave of industrialisation across Australia that accompanied the development of bauxite. State governments played an integral part in the development of internationally competitive resource industries. Politically, the development of these industries underwrote the hegemony of non-Labor parties in Western Australia for most of the 1960s and 1970s and in Queensland continuously from 1957 until the late 1980s. In New South Wales, by contrast, Labor governments presided over the establishment of the black coal export industry and its expansion during the "resources boom" from 1977 to 1982. In the Commonwealth of Australia, the second rush contributed to the political centre of gravity moving away from Victoria. Victoria was the colony most enriched by the nineteenth-century gold rush and it had led the Federation movement and the push for a protectionist tariff. In the last third of the twentieth century and the beginning of the twenty-first, as manufacturing declined and mining industries prospered, political power moved increasingly to New South Wales and mining States like Western Australia and Queensland.

The second rush did not produce a large influx of immigrants and the mining workforce was relatively small. But the indirect and long-term effects of the second rush were similar to those of the nineteenth-century gold rush. Wages in the mining sector became the pace-setters for other industries; ports and railways were built or expanded; new towns were constructed in remote Australia; regional centres like Gladstone, Kwinana and Geelong were boosted; and inputs and supplies were sourced by remote mining areas from elsewhere in Australia. The wave of economic development that accompanied the "minerals boom" of the 1960s and 1970s made for a more affluent and optimistic Australia, despite the end of the era of post-war prosperity globally in the mid-1970s. The era of political and social change in the late 1960s and early 1970s would not have been possible without the minerals boom.

As new towns were built in remote areas and harsh environments, mining companies needed to offer high wages and conditions to attract workers, particularly in times of full employment or near-full employment. Yet Despite attractive remuneration, the relations between companies and workers were often troubled. In New South Wales the Coal Industry Tribunal promoted better industrial relations than had existed in the coal industry in the interwar period. But this was not the case in the Pilbara iron ore industry. By the 1970s, frequent strikes and stoppages were of grave concern to Australian governments and the Japanese steel industry. The Robe River dispute of 1986 was symbolic of these poor industrial relations. After it, as State industrial relations laws changed, the mining industry and its workforce became amenable to the large-scale adoption of individual contracts. The ready accommodation of mining workers to such arrangements also reflects the historic sense of partnership between owners and mining workers, a bond reinforced by the common ground they shared: isolation and the need for mining industries, management and workers, to compete internationally. From the 1980s the mining industry moved away from the model of company-built towns accommodating their workforces to the fly-in-fly-out mode of operating (FIFO). This change has in turn led to some criticism of the mining industry both for the deleterious effect of FIFO on the mining towns built at such cost by mining companies and State governments in the 1960s and 1970s, and for the harmful social consequences of FIFO on mine-workers and their families.

The expansion of the mining industry into remote areas of Australia helped to foster a movement for national Aboriginal land rights that resulted in federal legislation for Aboriginal land rights in the Northern Territory in 1976. A movement to extend nationally the Northern Territory land rights regime, and a potential Aboriginal veto over new mining projects, provoked strong resistance from representatives of the mining industry, which developed as one of the most successful industry lobby-groups in the 1980s. The High Court's decision recognising native title in the 1992 Mabo decision temporarily divided the mining industry, but from the mid-1990s both the mining industry and Indigenous peoples found that the post-Mabo native title regime had greatly improved their relations. Mining activities also played a major part in the development of a national environmental movement in the 1970s and 1980s that greatly increased the regulatory burden and cost of mining projects.

The rise in the world price of oil in 1973 produced only a hiatus in the second rush – higher unemployment and inflation and a temporary drying up of investment in the mining industry. Reacting against an increase in foreign ownership of the mining sector, the Whitlam government introduced policies for greater federal regulation of the industry with mixed success. The Petroleum and Minerals Authority was stillborn, but controls over mineral exports encouraged the formation of Australian industry associations negotiating with overseas buyers, and in the period from 1973 to 1975 income earned from the sale of iron ore and coal increased significantly. The Foreign Investment Review Board remains the most enduring legacy of the Whitlam government's policies towards the mining sector. From 1977 mining industries in Australia received a significant boost due to the delayed effect of the rise in the world price of oil. More expensive oil led to a greater world demand for substitutes such as thermal coal, natural gas and uranium, and also for aluminium developed from low-coal-fired electricity plants. The "resources boom" from 1977 to 1982 saw steaming coal developed for export, the uranium export industry established, the North West Shelf natural gas project initiated and major developments in the aluminium industry.

Another world recession in 1982 and 1983 punctured the resources boom and prompted criticism that too much had been invested in mining industries such as iron ore and coal, which were not as profitable in the 1980s and 1990s as they had been in the 1960s and 1970s. Many Australians now hoped for great improvements from manufacturing and high technology exports in the last two decades of the twentieth century. They began to look at mining as a sector belonging to Australia's past in rather the same way that Australian policy-makers had done in the period of post-war reconstruction in the 1940s and 1950s. But generally, manufacturing and high-technology exports did not flourish. By contrast, particular sectors of the mining industry – coal, gold and alumina –

propped up Australia's deteriorating balance of payments with the rest of the world and helped thwart its descent to the "banana republic" status about which Paul Keating had warned in 1986.

Few advisers to government or companies in the mining industry foresaw in 2000 that Australia was about to embark on the biggest mining boom of the second rush and indeed one of the biggest booms in Australian history. It was inaugurated by China's entry into the WTO and the beginning of a new, material-intensive phase of its economic development between 2002 and 2012. China's economic growth made iron ore both Australia's major export earner and a bellwether of the fortunes of the Australian economy. The soaring iron ore price in 2011 was as much a sign of Australia's economic success as the plummeting value of iron ore exports after 2012 was a harbinger of more difficult times. While some analysts predicted that the China boom was a "commodities super cycle" that would last for generations, the reality was that that boom was the result of a particular and spectacular phase of China's material-intensive economic development that was extended by a fiscal stimulus package in response to the Global Financial Crisis. China's transition to a more modest economic growth brought with it significant problems of adjustment by mining industries and governments, which at the same time are responding to the longer-term challenge of climate change. The period from 1960 to 2012, it is argued, was the second great minerals rush in Australian history. The future will tell whether 2012 has marked just the end of the China boom, or something more fundamental to the future of the Australian economy.

Mining Gold on Kalgoorlie Gold Fields, 1987. NAA: A6135, K2/6/87/67

Prime Minister Bob Hawke with Chinese General Secretary Hu Yao Bang examining Channar iron ore mine, 1985. NAA: A6180, 15/4/85/1

Maps

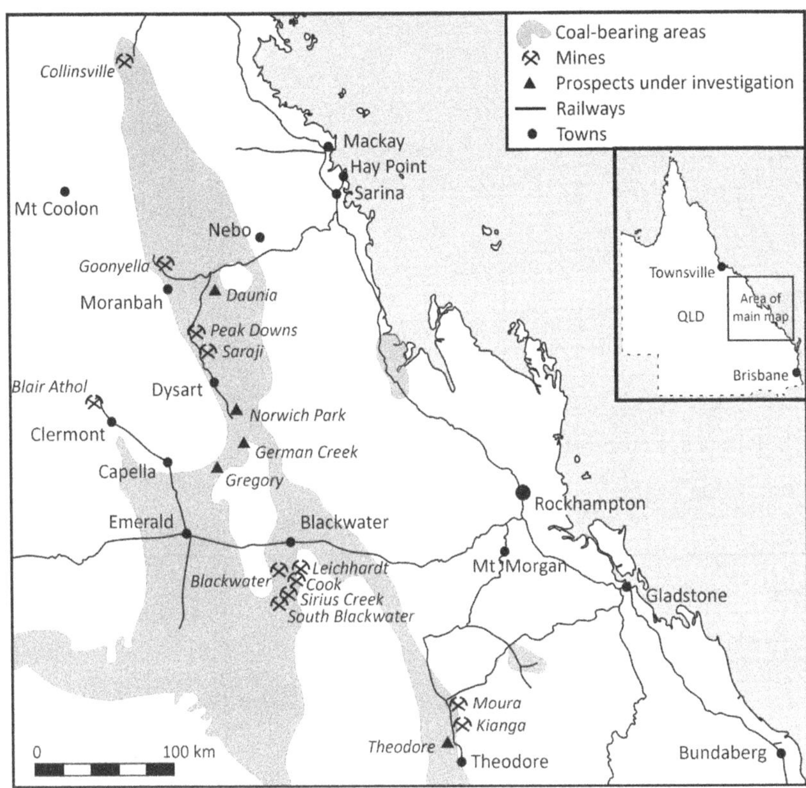

Queensland coal, mines and railways c. 1972

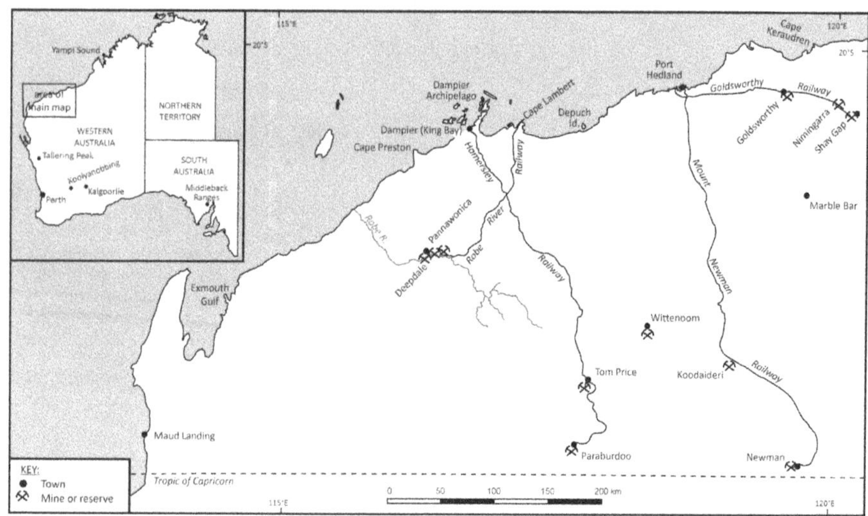

The Pilbara iron ore province c. 1972.

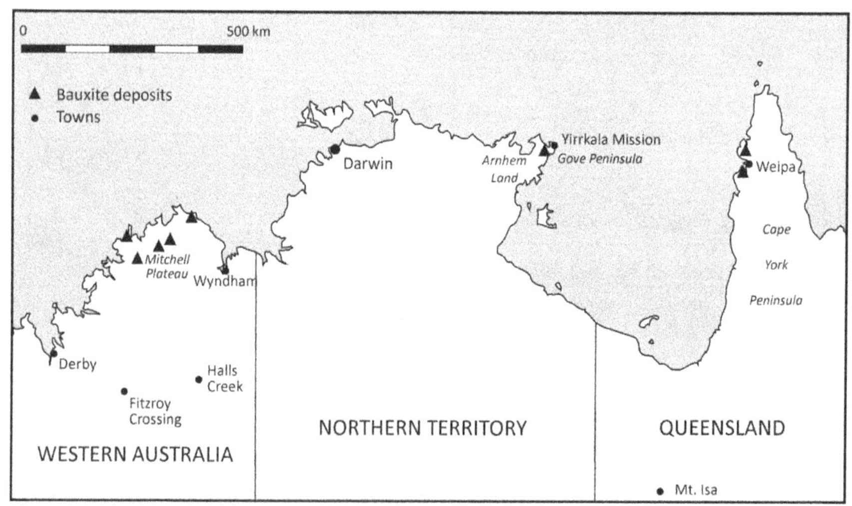

Bauxite deposits in northern Australia, c 1960

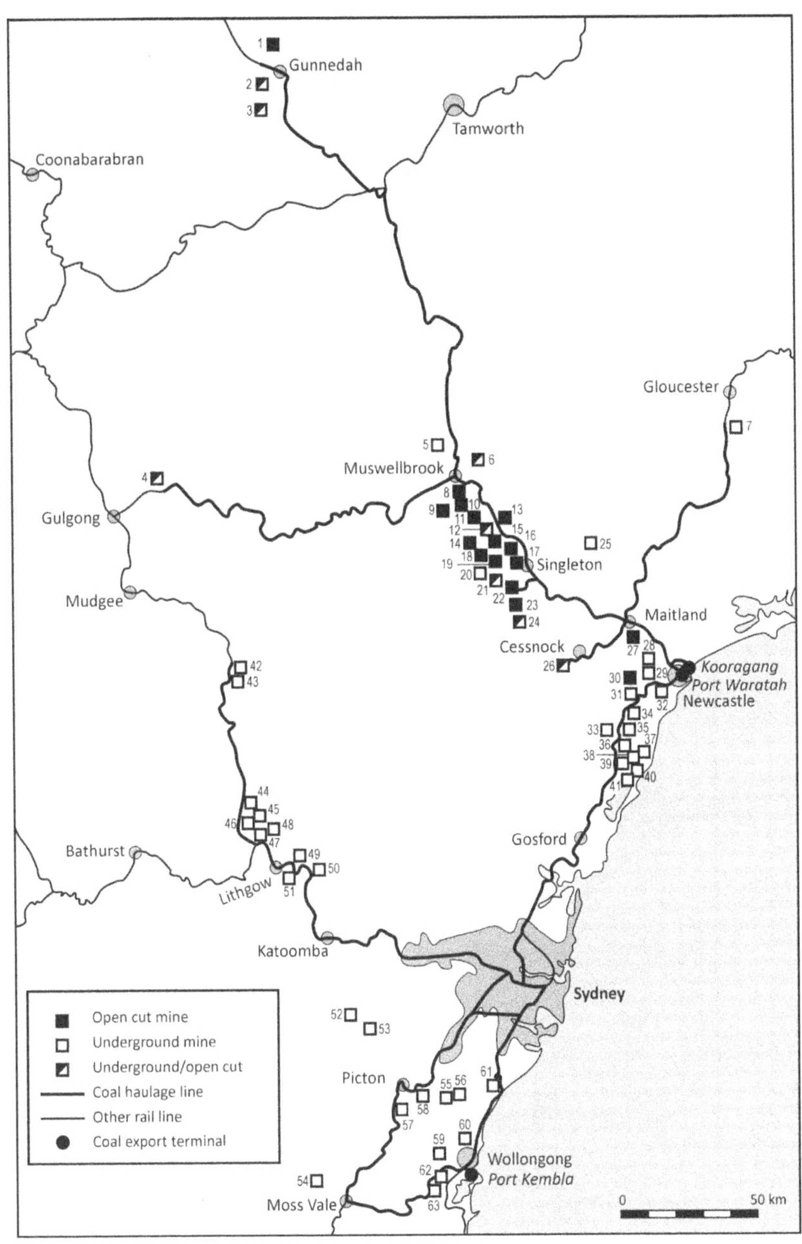

Coal in New South Wales in the 1990s

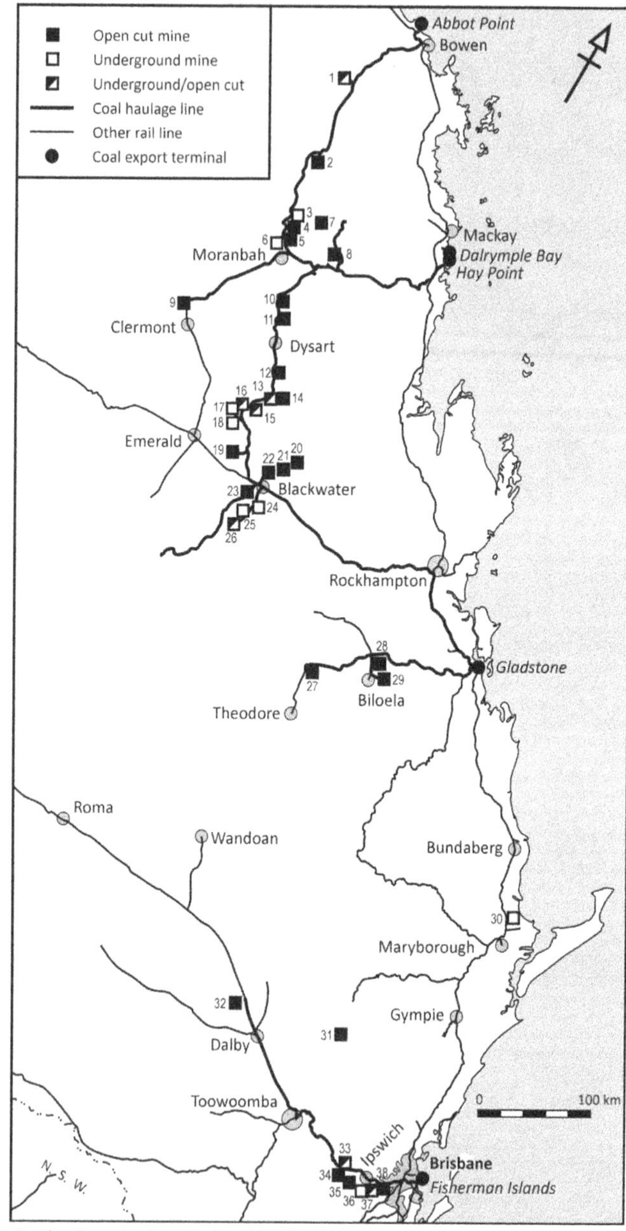

Coal mines in Queensland in the 1990s

Notes

[1] Ian W. McLean, *Why Australia Prospered: The Shifting Sources of Economic Growth*, Princeton University Press, Princeton, 2013, p. 80.
[2] Ibid.
[3] David Hill, *Gold: The Fever that Changed Australia*, William Heinemann, North Sydney, 2010, pp. xiii–xiv.
[4] C.R. Doran, "An Historical Perspective on Mining and Economic Change", in L.H. Cook and M.G. Porter (eds), *The Minerals Sector and the Australian Economy*, George Allen and Unwin, North Sydney, 1984, p. 39.
[5] Ibid.
[6] Ibid.
[7] Jurgen Österhammel, *The Transformation of the World: A Global History of the Nineteenth Century*, Princeton University Press, Princeton, 2014, pp. 733–6.
[8] McLean, op. cit., p. 90.
[9] Brian Pinkstone, *Global Connections: A History of Exports and the Australian Economy*, Australian Government Publishing Service, Canberra, 1992, p. 55.
[10] McLean, op. cit., p. 82.
[11] Ibid.
[12] Ibid., p. 47.
[13] Ibid., p. 91.
[14] Ibid., p. 83.
[15] Ibid., p. 82.
[16] Malcolm Knox, *Boom: The Underground History of Australia from Gold Rush to GFC*, Viking, Melbourne, 2013, pp. 98–107.
[17] Pinkstone, op. cit., p. 56.
[18] Ibid., p. 55.
[19] Ibid., p. 97.
[20] Peter Richmond, "The Origins and Development of the Collins House Group, 1915–1951", *Australian Economic History Review*, March 1987, pp. 3–29.
[21] Pinkston, op. cit., pp. 97–8; Max Griffiths, *Of Mines and Men: Australia's 20th Century Mining Miracle*, Kangaroo Press, East Roseville, 1998, p. 16; Arvi Parbo, *Down Under: Mineral Heritage in Australasia*, Australasian Institute of Mining and Metallurgy, Box Hill, 1992, pp. 222–3.
[22] Pinkstone, op. cit., pp. 119–20.
[23] Ibid., p. 120.
[24] Ibid., pp. 229–30.
[25] Geoffrey Blainey, *The Rush that Never Ended: A History of Australian Mining*, Melbourne University Press, Carlton, 1993, p. 327.
[26] K.K. Kennedy, *Mining Tsar: The Life and Times of Leslie Urquhart*, Allen & Unwin, North Sydney, 1986, Chapter 12.
[27] Doran, op. cit., p. 62.
[28] Ibid., pp. 62–3.
[29] McLean, op. cit., p. 222.
[30] Blainey, op cit.

[31] Geoffrey Blainey, *Mines in the Spinifex: The story of Mount Isa Mines*, Angus and Robertson, Sydney, 1960; Geoffrey Blainey, *The Peaks of Lyell*, Melbourne University Press, Melbourne, 1954; Geoffrey Blainey, *The Rise of Broken Hill*, Macmillan, Melbourne, 1968.
[32] Blainey,*The Rush that Never Ended: A History of Australian Mining*, Melbourne University Press, Carlton, 1993, p. vi.
[33] Knox, op. cit.
[34] Ibid., p. xiv.
[35] Jade Davenport, *Digging Deep: A History of Mining in South Africa*, Jonathan Ball Publishers, Jeppestown, 2013, p. 1.
[36] Ibid., p. 460.
[37] Doran, op. cit., p. 51
[38] Quoted in Harold Raggatt, *Mountains of Ore*, Landsdowne Presss, Melbourne, 1968, p. 14.
[39] Pinkstone, op. cit., pp. 379, 384. .
[40] A.H. Parbo, Address to Hailebury College Melbourne, 21 July 1986, Arvi Parbo Papers, Acc09/45, Box 20, National Library of Australia.
[41] Geoffrey Blainey, *White Gold: The Story of Alcoa of Australia*, Allen & Unwin, St Leonards, 1997.
[42] McLean, op. cit., pp. 221-2.
[43] John McIlwraith, Mesa Harvest: Robe River's First 25 Years, Robe River Iron Associates, Wickham, 1997, p. 34.
[44] Pinkstone, op. cit., p. 180.
[45] McLean, op. cit., p. 215.
[46] Robert C. Allen, *Global Economic History: A Very Short Introduction*, Oxford University Press, Oxford, 2011, pp. 135–6.
[47] David Pilling, *Bending Adversity: Japan and the Art of Survival*, Penguin, London, 2014, pp. 116-17.
[48] Pinkstone, op. cit, p. 388.
[49] Ibid, pp. 219-20.
[50] David Humphreys, *The Remaking of the Mining Industry*, Palgrave Macmillan, Basingstoke, 2015, pp. 20-21.
[51] McLean, op. cit., p. 228.

Chapter 1

[1] See Robin Gollan, *The Coalminers of New South Wales: A History of the Union*, Melbourne University Press, Carlton, 1963.
[2] Statement by B.W. Hartnell to the Joint Committee of the NSW Legislative Assembly and Legislative Council on the Coal Industry, August 1964, NSW Coal Proprietors' Records, Z224, Butlin Archives, ANU.
[3] Christopher Jay, *The Coal Masters: The History of Coal & Allied 1844–1994*, Focus Publishing, Double Bay, 1994, p. 160.
[4] Philip Deery, *Labour in Conflict: The 1949 Coal Strike*, Hale & Iremonger, Sydney, 1978; David Lee, "The 1949 Election: A Reinterpretation", *Australian Journal of Political Science*, Volume 29, No 3, 1994, pp. 501-19.
[5] Chris Fisher, *Coal and the State*, Methuen, Melbourne, 1987. p. 29.
[6] Paper on Coal exports from Australia, Department of National Development, n.d. 1970, National Archives of Australia (hereafter NAA): A1690, 1970/6140.
[7] Chris Fisher, op. cit., p. 186.

[8] Harold Raggatt, *Mountains of Ore*, Landsdowne Press, Melbourne, 1968, p.323.
[9] Paul Kelly, *The End of Certainty: Power, Politics, and Business in Australia*, Allen & Unwin, North Sydney, 1994.
[10] See Alan Rix, *Coming to Terms: The Politics of Australia's Trade with Japan 1945-57*, Allen & Unwin, North Sydney, 1986.
[11] Ian McLean, *Why Australia Prospered: The Shifting Sources of Economic Growth*, Princeton University Press, Princeton, 2013, p. 222.
[12] Statement by B.W. Hartnell to the Joint Committee of the NSW Legislative Assembly and Legislative Council on the Coal Industry, August 1964, NSW Coal Proprietors' Records, Z224, Butlin Archives, ANU.
[13] Fisher, op. cit., pp. 93–108.
[14] Stuart Macintyre, *Australia's Boldest Experiment: War and Reconstruction in the 1940s*, University of New South Wales Press, Sydney, 2015, pp. 284-87, pp. 428-34.
[15] Quoted in Fisher, op. cit., p. 102.
[16] Ibid., p. 118.
[17] Ibid., p. 88.
[18] W.G. McMinn, *A Constitutional History of Australia*, Oxford University Press, Melbourne, 1979, p. 192.
[19] Ibid.
[20] Quoted in Fisher, op. cit., p. 151.
[21] "Samuel Cochran" in John Ritchie (ed.), *Australian Dictionary of Biography*, Vol. 13, Melbourne University Press, Carlton, 1993, pp. 454-5.
[22] Ibid., p. 324,
[23] M.H. Ellis, *A Saga of Coal: The Newcastle Wallsend Coal Company's Centenary Volume*, Angus & Robertson, Sydney, 1969, p. 253.
[24] Harold Raggatt, *Mountains of Ore*, Landsdowne Press, Melbourne, 1968, p. 325.
[25] Fisher, op. cit., p. 170.
[26] Statement by B.W. Hartnell to the Joint Committee of the NSW Legislative Assembly and Legislative Council on the Coal Industry, August 1964, New South Wales Coal Proprietors' Records, Z224, Butlin Archives, ANU.
[27] Fisher, op. cit., p. 274.
[28] Fisher, op. cit., p. 275.
[29] Ibid., p. 382.
[30] Hugh Saddler, *Energy in Australia: Politics and Economics*, George Allen and Unwin, North Sydney, 1981, p. 105.
[31] Brian Robins, "Ludwig's Australian Connections", *Australian Financial Review*, 3 November 1978; Jay, op. cit., pp. 172-74.
[32] Raggatt, op. cit., p. 321; R.L. Whitmore, *Coal in Queensland: The First Fifty Years*, University of Queensland Press, St Lucia, 1981.
[33] Joan Priest, *The Thiess Story*, Boolarong Publication s, Ascot, 1981.
[34] Rix, op. cit., p. 151.
[35] Priest, op. cit.,, pp. 76-7.
[36] "Sir William Henry Spooner", in John Ritchie and Diane Langmore (eds), *Australian Dictionary of Biography, Vol. 16*, Melbourne University Press, Carlton, 2002, pp. 289–90.
[37] Quoted in ibid., p. 152.
[38] Letter from R.D. Innes, CRA, to Department of Trade, "Prospects of Japan Coal Trade with Mainland China and USSR", 27 April 1963, NAA: A1313, 1960/2473 part 2; Walter LaFeber, *The Clash: US–Japanese Relations Throughout History*, W.W. Norton & Company, New York, 1998, pp. 279–80.

[39] Anthony Cave Brown, *Wild Bill Donovan: The Last Hero*, Times Books, New York, 1982.

[40] Letter from Edward B. Nevin to General E.K. Smart, Australian Consul-General, New York, 20 November 1953, NAA: A1838, 736/2/3 part 1.

[41] Letter from Spooner to Casey, 18 August 1955, NAA: A1838, 736/2/3 part 1.

[42] Statement by Spooner, "Overseas Coal Exports – Proposed Trade Mission", 23 September 1954, NAA: A1838, 736/2/3 part 1.

[43] Rix, op. cit., p. 153

[44] Ibid and report by H.P. Reinbach on Marketing of Australian Coal, July 1953, NAA: A695, C219 part 3.

[45] Coal Exports from Australia, Department of National Development, n.d. 1970, NAA: A1690, 1970/6140.

[46] Memorandum from JG Murphy, Assistant Secretary, Joint Coal Board to Secretary, Department of External Affairs, 29 September 1954, NAA: A1838, 736/2/3 part 1.

[47] "Market for Blair Athol Coal", NAA: A1838, 736/2/3 part 1.

[48] Extract from Joint Coal Board Report, 1958–59, NSW Coal Proprietors, Z224, Box 64, C170 Sydney State Fair, Noel Butlin Archives, Australian National University.

[49] "Coal – Australian Supply Position" undated Department of Trade paper, 1958, NAA: A1313, 1960/2473 part 1.

[50] Minute from J.F. Atkins, Department of Trade, to Acting Secretary, Department of Trade, 26 August 1960, NAA: A1313, 1960/2473 part 1.

[51] Cablegram from Embassy in Tokyo to Department of External Affairs, 13 July 1955, NAA: A1838, 736/2/3 part 1.

[52] Christopher Jay, *The Coal Masters: The History of Coal & Allied 1844-1994*, Focus Publishing, Double Bay, 1994, pp. 175–6.

[53] Submission by Joint Coal Board to Senate Standing Committee on Foreign Affairs and Trade, June 1972, NSW Coal Proprietors' Records, Box 64, C170/1, Noel Butlin Archives, Australian National University.

[54] Letter from J.W. Austin, Chairman, Coal Exporter Division, NSW Combined Colliery Proprietors' Association, 2 September 1963, NAA: A1838, 736/2/3 part 1.

[55] "Robert James Heffron" in John Ritchie (ed.), *Australian Dictionary of Biography, Vol. 14*, Melbourne University Press, Carlton, 1996, pp. 427–9.

[56] Department of Trade, Review of Japanese Trade Agreement, 1960, Coal Background Paper, 1960, NAA: A1313, 1960/2473 part 1.

[57] Record of conversation between Spooner and representatives of the Japanese steel industry, 13 June 1960, NAA: A1313, 1960/2473 part 1.

[58] Ibid.

[59] Robert C., Allen, *Global Economic History: A Very Short Introduction*, Oxford University Press, New York, 2011. p. 136.

[60] Chalmers Johnson, *MITI and the Japanese Miracle: The Growth of Industrial Policy 1925–1975*, Stanford University Press, Stanford, 1982, p. 240.

[61] Statement by Spooner, "Increased Coal Export to Japan", 24 September 1962, NAA: A1838, 736/2/3 part 1.

[62] Extract from Joint Coal Board Annual Report, 1960–61, NSW Coal Proprietors Z224, Box 64, C170, Butlin Archives, ANU.

[63] Minute from Grenfell Ruddock to the Secretary of the Department of National Development, Harold Raggatt, 26 May 1954, NAA: A987 E1392 part 1. See generally Erik Eklund, *Steel Town: The Making and Breaking of Port Kembla*, Melbourne University Press, Carlton, 2002, pp. 153–4.

[64] Minute from Raggatt to Spooner, n.d. 1954, NAA: A987, E1392 part 1.

[65] "Sir Harold George Raggatt", in John Ritchie and Diane Langmore (eds), *Australian Dictionary of Biography, Vol. 16*, Melbourne University Press, Carlton, 2002, pp. 47–8.
[66] Rick Wilkinson, *Rocks to Riches: The Story of Australia's National Geological Survey*, Allen and Unwin, St Leonards, 1996, pp. 37-45.
[67] "Port Kembla Coal Loading Plant criticised", *Sydney Morning Herald*, 14 July 1959.
[68] Letter from James B. Shackell, Managing Director, Broken Hill Associated Smelters Pty Ltd to Menzies, 9 January 1961, NAA: A1209, 1961/1015.
[69] *Daily Telegraph*, 11 July 1959.
[70] Ibid.
[71] "Australia – the Virgin Land of Raw Materials" – Translation of undated Japanese newspaper article, [1961], NAA: A987, E1384.
[72] Letter from James B. Shackell, Managing Director, Broken Hill Associated Smelters Pty Ltd to Menzies, 9 January 1961, NAA: A1209, 1961/1015.
[73] Minute from Raggatt to Spooner, 19 April 1961, NAA: A987, E1392 part 1.
[74] "Port Kembla Coal Loading Plant criticised", *Sydney Morning Herald*, 14 July 1959.
[75] Draft Cabinet Submission, Port Facilities – Newcastle and Port Kembla, n.d. 1960, NAA: A987 E1392 part 1; Eklund, op. cit., pp. 154-5.
[76] Memorandum from F.L. McKay, Department of National Development, to Department of Labour and National Service, 25 August 1959, NAA: A987, E1392 part 1.
[77] Record of conversation between Spooner, officers of the Department of National Development and the New South Wales Premier (Heffron), n.d. 1961, NAA: A987, E1392 part 2.
[78] Cabinet Submission 981 from Spooner to Cabinet, "Port Facilities – Newcastle and Port Kembla, 25 January 1961, NAA: A987, E1392 part 1.
[79] Submission from Raggatt to Spooner, 7 December 1960, NAA: A987, E1392 part 1.
[80] Minute from Spooner to Raggatt, 18 January 1961, NAA: A987, E1392 part 1.
[81] Minute to Menzies, n.d. NAA: A463, 1965/5122.
[82] Cabinet submission 981 from Spooner to Cabinet, 5 January 1961, NAA: A987, E1392 part 1 NAA.
[83] Ibid.
[84] Letter from J.F. Nimmo, Acting Secretary, Prime Minister's Department, to Spooner, 24 February 1961, NAA: A987, E1392 part 1.
[85] Letter from Heffron to Menzies, 30 May 1961, NAA: A463, 1965/5122
[86] Ibid.
[87] Ibid.
[88] Minute from Spooner to Raggatt and Cochran, May 1961, NAA: NAA: A987, E1392 part 2.
[89] Johnson, op. cit., pp. 6, 230-41; La Feber, op. cit., Chapter 10.
[90] Jeffrey D. Wilson, *Governing Global Production: Resource Networks in the Asia-Pacific Steel Industry*, Palgrave Macmillan, Basingstoke, 2013, p. 41.
[91] Ibid., p. 38.
[92] Ibid., p. 41.
[93] Ibid., pp. 49-50.
[94] Ibid., p. 50.
[95] Ibid., p. 43.
[96] Johnson, op. cit., p. 56.
[97] Letter from Nagano to Spooner, 25 May 1961, NAA: A987, E1384.
[98] Ibid.
[99] Notes on Joint Coal Board meeting with Japanese Steel Mission, 29 June 1961, NAA: A987, E1392 part 3.

[100] Meeting of Spooner and officers of the Department of National Development with the Japanese Steel Industry Mission, 1 June 1961, NAA: A987 E1384.

[101] Meeting of the Japanese Steel Survey Mission with Joint Coal Board, 7 June 1961, NAA: 987, E1384.

[102] Japanese Steel Survey Mission – Conference held at Joint Coal Board offices on 29 June 1961, NAA: 987, E1392 part 3.

[103] "Japanese May Make Proposal over Port", *Sydney Morning Herald*, 22 June 1961; notes on Joint Coal Board meeting with Japanese Steel Mission, 29 June 1961, NAA: A987, E1392 part 3.

[104] Ibid.

[105] Letter from A.J. Day to McClintock, Department of Trade, 11 July 1961, NAA: A987, E1384.

[106] Notes on Joint Coal Board meeting with Coal Team of Japanese Steel Mission, 29 June 1961, NAA: A987, E1392 part 3.

[107] Ibid.

[108] Joint Coal Board Annual Report, 1961-62, NSW Coal Proprietors Z224, Box 64, C170, Butlin Archives ANU.

[109] Wilson, op. cit., p. 53.

[110] Letter from Menzies to Heffron, 10 August 1961, NAA: E1392 part 3.

[111] Ibid.

[112] Cablegram from Department of External Affairs to Embassy in Tokyo, 4 July 1962, NAA: A463, 1966/3717; Mikio Sumiya (ed.), *A History of Japanese Trade and Industry Policy*, Oxford University Press, Oxford, 2000, pp. pp. 478–80.

[113] Letter from Teruyoshi Tasaka to Sam Cochran, 28 November 1962, NAA A1313, 1960/2473 part 2.

[114] Minute by Maris King to the Secretary, Department of External Affairs, 7 January 1963, NAA: A1838, 759/1/21.

[115] Minute by Maris King, 18 January 1963, NAA: A1838, 759/1/21.

[116] Cablegram from McIntyre to the Department of External Affairs, 4 February 1963, NAA: A1838, 759/1/21.

[117] Letter from L.R. McIntyre to Arthur Tange, Secretary, Department of External Affairs, 7 May 1963, with record of conversation between Heffron and representatives of the Japanese steel industry, April 1963, NAA: A1313, 1970/2473 part 2.

[118] Notes by Australian Embassy in Tokyo on meeting of Heffron with Nagano and Inayama, n.d. 1961, NAA: A1313, 1960.2473 part 2.

[119] Cablegram from Embassy in Tokyo to Department of External Affairs, 30 November 1965, NAA: A1209, 1961/1015.

[120] Letter from Premier of New South Wales, Renshaw to Menzies, 14 February 1964, NAA: A463, 1965/5122.

[121] Export Trade in Coal: June 1964, NAA: A463, 1965/5122.

[122] Priest, op. cit., p. 139.

[123] Ibid.

[124] Ibid., pp. 139-40.

[125] Roger Stuart, 'Resources Development Policy: The Case of Queensland's Export Coal Industry', in Allan Patience (ed.), *The Bjelke-Petersen Premiership 1968-1983: Issues in Public Policy*, Longman Cheshire, Melbourne, 1985, pp. 57-8.

[126] Priest, op. cit., p. 142.

[127] Ibid., p, 143.

[128] Ibid., pp. 143-4.

[129] Ibid., p. 150.
[130] Ibid., pp. 150-2.
[131] Ibid., pp. 157-8.
[132] Minute from Raggatt to Spooner, Development of Kianga–Moura Coal Deposits Possible Commonwealth Assistance, 19 May 1961, NAA: A987, E414.
[133] Ibid., pp. 165-6.
[134] Note for file by Raggatt on Kianga–Moura Coal, 24 March 1961, NAA: A987, E414.
[135] Priest, op. cit., p. 169.
[136] Letter from Nicklin to Menzies, 13 September 1961, NAA: A987, E414.
[137] Ibid.
[138] Ibid., p. 170.
[139] Letter from Hiley to Spooner, 6 June 1961, NAA: A987, E414.
[140] Cameron Hazlehurst, *Gordon Chalk: A Political Life, Darling Downs Institute Press*, Towoomba, 1987, pp. 184-85; minute from Spooner to Raggatt, 3 October 1961, NAA: A987, E414.
[141] D. Frost, "The Revitalisation of Queensland Railways through Export Coal Shipments", *Journal of Transport History*, October 2014, pp. 131–47.
[142] Arvi Parbo, *Down Under: Mineral Heritage in Australasia*, Australian Institute of Mining and Metallurgy, Box Hill, 1992, p. 83.
[143] Priest, op. cit., p. 170.
[144] Trengove, op. cit., pp. 27-8.
[145] Ibid., p. 26.
[146] Jay, op. cit., p. 190.
[147] Trengove, op. cit., pp. 28-32.
[148] Ibid., p. 41.
[149] Ibid., p. 41.
[150] Trengove, op. cit., p. 44.
[151] Ibid., p. 47.
[152] Submission by Utah Development Company to Senate Select Committee on Foreign Ownership and Control, 11 May 1972, NSW Coal Proprietors Records, Z224, Box 64, C170, Noel Butlin Archives, Australian National University.
[153] Brian Galligan, *Utah and Queensland Coal: A Study in the Micro Political Economy of Modern Capitalism, and the State*, University of Queensland Press, St Lucia, 1989, p. 65.
[154] B.W. Hartnell, Submission By Joint Coal Board to Senate Standing Committee on Foreign Affairs and Trade, NSW Coal Proprietors, Box 64, C170/1, Noel Butlin Archives, Australian National University.
[155] Ibid.
[156] Ibid.
[157] Draft Cabinet submission by R.V. Swartz, Minister for National Development on Export of Black Coal, nd. 1971, NAA: A1690, 1970/399.
[158] B.W. Hartnell, Submission By Joint Coal Board to Senate Standing Committee on Foreign Affairs and Trade, NSW Coal Proprietors, Box 64, C170/1, Noel Butlin Archives, Australian National University.
[159] Ibid.
[160] Jay, op. cit., p.188.
[161] Ibid, p. 187.
[162] Ibid.
[163] Minute for Cabinet, Report of Cabinet Sub-Committee on Clutha Development Proposals, 17 March 1970, Wal Fife Papers, MS7626, Box 35, Clutha Negotiations File, National Library of Australia.

[164] Clutha Development, Proposals for an Agreement between the State of New South Wales and Clutha Development Pty for the establishment of New Ports and Railways for Coal Export Purposes, 17 March 1970, Fife Papers, MS 7626, Box 34, NLA.

[165] Ibid.

[166] Statement for Information of Cabinet, 22 January 1970, Wal Fife Papers, MS7626, Box 35, Clutha Negotiations File, NLA.

[167] Clutha Development, Proposals for an Agreement between the State of New South Wales and Clutha Development Pty for the establishment of New Ports and Railways for Coal Export Purposes, 17 March 1970, Fife Papers, MS 7626, Box 34, NLA.

[168] Treasury memorandum on Cabinet Minute submitted by the Minister for Mines for an agreement between the State of New South Wales and Clutha Development, 14 September 1970, Wall Fife Papers, MS7626, Box 35, Clutha Negotiations File, NLA.

[169] Robert Sorby, "Coal Board Calls for Common NSW–Qld Marketing Policy", *Australian Financial Review*, 26 November 1971.

[170] Submission by Joint Coal Board (A.C. Girard) to Senate Select Committee on Foreign Ownership and Control, Sydney, July 1973, NSW Coal Proprietors, Z224, Box 65, C170A, Noel Butlin Archives, Australian National University.

[171] Letter from B.W. Hartnell to Bott, 20 December 1971, NAA: A1690, 1970/399.

[172] Ibid.

Chapter 2

[1] F.G. Davidson and B.R. Stewardson, *Economics and Australian Industry*, Second Edition, Longman Cheshire, Melbourne, 1974, p. 93.

[2] Keith R. Smith, *The Great Challenge: The Saga of Yampi*, Keith R. Smith, Perth, 1978, p. 131.

[3] Malcolm Knox, *Boom: The Underground History of Australia from Gold Rush to GFC*, Viking, Melbourne, 2013, pp. 126–30.

[4] See E.M. Johnston-Liik, George Liik and R.M. Ward, *A Measure of Greatness: the Origins of the Australian Iron and Steel Industry*, Melbourne University Press, Carlton, 1998; Helen Hughes, *The Australian Iron and Steel Industry, 1948–1962*, Melbourne University Press, Carlton, 1964.

[5] Davidson and Stewardson, op. cit., p. 92.

[6] Ibid., pp. 82–3.

[7] Smith, op. cit., pp. 160–1.

[8] Ibid., p. 2.

[9] Ibid., p. 161.

[10] Ibid., p. 161.

[11] Minute from A.J. Maclachlan to Cabinet, 28 July 1937, in R.G. Neale (ed.), *Documents on Australian Foreign Policy, Volume 1 1937–38* (hereafter DAFP Vol. I), Australian Government Publishing Service, Canberra, 1975, p. 181.

[12] Francis Pike, *Hirohito's War: The Pacific War 1941–1945*, Bloomsbury, London, 2015, pp. 80-4.

[13] Memorandum from Longfield Lloyd to Secretary, Department of Commerce, 6 October 1937, DAFP Vol 1, pp. 234-5.

[14] Ibid., p. 235.

[15] "Yampi Sound – Development of Iron Ore Deposits" submitted by W.R. Hodgson to W.M. Hughes, 13 December 1937, DAFP Vol. 1, p. 240.

[16] Ibid.

[17] Jon White, "The Port Kembla Pig Iron Strike of 1938", *Labour History*, Vol. 37, November 1979, pp. 63-77.

[18] "Yampi Sound – Development of Iron Ore Deposits" submitted by W.R. Hodgson to W.M. Hughes, 13 December 1937, DAFP Vol. 1, p. 241.

[19] "Walter George Woolnough" in John Ritchie (ed.), *Australian Dictionary of Biography, Vol. 12*, Melbourne University Press, Carlton, 1986, pp. 572-3.

[20] Report by W.G. Woolnough, "Report on the Technical Aspects of Iron Ore Reserves in Western Australia", 1 March 1938, DAFP, Vol. 1, p. 352.

[21] Letter from Lyons to Torao Wakamatsu, Japanese Consul-General in Sydney, 17 May 1938, DAFP Vol 1, p. 342.

[22] Text of statement made by Lyons in House of Representatives on 19 May 1938, in cablegram from Lyons to Sir Earle Page, 18 May 1838, DAFP Vol. 1, pp. 347-8.

[23] Memorandum by J. McEwen, Minister for the Interior, 10 May 1940, in H. Kenway, H.J.W. Stokes and P.G. Edwards (eds), *Documents on Australian Foreign Policy, Volume III, January–June 1940*, Australian Government Publishing Service, Canberra, 1979, pp. 277–81.

[24] Richard West, *River of Tears, The Rise of the Rio Tinto Zinc Corporation Ltd*, Earth Island Corporation, London, 1972, p. 92; submission from W.H. Spooner to Cabinet, 14 June 1957, National Archives of Australia (hereafter NAA): A425 2003, C70/11048.

[25] Harold Raggatt, *Mountains of Ore*, Landsdowne Press, Melbourne, 1968, p. 109.

[26] See F.K. Crowley, *Australia's Western Third: A History of Western Australia*, Heinemann, Melbourne, 1960.

[27] Raggatt, op. cit., pp. 109-10.

[28] Jeffrey D. Wilson, *Governing Global Production: Resource Networks in the Asia Pacific Steel Industry*, Palgrave Macmillan, Basingstoke, 2013, p. 65.

[29] "Sir David Brand" in John Ritchie (ed.), *Australian Dictionary of Biography* (ADB), vol. 13, Melbourne University Press, Carlton, 1989, pp. 249–50.

[30] Humphrey McQueen, *Gone Tomorrow: Australia in the 80s*, Angus & Robertson, Sydney, 1982, p. 80.

[31] See Ronda Jamison, *Charles Court: I Love This Place*, St George Books, Osborne Park, 2011.

[32] Letter from Spooner to Menzies, 23 September 1959, NAA: A1209, 1961/36 part 1.

[33] Letter from Charles Court to Lindesay Clark, 18 September 1959, State Records Office of Western Australia (hereafter SROWA) item 298/69.

[34] Ibid., "Press Announcement by Brand", 18 September 1959.

[35] Ibid., letter from Brand to Menzies, 11 September 1959.

[36] *Australian Financial Review*, 15 October 1959.

[37] Letter from Brand to Menzies, 23 September 1959, NAA: A1209, 1961/36 part 1.

[38] Cable from A.W. Fadden to Dumas, Department of Industrial Development, Perth, n.d. October 1959, SROWA, item 298/69.

[39] Statement of Menzies in the House of Representatives, 6 October 1959, NAA: A1209, 1961/36 part 1; record of discussion at the Imperial Hotel, Tokyo, on 13 June 1960 between Spooner and representatives of the Japanese steel industry, NAA: M2576, 5.

[40] Department of National Development Review of Australian Iron Ore Resources, March 1960, NAA: A425, C70/11048.

[41] Raggatt, op. cit., pp. 110-11.

[42] Cabinet decision 722, 31 March 1960, NAA: A4940, C3063 part 1; David Lee, "Reluctant Relaxation: The end of the Iron Ore Export Embargo and the Origins of Australia's Mining Boom, 1960–1966", *History Australia*, Vol. 10, December 2013, pp. 154-55.

[43] "Sir Colin Syme" in Melanie Nolan (ed.), *Australian Dictionary of Biography*, vol. 18, Melbourne University Press, Carlton, 2012, pp. 481–2.

[44] Letter from Syme to Spooner, 2 April 1960, NAA: A4940, C3063 part 1.

[45] Ibid.

[46] Letter from Spooner to McEwen, May 1960, NAA: A4940, C3063 part 1.

[47] Minute from E.J.B. Foxcroft to McEwen, 9 May 1960, NAA: A4940, C3063 part 1.

[48] Peter Golding, *Black Jack McEwen: Political Gladiator*, Melbourne University Press, Carlton, 1991.

[49] Speech by the Minister for Trade, John McEwen, 8 March 1961 in J.G. Crawford (ed.), *Australian Trade Policy 1942-1966: A Documentary History*, Australian National University Press, Canberra, 1968, pp. 539–42.

[50] Statement by the Minister for Trade (J. McEwen) in Canberra on 22 February 1960 in Crawford (ed.), op. cit., p. 522.

[51] Speech by the Prime Minister (R.G. Menzies) in the House of Representatives, 16 November 1960 in Crawford (ed.), op. cit., p. 549.

[52] Statement by the Minister for Trade (J. McEwen) in the House of Representatives, 17 August 1961 in Crawford (ed.), op. cit., pp. 300–02.

[53] Geoffrey Blainey, *The Rush that Never Ended: A History of Australian Mining*, Melbourne University Press, Carlton, 2003; Ian W. McLean, *Why Australia Prospered: The Shifting Sources of Economic Growth*, Princeton University Press, Princeton, 2013, pp. 229–30. In 1951 minerals and fuels represented just one per cent of total exports.

[54] Malcolm Knox, *Boom: The Underground History of Australia from Gold Rush to GFC*, Viking, Melbourne, 2013, p. 146.

[55] Letter from Brand to Menzies, 20 October 1960, NAA: A1209, 1961/36 part 1.

[56] Ibid.

[57] Minute from Bunting to Menzies, 16 November 1960, NAA: A4940, C3063 part 1.

[58] Ibid.

[59] Ibid.

[60] Cabinet decision 1136, 24 November 1960, NAA: A4940, C3063 part 1.

[61] Ibid.

[62] Davidson and Stewardson, op. cit., p. 94.

[63] Letter from Arthur Griffith, Western Australian Minister for Mines, to Spooner, 22 May 1961, NAA: A4940, C3063 part 2.

[64] The management of Hamersley Iron was led by CRA, Mount Newman after 1966 was led by BHP, Consolidated Goldfields Australia contributed Australian management to the Goldsworthy Associates, and though Robe River's management was initially led by Cleveland Cliffs, it later passed into the hands of Australian mining companies, Peko-Wallsend and North Limited.

[65] Submission from Spooner to Cabinet, 20 May 1963, NAA: A4940, C3063 part 2.

[66] Ibid.

[67] Memorandum from F.S. Anderson to Sir Maurice Mawby, 20 June 1963, folder 62, Sir Maurice Mawby Papers, University of Melbourne Archives.

[68] Submission from Spooner to Cabinet, 20 May 1963, NAA: A4940, C3063 part 2.

[69] Cabinet decision 830, 5 June 1963, NAA: A4940, C3063 part 2; Raggatt, op. cit., p. 118.

[70] Manners, Gerald, *The Changing World Market for Iron Ore, 1950-1980: An Economic Geography*, Johns Hopkins University Press, Baltimore, 1971, p. 229.

[71] Robert Duffield, *Rogue Bull: The Story of Lang Hancock King of the Pilbara*, Collins, Sydney 1979; Neil Phillipson, *Hancock: Man of Iron*, Wren Publishing, Melbourne, 1974.

[72] Phillipson, op. cit., pp. 19–20

[73] Duffield, op. cit., pp. 67–8.

[74] Phillipson, op. cit., p. 75.

[75] See also Alan Trengove, *Adventure in Iron: Hamersley's First Decade*, Stockwell Press, Mont Albert, 1976, p. 18

[76] This information was provided to the author by Chris McSweeney, a former officer of the WA Department of Industrial Development.

[77] Phillipson, op. cit., 1974, pp. 73–81.

[78] "Mountains of Iron Ore" from CRA Gazette April 1992.

[79] Duffield, op. cit., p. 22.

[80] West, op. cit., p. 91.

[81] Noel Olive, *Enough is Enough: A History of the Pilbara Mob*, Fremantle Arts Centre Press, Fremantle, 2007.

[82] Letter from Hancock and Wright to Department of Industrial Development, Perth, 6 November 1959, SROWA, item 298/69.

[83] Avery, David *Not on Queen Victoria's Birthday: The Story of the Rio Tinto Mines*, Collins, London, 1974, pp. 139–59.

[84] Peter Thompson and Robert Macklin, *The Big Fella: The Rise and Rise of BHP Billiton*, William Heinemann Australia, Sydney, 2009, p. 343.

[85] R.H. Harding, *Wholeheartedly and At Once: A History of the First Operation of Mary Kathleen Uranium Ltd*, CRA Limited, Melbourne 1992, p. 24.

[86] Avery, op. cit., pp. 395-408.

[87] "Application for Temporary Reserves for Iron Ore by Wright Prospecting Pty, Hancock Prospecting Pty Ltd and Rio Tinto Southern Pty Ltd", 15 May 1961, SROWA, item 298/69.

[88] Phillipson, op. cit., pp. 66-7.

[89] Bruno Campana, "Hamersley and its Men"; a Contribution to the Exploration History of the Hamersley Iron Ore Province Western Australia", 16 July 1967, copy provided by courtesy of Chris McSweeney.

[90] Letter from John Hohnen, Rio Tinto to S. Idoh, Maubeni-Idoh Co. 11 April 1962, Maurice Mawby Papers, folder 60, University of Melbourne Archives; "Application for Temporary Reserves for Iron Ore by Wright Prospecting Pty, Hancock Prospecting Pty Ltd and Rio Tinto Southern Pty Ltd", 15 May 1961; letter from Blake Pelly, Rio Tinto, to Arthur Griffith, 10 May 1961 SROWA, item 298/69.

[91] Letter from Val Duncan to Arthur Griffith, 4 November 1961, SROWA, item 298/69.

[92] Ibid.

[93] Ibid.

[94] Ibid.

[95] Minute from Griffith to Premier in Cabinet, 8 December 1961, WA SRO, 317/72 Volume 3.

[96] Quoted in Jamieson, op. cit., p. 211.

[97] See generally Peter Yule, *William Laurence Baillieu: Founder of Australia's Greatest Business Empire*, Hardie Grant Books, Richmond, 2012.

[98] "William Sydney Robinson" in Geoffrey Serle (ed.), *Australian Dictionary of Biography, Vol. 11*, Melbourne University Press, Carlton, 1988, pp. 428-33.

[99] "Sir Maurice Alan Edgar Mawby" in John Ritchie (ed.), *Australian Dictionary of Biography, Vol. 15*, Melbourne University Press, Carlton, 2000, pp. 333–5.

[100] Quoted in Kosmas Tsokhas, *Beyond Dependence: Companies, Labour Processes and Australian Mining*, Oxford University Press, Melbourne, 1986, p. 41.

[101] Raggatt, op. cit., p. vii.

[102] Letter from Arthur Griffith to Blake Pelly, 30 March 1962, SROWA, 317/72 Volume 3.

[103] Letter from Mawby to Brand, 12 November 1962, Maurice Mawby Papers, folder 60, University of Melbourne Archives; Jamieson 2006, p. 212; "Mountains of Iron Ore", *CRA Gazette* April 1992.

[104] Alan Trengove, *Discovery: Stories of Modern Mineral Exploration*, Stockwell Press, Clayton, 1979, pp. 118–20; memorandum from Anderson to Mawby, 28 August 1962, "Negotiations with Hancock and Wright", Maurice Mawby Papers, University of Melbourne Archives. CRA yielded to the claim of Hancock and Wright that they had introduced Rio Tinto to the whole Pilbara, West Pilbara and Ashburton area and thus entitled to royalties even outside the temporary Hancock/Wright/Rio Tinto temporary reserves.

[105] Letter from Mawby to Roy Wright, 9 October 1962, "Negotiations with Hancock and Wright", Maurice Mawby Papers, University of Melbourne Archives.

[106] John McIlwraith, *The First 500 Million: The Mt Newman Story*, Iron Ore BHP-Utah Minerals Division, Perth, 1988, pp. 14–19.

[107] Phillipson, op. cit., p. 101.

[108] Memorandum from Anderson to Mawby, 5 July 1963, folder 62, Sir Maurice Mawby Papers, University of Melbourne Archives.

[109] McIlwraith, op. cit., p. 18.

[110] Stephen M. Voynick, *Climax: the History of Colorado's Climax Molybdenum Mine*, Mountain Press Club Co., Missoula, 1996.

[111] McIlwraith op. cit., pp.14–19.

[112] "Sir Robert David Garrick Agnew" in Diane Langmore (ed.), *Australian Dictionary of Biography*, Vol. 17, Melbourne University Press Carlton, 2007, pp. 5–6.

[113] John McIlwraith, *Mesa Harvest: Robe River's First Years, Robe River Iron Associates*, Wickham, 1997, p. 12.

[114] Ibid., p. 11.

[115] Ibid., p. 12.

[116] Terry Reynolds and Virginia P. Dawson, *Iron Will: Cleveland Cliffs and the Mining of Iron Ore, 1847-2006*, Wayne State University Press, Detroit, 2011, p. 201.

[117] Letter from Mawby to R.C. Atherton, 26 March 1952, folder 02, Sir Maurice Mawby Papers, University of Melbourne Archives.

[118] "Frank Struan Anderson" in John Ritchie (ed.), *Australian Dictionary of Biography, Vol. 13*, Melbourne University Press, Catlton, 1999, pp. 47–8.

[119] Letter from Mawby to Hancock and Wright, 1 October 1962, "Relations with Hancock and Wright", Maurice Mawby Papers, University of Melbourne Archives.

[120] Letter from Anderson to Mawby, 28 August 1962, Sir Maurice Mawby Papers, University of Melbourne Archives.

[121] Memorandum from Anderson to Mawby, 5 July 1963, folder 62, Sir Maurice Mawby Papers, University of Melbourne Archives.

[122] Ibid., Memorandum from F.S. Anderson to Maurice Mawby, 20 June 1963.

[123] Alan Trengove, *Adventure in Iron: Hamersley's First Decade*, Stockwell Press, Mont Albert, 1976, p. 50.

[124] Ibid., p. 51.

[125] Trengove, op. cit., pp. 105–6.

[126] Ibid., Letter from Mawby to Duncan, 30 August 1962, folder 62, Sir Maurice Mawby Papers, University of Melbourne Archives.

[127] Trengove, op. cit., p. 51.

[128] Letter from Mawby to A.M. Baer, 7 February 1964, folder 07. Sir Maurice Mawby Papers, University of Melbourne Archives.

[129] Note by Court of discussion with Sir Ian McLeannan, 4 December 1963, SROWA, item 1963/0049.
[130] Record of discussions between Struan Anderson and Charles Court, 15 August 1964, folder 64, Sir Maurice Mawby Papers, University of Melbourne Archives.
[131] Letter from Mawby to Menzies, 13 October 1964, NAA: A4940, C3063 part 2.
[132] Ibid.
[133] Ibid.
[134] Telegram from Menzies to Brand, 13 November 1964, NAA: A1209, 1961/36 part 2.
[135] Jerry Shields, *The Invisible Billionaire, Daniel Ludwig*, Houghton Mifflin, Boston, 1986.
[136] Letter from V.W. Wagner, Executive Vice President and Treasurer, National Bulk Carriers, to Brand, 10 July 1964, SROWA, series 3121, item 15.
[137] Ibid.
[138] Ibid , p. 63.
[139] Max Griffiths, *Of Mines and Men: Australia's 2oth Century Mining Miracle*, Kangaroo Press, East Roseville, 1998, p. 61.
[140] E.T. Biggs, "CRA relations with Marubeni-Idoh", n.d. 1962, folder 60, Maurice Mawby Papers, University of Melbourne Archives.
[141] Letter from Mawby to John Hohnen, 11 April 1962, folder 60, Maurice Mawby Papers, University of Melbourne Archives.
[142] Alan Trengove, *Adventure in Iron: Hamersley's First Decade*, Stockwell Press, Mont Albert, 1976, p. 70.
[143] Griffiths, op. cit., p. 62.
[144] Trengove, op. cit., p. 70.
[145] Ibid., Chapter 6.

Chapter 3

[1] Jeffrey D. Wilson, *Governing Global Production: Resource Networks in the Asia Pacific Steel Industry*, Palgrave Macmillan, Basingstoke, p. 49. Gerald Manners, *The Changing World Market for Iron Ore 1950-1980: An Economic Geography*, The Johns Hopkins Press, Baltimore, 1971, pp. 317-18.
[2] Leon Glezer, "Sir Ian Potter and his Generation" in R.T. Appleyard and C.B. Schedvin (eds), *Australian Financiers: Biographical Essays*, Macmillan, South Melbourne, 1988, pp. 401-27.
[3] Peter Yule, *Ian Potter: Financier, Philanthropist and Patron of the Arts*, Miegunyah Press, Carlton, 2006, pp. 247-8.
[4] Grant Fleming, David Merrett and Simon Ville, *The Big End of Town: Big Business and Corporate Leadership in Twentieth Century Australia*, Cambridge University Press, Cambridge, 2004, pp. 104-5.
[5] Robert Duffield, *Rogue Bull: The Story of Lang Hancock, King of the Pilbara*, Collins, Sydney, 1979, pp. 91-2.
[6] Letter from John Payne, Director of the Mount Newman Iron Ore Company, to John McEwen, 26 February 1965, National Archives of Australia (hereafter NAA) NAA; M2568, 81.
[7] Cabinet decision 620, 11 November 1964, NAA: A4940, C3063 part 2.
[8] Letter from Payne to McEwen, 26 February 1965, NAA: M2568, 81, NAA.
[9] Wilson, op. cit., pp. 66–9.
[10] Submission 752 from Fairbairn to Cabinet, 'Price of Iron Ore Exports-the Mount Newman Case', 23 April 1965, NAA: A4940, C3063 part 2.

11 Ibid.
12 Letter from Clark to Holt, 12 March 1965, NAA: A4940, C3063 part 2; G. Lindesay Clark, *Built on Gold: Recollections of Western Mining*, Hill of Content, Melbourne, 1983, pp. 171–7.
13 Minute from J.H. Garrett, First Assistant Secretary, Treasury, to Harold Holt, 29 March 1965, NAA: M2568, 152.
14 Letter from Sir Gordon Lindsay Clark to Harold Holt, 12 March 1965, NAA: A4940, C3063 part 2.
15 Submission 752 from Fairbairn to Cabinet, "Price of Iron Ore Exports" – the Mount Newman Case', 23 April 1965, NAA: A4940, C3063 part 2.
16 Letter from A.T. Carmody, Acting Secretary, Department of Trade and Industry, to Sir John Bunting, 8 April 1965, NAA: M2568, 152.
17 Cabinet decision 857 without memorandum "Iron Ore Exports", 8 April 1965, NAA: A4940, C3063 part 2.
18 Submission 752 from Fairbairn to Cabinet, "Price of Iron Ore Exports – the Mount Newman Case", 23 April 1965, NAA: A4940, C3063 part 2.
19 Ibid.
20 Cabinet decision 958, 18 May 1965, NAA: A4940, C3063 part 3.
21 CSR paper left with Menzies, "Mt Newman contract price", 25 May 1965, NAA: A4940, C3063 part 3.
22 Note for file by Peter Lawler, Deputy Secretary of the Prime Minister's Department of discussion between Menzies and Vernon, 4 June 1965, NAA: A4940, C3063 part 3.
23 Ibid.
24 Cabinet decision 989, 26 May 1965, NAA: A4940, C3063 part 3.
25 Ibid.
26 Ibid.
27 "Prices for ore under fire", *Australian Financial Review*, 6 July 1965; memorandum from Gordon Jackson to Sir James Vernon, 6 July 1967, Sir Gordon Jackson Papers, MS 8353, Series 6, Box 9, National Library of Australia.
28 Minute by Charles Court, 3 December 1962, State Records Office of Western Australia (hereafter SROWA), item 1963/0049; "History of Development" January 1968, SROWA item 1970/0292.
29 Ibid., memorandum by Court for Brand and Griffith, 27 October 1963.
30 Ibid.
31 Ibid., "Notes given by Mr W.E. Dohnal on Estimated Economic Impact of Cliffs Pelletisation Project", 13 January 1965.
32 Ibid., memorandum re Robe River (Cliffs) Pelletization Project and Composition of Limonite and Hematite ores, n.d. [1964].
33 Ibid., letter from Court to Dohnal, 10 November 1964. Court wrote that: "Also I would not hesitate to mention to [Harold Raggatt] that the export of pellets or similar material will also mean a greater contribution to the national economy and also ensure quicker industrial development in this fairly undeveloped part of the North West of Western Australia."
34 Ibid., notes by Bill Dohnal for Court, n.d. October 1964.
35 Ibid., note of discussions between Dohnal, Court, Brand and the Western Australian Under Secretary of Mines, n.d. 1964.
36 Ibid., Cliffs International Inc. Summary of Proposal for Western Australian Government, 28 October 1964.
37 Ibid., letter from Dohnal to Court, 3 December 1964.

38 Letter from Dohnal to Court, 3 March 1965, SROWA item 1970/0292.
39 Ibid., letter from Dohnal to Court, 5 March 1965.
40 Ibid., minute from Court to Brand and Griffith, 15 March 1965.
41 Ibid., letter from Court to Dohnal, 16 March 1965.
42 Ibid.
43 Raggatt, op. cit., p. 126.
44 Submission 613 from Fairbairn to Cabinet "Price of Iron Ore Exports – Hamersley Iron", 24 August 1965, NAA: A1209, 1961/36 part 3.
45 Ibid.
46 Ibid.
47 Letter from R.T. Madigan, Managing Director, Hamersley Iron, to Secretary, Department of National Development, 18 October 1965, NAA: A4940, C3063 part 3.
48 Ibid.
49 "Commercial Aspects of the Development of Australia's Iron Ore Resources", 10 November 1965, NAA: A1209, 1961/36.
50 Submission 1134 from Fairbairn to Cabinet, "Hamersley Iron – Further Export of Iron Ore Pellets", 15 November 1965, NAA: A4940, C3063 part 3.
51 "Govt rejects Japanese iron ore deal", *Australian Financial Review*, 14 January 1966.
52 Letter from Mawby to Menzies, 1 December 1965, NAA: A1209, 1961/36 part 3.
53 Ibid.
54 Ibid.
55 "Pellet plant can go ahead", *The Australian*, 30 March 1966; "Hamersley to go ahead with $45 million plant", The *Age*, 30 March 1966.
56 "Mt Newman Iron Ore Project", 25 May 1965, NAA: A4940, C3063 part 3.
57 Ibid.
58 Chronology of Hamersley–Newman Relationship, n.d. Mawby Papers, folder 62, University of Melbourne Archives.
59 Letter from Eugene Trefethen, Kaiser Steel, to Val Duncan, Chairman, Rio Tinto Zinc, 24 Januiary 1966, Mawby Papers, folder 58, University of Melbourne Archives.
60 Chronology of Hamersley–Newman Relationship, n.d. Mawby Papers, folder 62, University of Melbourne Archives.
61 Letter from Vernon to Mawby, 17 February 1966, Mawby Papers, folder 62, University of Melbourne Archives.
62 Record of telephone conversation between McEwen and Frank Coolbaugh, 8 March 1966 and record of conversation between McEwen and Val Duncan, 9 March 1966, NAA: M58, 205.
63 Telegram from Duncan to Lord Baillieu, 6 March 1966, folder 59 Mawby Papers, University of Melbourne Archives.
64 "Mt Newman plans now left the three", *Australian Financial Review*, 23 August 1966.
65 Letter from Dohnal to Court, 28 November 1966, SROWA, item 1970/0292.
66 "Mt Newman iron ore plan in balance", *Australian Financial Review*, 25 September 1966.
67 David Lee, "The Emergence of Iron Ore Giants: Hamersley Iron and the Mount Newman Company", *Journal of Australasian Mining History*, Vol. 11, no. 1, 2013, pp. 61–77.
68 Ronda Jamieson, *Charles Court: I Love this Place, St George Books*, Osborne Park, 2011, pp. 173-4.
69 "Mt Newman deal near signature", *Australian Financial Review*, 7 June 1966.
70 Submission 555 from Fairbairn to Cabinet, "Proposed Revision of the Mt Newman Contract", 24 October 1966, NAA: A1209, 1961/36 part 3.

71 Cablegram from Australian Embassy, Tokyo, to Department of National Development, 21 October 1966, NAA: A1209, 1961/36 part 4; "Behind the iron ore offers", *Australian Financial Review*, 27 October 1966; cable from A.J. Rew to Duncan and cable from Trefethen to Mawby, 20 October 1966, folder 62, Mawby Papers, University of Melbourne Archives.

72 Minute from A.J. Campbell, Acting Secretary, Department of Trade and Industry, to McEwen, 25 October 1966, NAA: A4940, C3063 part 4.

73 Cabinet decision 658, 25 October 1966, NAA: A4940, C3063 part 4.

74 *Australian Financial Review*, 3 April 1967.

75 Yule, op. cit., p. 248.

76 Kaiser Industries memorandum, "Australia is losing Major Tonnages of Pellets and Ore to Other Countries as a Result of Lack of Flexibility in Administering Guideline Pricing", 29 May 1967, NAA: A1209, 1967/7571.

77 Ibid.

78 Ibid.

79 Minute by Court, 1 July 1967, SROWA 1970/0292. Court was told by a visiting member of Yawata Steel, Yoshihara Iimura, in June 1967 that: "The Brazilians are laughing their heads off. He went on to say that since the Commonwealth introduced guideline prices, Japanese Steel Mills had written 10 million tons of additional contracts per annum and not any of it had gone to Australia"; also minute by Court, 17 July 1967.

80 Minute from Lawler to Holt, 1 August 1967, NAA: A1209, 1967/7571.

81 Statement for press by David Fairbairn on "Iron Ore Export Policy", 25 August 1967, NAA: A1209, 1967/7571; "Robe River versus Hamersley: Guideline switch Confuses Ore Groups", *Australian Financial Review*, 28 August 1967.

82 Cabinet decision 25, 9 December 1969, NAA: A5869, 28.

83 Alan Trengove, *Adventure in Iron: Hamersley's First Decade*, Stockwell Press, Clayton, p. 108.

84 "WA Iron Ore Export Orders Now Top $5000 Million", *Australian Financial Review*, 18 November 1969.

85 Donald W. Barnett, *Minerals and Energy in Australia*, Cassell, Stanmore, pp. 187-9.

86 Letter from Dohnal to Court, 1 November 1966, SROWA, item 1970/0292.

87 Ibid., letter from Court to Dohnal, 15 November 1966.

88 Ibid., letter from Cliffs Western Australia to Japanese Steel Mills, 11 November 1966.

89 Ibid., letter from Dohnal to Court, 28 November 1966.

90 Ibid., minute by Court, 18 December 1966.

91 Ibid., press statement by Court, 3 September 1967.

92 Ibid., minute by Court, 3 February 1967.

93 Ibid., cablegram from Brand to Stuart Harrison, President, Cleveland Cliffs, 11 May 1967.

94 Ibid., letter from Court to Dohnal, 23 May 1967.

95 Ibid., minute by Court, 1 July 1967.

96 See Jerry Shields, *The Invisible Billionaire, Daniel Ludwig*, Houghton Mifflin, Boston, 1986.

97 Minute by Court, 15 August 1967, SROWA, item 1970/0292.

98 Ibid., minute by Court 24 August 1967; and press statement by Court, 3 September 1967. "Cleveland-Cliffs seeks to sell ore to Japan", *Wall Street Journal*, 8 September 1967.

99 Minute by Court, 30 September 1967; and minute by Court 26 September 1967 SROWA, item 1970/0292.

100 Ibid., letter from Court to Tanabe, 17 December 1967.
101 Ibid., letter from Court to Fairbairn, 4 December 1967.
102 Ibid., telex from Mekata (Melbourne) to Court, December 1967.
103 Ibid., letter from Y. Iimura to Dohnal, 5 February 1968.
104 Ibid., letter from Court to Dohnal, 14 February 1968.
105 John McIlwraith, *Mesa Harvest: Robe River's First 25 Years*, Robe River Iron Associates, Wickham, p. 30.
106 Ibid., p. 34.
107 Trengove, op. cit., pp. 24-5.
108 'The nation's heaviest track took only 10 months to build', *Australian Financial Review*, 8 November 1966; John Joyce and Allan Tilley, *Railways in the Pilbara*, J & A Publications, Wembley, 1980, pp. 41-2.
109 Trengove op. cit., p. 157.
110 Ibid., p. 165.
111 John McIlwraith, *The First 500 Million: The Mt Newman Story*, Iron Ore BHP-Utah Minerals International, Perth, 1998, p. 37
112 Ibid., p. 35.
113 Ibid.,, pp. 35-7.
114 Memorandum from F.S. Anderson to Maurice Mawby, 20 June 1963, folder 62, Maurice Mawby Papers, University of Melbourne Archives.
115 Trengove, op. cit., 1976, pp. 57-9.
116 Ibid., pp. 82-5.
117 Ibid.
118 Letter from Maurice Mawby to Peter Howson, 12 August 1968, Mawby Papers, folder 143.
119 Hamersley Iron, *Early Mining in the Pilbara: the 20th Year of Hamersley Iron*, Hamersley Iron, Perth, 1986, p. 5.
120 Ibid., pp. 168-70.
121 Murray Shaw, *Moving Mountains: The Evolution of Port Hedland Harbour*, Hesperian Press, Victoria Park, 2006, p. x.
122 Ibid., p. 30.
123 Ibid., p. 43.
124 Ibid., pp. 35-37, and p. 50.
125 Ibid., p.58.
126 Ibid., pp. 59-60.
127 Letter from Dohnal to Court, 6 January 1967, SROWA item 1970/0292.
128 John McIlwraith, *Mesa Harvest: Robe River's First 25 Years*, Robe River Iron Ore Associates, Wickham, 1997, p. 26.
129 Ibid., pp. 28–9.
130 Trengove, op. cit., p. 140.
131 Ibid., p. 142.
132 Ibid., , p. 131.
133 Ibid.,, p. 122.
134 Ibid., p. 132.
135 D.N. F. Dufty, *Industrial Relations in the Pilbara Iron Ore Industry*, Western Australian Institute of Technology, Perth, 1984, p. 56.
136 Ibid., p. 42.
137 Trengove, op. cit., p. 138.
138 Ibid., p. 123.

139 John McIlwraith, *The First 500 Million: The Mt Newman Story*, Iron Ore BHP-Utah Minerals International, Perth, 1988, p. 52.
140 Dufty, op. cit., p. 55.
141 Ibid., p. 59.
142 Ibid., p. 284.
143 Ibid., p. 287.
144 Trengove, op. cit., p. 146.
145 Ibid., p. 124.
146 Ibid., p. 125.
147 Dufty, op. cit., p. 286.
148 Brian Pinkstone, *Global Connections: A History of Exports and the Australian Economy*, Australian Government Publishing Service, Canberra, 1992, p. 179.
149 Minute by J.H. Lord, Director, Geological Survey to Minister for Mines, 17 May 1971, SROWA, series 164, item 1972/509 vol. 2.
150 Undated notes by Court for incoming Labor government, [1971], SROWA, series 164, item 1972/509 vol. 1.
151 Minute by J.H. Lord, Director, Geological Survey to Minister for Mines, 17 May 1971, SROWA, series 164, item 1972/509 vol. 2.
152 Ibid.
153 *Iron Ore (Hanwright) Agreement Act* 1967.
154 Minute by Court, 4 August 1967, SROWA, item 1970/0292.
155 Phillipson, op. cit., p. 153.
156 Ibid., p. 156.
157 Trengove, op. cit., pp. 104–5.
158 *Iron Ore (Hanwright) Agreement Amendment Act* 1968.
159 Phillipson, op. cit., p. 157.
160 Trengove, op. cit., p. 105.
161 Phillipson, op. cit., p. 160; Jamieson, op. cit., p. 214
162 Phillipson, op. cit., p. 145
163 Quoted in Jamieson, op. cit., p. 216.
164 Charles Court, "Report on Iron Ore, Iron and Steel Policy", 28 August 1970, SROWA, series 164, item 1972/509 Vol. 1.
165 Ibid.
166 Memorandum by David Oxer for Don May, Minister for Mines, on "Public Relations Opinion on the Inherited Hancock and Wright Conflict", 27 February 1972, SROWA, consignment 7162, item 182.
167 Letter from Hancock to Court, 1 December 1969, SROWA, consignment 7156, item 19.
168 Phillipson, op. cit., p. 169.
169 Letter from William J. Verity Jr to Court, 9 September 1970, SROWA, series 3121, item 18.
170 Cable from Court to Armco Steel Corporation, 3 December 1970, SROWA, series 164, item 1972/509 vol. 1.
172 I am indebted to David Moore for this information.
172 Ibid., p. 179.
173 Press Release by Premier's Department, 26 June 1971, SROWA, series 164, item 1972/509 vol. 2.

Chapter 4

1 Max Griffiths, *Of Mines and Men: Australia's 20th Century Mining Miracle 1945–1985*, Kangaroo Press, East Roseville, 1998, p. 50.

2 Kosmas Tsokhas, *Beyond Dependence: Companies, Labour Processes and Australian Mining*, Oxford University Press, Melbourne, 1986.

3 Harold Raggatt, *Mountains of Ore*, Landsdowne Press, Melbourne, p. 77.

4 George David Smith, *From Monopoly to Competition: the Transformations of Alcoa, 1988–1986*, Cambridge University Press, Cambridge, 1988; Charles C. Carr, *Alcoa: An American Enterprise*, Rinehart & Co., New York, 1952.

5 Smith, op. cit., p. 144.

6 Ibid., p. 1.

7 Merton J. Peck, *Competition in the Aluminium Industry, 1945–1958*, Cambridge, MA, 1961, chapter 9; Humphrey McQueen, *Gone Tomorrow: Australia in the 80s*, Angus and Robertson, Sydney, 1982, pp. 47–8.

8 Smith, op. cit., p. 241; Sheller, op. cit., p. 51.

9 Sheller, op. cit., p. 51.

10 Ibid., pp. 151–57.

11 S. Brubaker, *Trends in the World Aluminium Industry*, John Hopkins University Press for Resources for the Future, Baltimore, 1967, p. 101.

12 *The History of the British Aluminium Company Limited, 1894-1955*, The British Aluminium Company, London, c. 1956.

13 Comalco Limited, *The Development of the Bauxite, Alumina and Aluminium Industries*, A Submission to the Senate Standing Committee on Natural Resources, Melbourne, 1981; T.K. McDonald, *The Aluminium Industry: An Australian Perspective*, Comalco, Canberra, 1984, pp. 4–6.

14 Alcan Australia Limited, *40 Years on: A History of Alcan in Australia*, Granville, 1981, p. 7.

15 *Commonwealth Parliamentary Debates* (CPD), House of Representatives (HR), 25 June 1941, p. 422; Brian Carroll, *Potlines and People: A History of the Bell Bay Smelter*, Comalco Ltd, Burwood, 1980, pp. 1–3.

16 *Commonwealth Parliamentary Debates* (CPD), House of Representatives, 13 October 1943, p. 501.

17 Raggatt, op. cit., p. 77. Annual production of bauxite in Australia ranged between 3467 tons in 1950 to 7,700 tons in 1959 about half of which came from Victoria and the other half from New South Wales.

18 Australian Aluminium Production Commission, *Annual Report 1945–46*, Canberra, 1956, p. 4.

19 Carroll, op. cit.

20 "Sir Maurice Alan Edgar Mawby" in John Ritchie (ed.), *Australian Dictionary of Biography, Vol. 15*, Melbourne University Press, Carlton, 2000, pp. 333–5.

21 Notes from Howard Beale to Arthur Fadden, "Development of Bauxite Resources in Queensland and the Northern Territory", folder 55, Mawby Papers, University of Melbourne Archives. Tsokhas, op. cit., pp. 40–1.

22 Rick Wilkinson, *Rocks to Riches: The Story of Australia's National Geological Survey*, Allen & Unwin, Sydney, 1986, pp. 81–2.

23 Janette Ryan, *The Development of the Australian Aluminium Industry 1944–1955*, Occasional Paper No. 6, Transnational Corporations Research Project, Sydney, 1984, pp. 7–15.

24 Sheller, op. cit., p. 16.

25 Ibid., p. 18.
26 Smith, op. cit., pp. 367-8.
27 Aide Memoire, New Guinea Resources Prospecting Company – Gove Bauxite, 16 January 1958, folder, Mawby Papers, University of Melbourne Archives.
28 David Lee, "The Development of Bauxite at Gove, 1955-1975", *Journal of Australasian Mining History*, Vol. 12, October 2014, pp. 131-47.
29 Raggatt, op. cit., pp. 80-1.
30 H. Evans, "Preliminary Report on Bauxite Deposits, Albatross Bay, Cape York Peninsula", 11 January 1956, folder 10, Mawby Papers, University of Melbourne Archives.
31 L.B. Robinson to Maurice Mawby, 28 June 1957, folder 0103 Mawby Papers, University of Melbourne Archives.
32 See David Lee, "Comalco's Development of Queensland Bauxite, 1955-1975", *Australian Journal of Politics and History*, Volume 62, Number 1, 2016, pp. 44-58.
33 Undated notes on Commonwealth Aluminium Corporation, n.d. 1957, folder 181, Mawby Papers, University of Melbourne Archives.
34 "Sir Donald James Hibberd" in Diane Langmore (ed.), *Australian Dictionary of Biography, Vol. 17*, Melbourne University Press, Carlton, 2007, pp. 526-7.
35 Letter from Mawby to Gair, 10 January 1957 and letter from Gair to Mawby, 27 February 1957, folder 181, Mawby Papers, University of Melbourne Archives.
36 See K.D. McDonald, "The Negotiation and Enforcement of Agreements with State Governments Relating to the Development of Mineral Ventures", *Australian Mining and Petroleum Law Journal*, vol. 1, 1 1977, pp. 29-47; Anne Fitzgerald, *Mining Agreements: Negotiated Frameworks in the Australian Minerals Sector*, Prospect Books, Chatswood, 2010; G. Lewis, "Queensland Nationalism and Australian Capitalism" in E.L. Wheelwright and K. Buckley (eds), *Essays in the Political Economy of Australian Capitalism*, Vol. 2, ANZ Book Company, Sydney, 1978, p. 122.
37 Richard West, *River of Tears: The Rise of the Rio Tinto-Zinc Corporation*, Earth Island Limited, London, 1972, p. 100.
38 Notes by D. Hibberd of interview with Queensland Minister for Health and Home Affairs (Noble) and Minister for Mines (Evans) on 30 August 1957, folder 181, Mawby Papers, University of Melbourne Archives.
39 R.B. McKern, *Multinational Enterprises and Natural Resources*, McGraw Hill, Adelaide, 1976, p. 220.
40 Note by Hibberd of telephone message from C.A. Byrne, 21 October 1957, folder 181, Mawby Papers, University of Melbourne Archives.
41 West, op. cit., p. 100.
42 Griffiths, op. cit., p. 47; Knox, op. cit., pp. 154-5.
43 Quoted in West, op. cit., p. 100-1.
44 Submission 808 from Hasluck to Cabinet Committee, 19 August 1960, National Archives of Australia (hereafter NAA): A4940, C2323 part 1.
45 Smith, op. cit., pp. 322-6; Joseph Wechsberg, *The Merchant Bankers*, Little Brown and Company, New York, 1968, pp. 128–62.
46 Letter from Mawby to A.M. Baer, 5 June 1959, Mawby Papers, folder 08 and cable from Mawby to L.B. Robinson, 12 April 1960, Mawby Papers, folder 31, University of Melbourne Archives.
47 Note by Hibberd dated 25 September 1959 of a meeting between Consolidated Zinc and the British Aluminium Company on 24 September 1959, folder 11, Mawby Papers, University of Melbourne Archives; letter from Hibberd to Mawby, 5 July 1960, folder 180, Mawby Papers, University of Melbourne Archives.

48 Hibberd, "Bell Bay – General Policy Considerations", 16 November 1959, folder 11, Mawby Papers, University of Melbourne Archives; letter from Spooner to Robinson, 14 November 1959, folder 11, Mawby Papers, University of Melbourne Archives.

49 Letter from Hibberd to Mawby, 4 May 1959, folder 11, Mawby Papers, University of Melbourne Archives; note by Hibberd dated 25 September 1959 of meeting between Consolidated Zinc and British Aluminium at Norfolk House on 24 September 1959, folder 11, Mawby Papers, University of Melbourne Archives.

50 Cable from Hibberd to Robinson, 9 May 1960 and A.V. Smith's "Note Governing Broad Essentials Regarding Bell Bay", n.d. 1960 and record of meeting with Senator Spooner on Bell Bay, 27 June 1960, folder 11, Mawby Papers, University of Melbourne Archives.

51 Letter from Mawby to A.V. Smith, 13 May 1960, folder 11, Mawby Papers, University of Melbourne Archives.

52 Notes by Prime Minister's Department on Hasluck's submission 808, 22 August 1960, NAA: A4940, C2323 part 1.

53 Alan Rix, *Coming to Terms: The Politics of Australia's Trade with Japan 1945-57*, Allen and Unwin, North Sydney, 1986, p. 150.

54 Submission 528 from Spooner to Cabinet, January 1963, NA: A4940, C2323 part 1.

55 Notes by A.W. McCasker, Prime Minister's Department, February 1963, NAA: A4940, C2323 part 1.

56 Cabinet Decision 650, 13 February 1963, NAA: A4940, C2323 part 1.

57 Submission 643 from Hasluck and Spooner to Cabinet, 19 April 1963, NAA: A4940, C2323 part 1.

58 Cabinet decision 781, 8 May 1963, NAA: A4940, C2323 part 1.

59 Bain Attwood, *Rights for Aborigines*, Allen & Unwin, Crows Nest, 2003, pp. 215-19.

60 Ibid., p. 221.

61 Ibid., pp. 221-3.

62 Ibid., p. 222.

63 Ibid., pp. 226-36.

64 KACC proposal for variation of Comalco Association, 26 March 1962, folder 180, Mawby Papers, University of Melbourne Archives.

65 Letter from R.C. Atherton to John Cocks, 4 December 1972, folder 95, Mawby Papers, University of Melbourne Archives; see also Sheller, op. cit., pp. 56–7.

66 Cable from Mawby to Duncan and E.E. Trefethen, 2 May 1963, folder 82, Mawby Papers, University of Melbourne Archives.

67 Memorandum from Sid Christie to Mawby, 15 January 1962, folder 16, Mawby Papers, University of Melbourne Archives.

68 Letter from Clarrie Byrne to Maurice Mawby, 4 October 1960, folder 012 Mawby Papers, University of Melbourne Archives.

69 Comalco, Projected Development and Estimated Future Profitability, 12 June 1964, folder 32, Mawby Papers, University of Melbourne Archives.

70 Comalco Limited, Submission to the Senate Standing Committee on Natural Resources, *The Development of the Bauxite, Alumina and Aluminium Industries*, Comalco, Melbourne, 1981, p. 106.

71 Comalco, Projected Development and Estimated Future Profitability, 12 June 1964, folder 32, Mawby Papers, University of Melbourne Archives.

72 McKern, op. cit., p. 221.

73 Cable from Duncan to Trefethen, 13 February 1963, folder 82, Mawby Papers, University of Melbourne Archives.

74 Cable from Duncan to Mawby, 25 April 1963 and cable from Mawby to Duncan and

Trefethen, 2 May 1963, folder 82, Mawby Papers, University of Melbourne Archives.

75 Cable from R.C. Atherton to J.N.V. Duncan, 13 February 1963, folder 82, Mawby Papers, University of Melbourne Archives.

76 Cable from Hibberd to Atherton, 21 February 1963, folder 82, Mawby Papers, University of Melbourne Archives.

77 Cable from Hibberd to Atherton and Clayton, 29 March 1963 and cable from Mawby to D.A. Rhoades, folder 82, 29 May 1963, Mawby Papers, University of Melbourne Archives.

78 Ian McLean, *Why Australia Prospered: The Shifting Sources of Economic Growth*, Princeton University Press, Princeton, 2013, p. 84.

79 John Fogarty and Noel Wotton, *Breakthrough: The Establishment and Development of Queensland Alumina*, Hargreen Publishing Company, North Melbourne, 1992, p. 7.

80 McDonald, op. cit., p. 16.

81 Letter from Maurice Mawby to Alfred Baer, 4 December 1968, folder 010 Mawby Papers, University of Melbourne Archives.

82 Geoffrey Blainey, *The Rush That Never Ended: A History of Australian Mining*, Melbourne University Press, Carlton, 2003, p. 333.

83 Alcan, op. cit., pp. 22-3.

84 Comalco, *The Bluff Smelter*, Comalco, Auckland, 1971; J.C. Cooper, "Japanese Aluminium Smelter Location Policy 1975–85", *Geography*, Vol. 64, 3 November 1979, pp. 334-38.

85 Geoffrey Blainey, *White Gold: The Story of Alcoa of Australia*, Allen & Unwin, St Leonards, 1997, pp. 28-30.

86 Raggatt, op. cit., p. 95.

87 Blainey, op. cit., pp. 22-3.

88 Ibid., pp. 40-2.

89 Ibid., p. 42.

90 Ibid.

91 Griffiths, op. cit., p. 49.

92 Smith, op. cit., p. 326.

93 Lindesay Clark, *Built on Gold: Recollections of Western Mining*, Hill of Content, Melbourne, 1983, pp. 186-8.

94 Blainey, op. cit., p. 57.

95 Ibid.

96 Griffiths, op. cit., p. 50.

97 Smith, op. cit., pp. 326-9.

98 Ibid., p. 328.

99 Blainey, op. cit., pp. 648.

100 Ronda Jamieson, *Charles Court: I Love this Place*, St George Books, Osborne Park, 2011, pp. 148-50.

101 Smith, op. cit., p. 328.

102 Ibid., pp. 328-9.

103 Memorandum by Arvi Parbo on conversation with Mitsubishi, 5 March 1974, Arvi Parbo Papers, Acc 09/45 Box 6, National Library of Australia.

104 Blainey, op. cit., pp. 88-108.

105 Ibid., pp. 116-17.

106 Ibid., p. 134.

107 Ibid., p. 124.

108 Ibid., pp. 134-6.

109 Ibid., p. 136.
110 Quoted in Donald Horne, *Time of Hope: Australia 1966–72*, Angus & Robertson, Sydney, 1980, p. 103.
111 Letter from Menzies to CSR and BHP, 28 August 1964, NAA: A4940, C2323 part 2.
112 Cabinet submission 233 from Barnes and Spooner, 28 May 1964, NAA: A4940, C2323 part 2.
113 Notes by J. Colwell on submissions by Fairbairn and Barnes, 16 August 1965, NAA: 4940, C2323 part 2.
114 Notes by J. Colwell, Prime Minister's Department on submissions by Fairbairn and Barnes, 16 August 1965; cabinet decision, 7 September 1965, NAA: A4940, C2323 pt 2.
115 Comments provided to Treasurer on Barnes submission 329, 11 August 1967, A571, 1966/3337 part 6.
116 Grant Fleming, David Merrett and Simon Ville, *The Big End of Town: Big Business and Corporate Leadership*, Cambridge University Press, Melbourne, 2004, pp. 104–5 and pp. 174-6.
117 Dennis O'Brien, "The Vision at Gove", *Bulletin*, 16 November 1968.
118 Ibid.
119 Department of Territories note on Nabalco's interview with Barnes on 7 December 1967, A1209, 1967/7863 part 1, NAA. Griffin and Vernon were present with directors of Alusuisse.
120 Nabalco aide-memoire, 5 December 1967, A1209, 1967/7863 part 1, NAA.
121 Ibid.
122 Department of Treasury minute to Conron, n.d. 1968, A571, 1966/3337 part 6, NAA.
123 Letter from Barnes to Gorton, A1209, 1967/7863 part 1, NAA.
124 Statement by E.G. Whitlam, Leader of the Opposition, 10 December 1968, NAA: A1209, 1967/7863 part 2.
125 Ibid.
126 Announcement by Barnes on the agreement to develop Gove bauxite, NAA: A1209, A452, 1968/1204.
127 J.C. Altman, *Aborigines and Mining Royalties in the Northern Territory*, Australian Institute of Aboriginal Studies, Canberra, 1983, pp. 4–9; Paul Kauffman, *Wik, Mining and Aborigines*, Allen & Unwin, St Leonards, 1998, pp. 42–3.
128 Altman, op. cit., pp. 17–20.
129 Highly confidential note by CSR on Australian/Swiss Relationships over Gove Alumina Project, NAA: A1209, 1967/7863 part 1.
130 Notes by the Australian partners on Gove project for discussion with the Commonwealth, 27 May 1968, NAA: A1209, 1967/7863 part 2.
131 CSR note of discussions between CSR and Alusuisse in Zurich, 14 May 1968, NAA: A571, 1966/3337 part 7.
132 Record of meeting between representatives of Nabalco, the Minister for the Interior, and the secretaries of the Department of National Development and the Department of the Interior on 6 June 1968, NAA: A1209, 1967/7863 part 1.
133 CSR note of discussions between CSR and Alusuisse in Zurich, 14 May 1968, A571, 1966/3337 part 7, NAA.
134 Letter from Nixon to Griffin, 28 May 1968, NAA: A1209, 1967/7863 part 1.
135 Ibid., Letter from Griffin to R.S. Swift, 22 July 1968.
136 Ibid.
137 Ibid. See also letter from E.R. Meyer (Chairman) and P.H. Muller, Director, of Alusuisse, to Nixon, 20 August 1968, A1209, 1967/7863 part 1, NAA.

[138] Ibid., notes by CSR for discussion with Commonwealth government, 21 June 1968.
[139] Submission to Minister for Interior on Gove Alumina Project, August 1968, A1209, 1967/7863 part 1, NAA.
[140] Draft cabinet submission by Nixon, October 1968, A1209, A1967/7863 part 2.
[141] Malcolm Knox, *Boom: The Underground History of Australia from Gold Rush to GFC*, Viking, Melbourne, 2013, pp. 156–7.
[142] Letter from W.E.H. Stanner, Professor of Anthropology, Australian National University, to W.C. Wentworth, Minister-in-charge of Aboriginal Affairs, 1 April 1969, A432, 1968/649 part 7.
[143] Statement by McMahon, Official Opening of Nabalco, 1 July 1972, A1209, 1976/520, NAA.
[144] Sheller, op. cit., pp. 16-17.
[145] Griffiths, op. cit., p. 50.

Chapter 5

[1] Alan Trengove, *Discovery: Stories of Modern Mineral Exploration*, Stockwell Press, Mont Albert, pp. 86-7. From 1892 to 1977 more than 1800 wells were drilled and about $1200 million spent by the petroleum industry, "but most of the effort and expense produced nothing but geological information".
[2] Ibid., p. 88.
[3] Ibid., p. 83.
[4] Ibid., p. 88.
[5] Harold Raggatt, *Mountains of Ore*, Landsdowne Press, Melbourne, 1968, p. 231.
[6] Max Griffiths, *Of Mines and Men: Australia's 20th Century Mining Miracle 1945-1985*, Kangaroo Press, East Roseville, 1998, p. 74.
[7] Trengove, op. cit., pp. 83-4.
[8] Raggatt, op. cit., p. 295.
[9] Arvi Parbo, *Down Under: Mineral Heritage in Australasia*, Australasian Institute of Mining and Metallurgy, Box Hill, 1992, p. 176.
[10] Trengove, op. cit., pp. 84-5.
[11] Peter Thompson and Robert Macklin, *The Big Fella: The Rise and Rise of BHP Billiton,* William Heinemann Australia, Sydney, 2009, p. 95.
[12] Raggatt, op. cit., pp. 250-1.
[13] Griffiths, op. cit., p. 74.
[14] Raggatt, op. cit., p. 216.
[15] Thompson and Macklin, op. cit., p. 102.
[16] Trengove, op. cit. p. 98.
[17] Ibid., pp. 98-9.
[18] Hugh Saddler, *Energy in Australia: Politics and Economics*, George Allen & Unwin, North Sydney, p. 92.
[19] Ibid., p. 90.
[20] Trengove, op. cit., p. 107.
[21] Thompson and Macklin, op. cit., p. 109.
[22] Ibid., p. 80.
[23] Raggatt, op. cit., pp. 268-9.
[24] Saddler, op. cit., p. 99.
[25] Griffiths, op. cit., p. 82.

26 Brian Pinkstone, *Global Connections: A History of Exports and the Australian Economy*, Australian Government Publishing Service, Canberra, 1992, p. 178.
27 Raggatt, op. cit., p. 191.
28 Ibid., p. 191.
29 Ibid., pp. 195-7.
30 Ibid., p. 204; Parbo, op. cit., pp. 132-3.
31 Ciaran O'Faircheallaigh, *Mining and Development: Foreign-Financed Mines in Ireland, Papua New Guinea and Zambia*, Croom Helm, Sydney, 1984, p. 216.
32 Raggatt, op. cit., p. 209.
33 Ibid.
34 Submission 187 from Barnes to Cabinet, 29 March 1967, NAA: A4940, C4491.
35 Ibid.
36 O'Faircheallaigh, op. cit., p. 234.
37 Ibid., p. 137.
38 Raggatt, op. cit., p. 138.
39 Ibid., pp. 139-40.
40 Trengove, op. cit., p. 72.
41 Ibid., p. 75.
42 Ibid., p. 75.
43 Paul Kauffman, *Wik, Mining and Aborigines*, Allen & Unwin, St Leonards, 1998, pp. 48-50.
44 Trengove, op. cit., p. 76.
45 Raggatt, op. cit., p. 151.
46 Ibid.
47 Griffiths, op. cit., pp. 54-5.
48 Pinkstone, op. cit., p. 179.
49 Raggatt, op. cit., p. 142.
50 Eric Eklund, *Mining Towns: Making a Living, Making A Life*, University of New South Wales Press, Sydney, 2012, p. 211.
51 Letter from Arvi Parbo to Rex Connor, Minister for Minerals and Energy, 31 August 1973, Arvi Parbo Papers, Acc 09/45 Box 6, National Library of Australia.
52 Trengove, op. cit., p. 146.
53 Raggatt, op. cit., pp. 142-3.
54 Lindesay Clark, *Built on Gold: Recollections of Western Mining*, Hill of Content, Melbourne, 1983, p. 203.
55 Trengove, op. cit., p. 155.
56 Ibid., pp. 154-5.
57 Raggatt, op. cit., p. 147.
58 Clark, op. cit., pp. 207-8.
59 Ken Spillman, *A Rich Endowment: Government and Mining in Western Australia 1829–1994*, University of Western Australia Press, Nedlands, 1993, pp. 197-8.
60 Alice Cawte, *Atomic Australia 1944-1990*, University of New South Wales Press, Kensington, 1992, p. 136; Geoffrey Blainey, *The Rush that Never Ended: A History of Australian Mining*, Melbourne University Press, Carlton, 2003, pp. 374-5.
61 Trevor Sykes, *The Money Miners: Australia's Mining Boom 1969-1970*, Wildcat Press, Sydney, 1978, p. 210.
62 Cawte, op. cit., pp. 136-7.
63 Blainey, op. cit., p. 374.
64 Quoted in Donald Horne, *Time of Hope: Australia 1966-72*, Angus & Robertson, Sydney, 1980, p. 101.

65 Malcolm Knox, *Boom: The Underground History of Australia, from Gold Rush to GFC*, Viking, Melbourne, p. 195.
66 Donald W. Barnett, *Minerals and Energy in Australia*, Methuen Australia, North Ryde, 1979, pp. 4-5.
67 Ibid., p.3.
68 "Minerals boom is opportune for Australia", *Sydney Morning Herald*, 27 October 1969.
69 Pinkstone, op. cit., p. 381.
70 McLean, op. cit., p. 216.
71 Pinkstone, op. cit., p. 183.
72 Ibid., p. 388.
73 Ibid., op. cit., p. 388.
74 Tom Conley, *The Vulnerable Country: Australia and the Global Economy*, University of New South Wales Press, Kensington, 2009, p. 104.
75 Pinkstone, op. cit., p. 388.
76 Reserve Bank of Australia, *Australian Economic Statistics 1949–50 to 1996–1997, Occasional Paper No 8.*
77 Barrie Dyster and David Meredith, *Australia in the Global Economy: Continuity and Change*, Second Edition, Cambridge University Press, Port Melbourne, 2012, p. 223; Pinkstone, op. cit., p. 169.
78 Ibid.
79 Barnett, op. cit., p. 3; Pinkstone, op cit., p. 388.
80 Dyster and Meredith, op. cit., p. 200.
81 Barnett, op. cit., p. 5.
82 Donald Horne, *The Lucky Country*, Penguin, Melbourne, 2009 (first published 1964).
83 Ibid., p. 237.
84 Donald Horne, *The Next Australia*, Angus and Robertson, Sydney, 1970, p. 36.
85 Trevor Dawson-Grove, "Nickel excites world interest", *Sydney Morning Herald*, 27 October 1969.
86 Phyllis Rosendale, "Australia in the Global Economy" in P.J. Boyce and J.R. Angel (eds), *Independence and Alliance: Australia in World Affairs 1976-80*, George Allen & Unwin, North Sydney, 1983, p. 110.
87 Pinkstone, op. cit., p. 180.
88 Ibid., pp. 180-1.
89 Letter from A.M. Baer to James Callaghan, British Chancellor of the Exchequer, 18 May 1966, folder 010, Mawby Papers, University of Melbourne Archives.
90 Cablegram from Downer to McMahon, 30 May 1968 in S.R. Ashton, Carl Bridge and Stuart Ward (eds), *Documents on Australian Foreign Policy: Australia and the United Kingdom 1960-1975*, Department of Foreign Affairs and Trade, Canberra, 2010, pp. 658-9.
91 Ibid., letter from McMahon to Jenkins, 5 July 1968, pp. 660-3.
92 Ibid., despatch from Johnstone to Bowden, 5 June 1967, p. 614.
93 Ibid., p. 610.
94 Submission No, 548 from McMahon to Cabinet, 20 November 1967 in Ashton, Bridge and Ward, op. cit., pp. 640–44.
95 Press statement by Holt, 12 December 1967 in ibid., pp. 651-3.
96 Ibid.
97 Paper on Australian dollar parity, n.d. 1971, folder 143, Mawby Papers, University of Melbourne Archives.
98 Kosmas Tsokhas, *A Class Apart? Businessmen and Australian Politics 1960–1980*, Oxford University Press, Melbourne, 19854, p. 58.

99 Ibid., pp. 58-9.
100 Tsokhas, op. cit., p. 60.
101 Ibid., pp. 61-2.
102 Knox, op. cit., pp. 190-1.
103 Sykes, op. cit., 86.
104 Knox, op. cit., 194.
105 Sykes, op. cit., p. 90.
106 Ibid., pp. 91-2.
107 Ibid., pp.116-17.
108 Ibid., p. 141.
109 *Economist*, quoted in ibid., p. 144.
110 Ibid., pp. 200-2.
111 Ibid., pp. 217-18.
112 Cawte, op. cit., p. 137.
113 Ibid., pp. 253-4.
114 Frank Crowley, *Tough Times: Australia in the Seventies*, William Heinemann Australia, Richmond, 1986, p. 17-18.
115 Ibid., p. 236.
116 Horne, op. cit., p. 106
117 Humphrey McQueen, *Gone Tomorrow: Australia in the 80s*, Angus and Robertson, Sydney, 1982. p. 303
118 Sykes, op. cit., p. 240.
119 Ibid., p. 241.
120 Ibid.
121 Sykes quoted in Crowley, op. cit., p. 18.
122 Ibid.
123 Richard West, *River of Tears: The Rise of the Rio Tinto-Zinc Corporation Ltd.*, Earth Island Limited, London, 1972, pp. 101-6.
124 Ibid., p. 107.
125 Donald Horne, *Time of Hope: Australia 1966-72*, Angus & Robertson, Sydney, 1980, p. 103.
126 Ibid.

Chapter 6

1 See Peter Golding, *Black Jack McEwen: Political Gladiator* Melbourne University Press, Carlton, 1996.
2 C.J. Lloyd, "Reginald Francis Xavier 'Rex' Connor", in John Ritchie (ed.), *Australian Dictionary of Biography, Volume 13*, Melbourne University Press, Carlton, 1993, pp. 486–8.
3 Ibid.
4 Jenny Hocking, *Gough Whitlam: A Moment in History*, Miegunyah Press, Carlton, 2008, pp. 276, 295.
5 For Whitlam's relationship with Connor during the Whitlam Government see Jenny Hocking, *Gough Whitlam: His Time: the Biography. Volume II* Miegunyah Press, Carlton, 2012, passim.
6 Michael Sexton, *Illusions of Power: the Fate of a Reform Government* George Allen and Unwin, North Sydney, 1979, p. 97.
7 Quoted in Griffiths, op. cit., p. 104.

8 "Foreign Explorers Await Cabinet Decision", *Australian Financial Review* (hereafter *AFR*), 13 March 1973.

9 Alex Millmow, "Australia and the Keynesian Revolution" in Samuel Furphy (ed.), *The Seven Dwarfs: The Age of the Mandarins* ANU Press, Canberra, 2015, pp. 76–7.

10 Editorial, "The Miners and the Dollar", *AFR*, 3 January 1973; Trevor Sykes, "Two Mines Face Currency Crisis", *AFR*, 19 February 1973.

11 *AFR*, 14 February 1973.

12 Article on Connor, *AFR*, 11 January 1973.

13 Maximilian Walsh, "Miners Anger Govt: Secretive Deal Threatens Uranium Contract", *AFR*, 4 May 1973.

14 Brian Toohey, "Australian Coal Marketing Methods under Fire Again", *AFR*, 2 March 1973.

15 Letter from B.W. Hartnell to Connor, 6 February 1973 and paper by Hartnell, "Commonwealth Control of Coal Mining Industry", 19 December 1972, Papers of Rex Connor, Series 2, Box 8, D61 Joint Coal Board Controls part 2, University of Wollongong.

16 Submission from Connor to Cabinet, "Export of Black Coal", 2 January 1973, Papers of Rex Connor, Series 2, Box 9, Export Controls Part 1, University of Wollongong.

17 Ibid.

18 Ibid.

19 Ibid.

20 Handwritten notes by Connor on reasons for export controls, n.d. 1973, Papers of Rex Connor, Series 2, Box 8, D61, University of Wollongong.

21 *AFR*, 2 March 1973; Robert Haupt, "The AIDC to be Development Spearhead", *AFR*, 7 March 1973.

22 "AIDC Bill Contrary to Free Enterprise", *AFR*, 18 October 1973.

23 Report of speech by Connor at Twelfth General Meeting of the American Chamber of Commerce in Melbourne, *AFR*, 26 March 1973.

24 Letter from Connor to Whitlam, 15 August 1973, Papers of Rex Connor, Series 2 Box 45, D61/2/534, University of Wollongong.

25 Gary Smith, "Minerals and Energy" in Allan Patience and Brian Head (eds), *From Whitlam to Fraser: Reform and Reaction in Australian Politics*, Oxford University Press, Melbourne, 1979, p. 238.

26 Fred Brenchley, "New Mineral Rules – Whitlam Gives the Details", *AFR*, 30 October 1973.

27 Jeffrey D. Wilson, *Governing Global Production: Resources Networks in the Asia–Pacific Steel Industry*, Palgrave Macmillan, Basingstoke, p. 88.

28 Letter from Connor to Whitlam, 11 April 1973, Papers of Rex Connor, Series 2 Box 45, D61/2/534, University of Wollongong.

29 Brian Toohey, "State Ownership Italian Style Could be Model for Petroleum Authority", *AFR*, 10 December 1973.

30 Submission 488 from Connor to Cabinet, 27 June 1973, Series 2 Box 45, D61/2/534, Papers of Rex Connor, University of Wollongong.

31 Ibid.

32 Cabinet decision 527, 16 April 1973, Papers of Rex Connor, Series 2 Box 45, D61/2/534, University of Wollongong.

33 Letter from Connor to Crean, 22 July 1974, Papers of Rex Connor, Series 2 Box 45, D61/2/534, University of Wollongong.

34 Letter from Crean to Whitlam, 1 March 1974, Papers of Rex Connor, Series 2 Box 45,

D61/2/534, University of Wollongong.

35 Comments on the Petroleum and Minerals Authority Bill, n.d. Papers of Rex Connor, Series 2 Box 45, D61/2/534, University of Wollongong, Wollongong.

36 Ibid.

37 Minute, Bob Sorby to Connor, 17 February 1974, Papers of Rex Connor, Series 2 Box 45, D61/2/534, University of Wollongong.

38 Letter from Dunstan to Whitlam, 26 November 1973, Papers of Rex Connor, Series 2 Box 45, D61/2/534, University of Wollongong.

39 Letter from Tonkin to Whitlam, 17 May 1973, Papers of Rex Connor, Series 2 Box 45, D61/2/534, University of Wollongong.

40 Sarah Burnside, "Mineral Booms, Taxation and the National Interest", *History Australia* 10, no. 3, 2013, p. 180.

41 George Williams, "The Whitlam Government and Constitutional Reform" in Jenny Hocking and Colleen Lewis (eds), *It's Time Again: Whitlam and Modern Labor*, Melbourne Publishing Group, Armadale, 2003), pp. 205-6.

42 Griffiths, op. cit., pp. 105-6.

43 Wilson, op. cit., 90.

44 Ibid., p. 90.

45 Ibid.

46 John McIlwraith, "Iron Ore Miners Calculate Solid Gains on New Japanese Prices", *AFR*, 17 September 1974.

47 Brian Pinkstone, *Global Connections: A History of Exports and the Australian Economy*, Australian Government Publishing Service, Canberra, 1992, p. 214.

48 Ibid.

49 "Coal Industry's United Front to Overseas buyers", *AFR*, 2 July 1973.

50 Discussions between S. Nagano, Chairman, Nippon Steel Corporation, Inayama, Representative Director and President, Nippon Steel Corporation, S. Tanabe, Managing Director, Nippon Steel Corporation, and B.W. Hartnell, Chairman, Joint Coal Board, in Tokyo, 26 July 1972, Papers of Rex Connor, Series 2, D61, Box 8, University of Wollongong.

51 Minute, R.J. Gray to Connor, 11 March 1974, Papers of Rex Connor, D61, Series 2, Box 8, D61/2/100, University of Wollongong.

52 Transcript of ABC radio broadcast, AM, 28 March 1974, Papers of Rex Connor, D61, Series 2, Box 8, D61/2/100, University of Wollongong.

53 Minute from R.J. Gray to Connor, 11 March 1974, Papers of Rex Connor, D61, Series 2, Box 8, D61/2/100, University of Wollongong, Wollongong.

54 Minute from R.J. Gray to Connor, 29 June 1974, Papers of Rex Connor, Series 2, Box 8, D61/2/91, University of Wollongong.

55 Ibid.

56 "Building Energy Plants or Just Monuments", *AFR*, 14 May 1974.

57 Ibid.

58 Department of Minerals and Energy Brief for discussions with Australian Coal Association – Prices for April 1975, 10 March 1975, Papers of Rex Connor, Series 2, Box 57, D61/2/668, University of Wollongong.

59 Department of Minerals and Energy Brief for discussions with Australian Coal Association – Prices for April 1975, 10 March 1975, Papers of Rex Connor, Series 2, Box 57, D61/2/668, University of Wollongong.

60 Memorandum by G.F. Clarke, Counsellor (Minerals), Tokyo, 26 March 1975 conveying telex from Ryan to Hewitt drafted in the Australian Embassy, Tokyo, Papers of Rex

Connor, Series 2, Box 57, D61/2/668, University of Wollongong.

61 Ibid.

62 Note for File by Hewitt, 1 April 1975, Papers of Rex Connor, Series 2, Box 57, D61/2/668, University of Wollongong.

63 "Lots of Holes in Coal Agreement", *AFR*, 17 July 1975.

64 Pinkstone, op. cit., pp. 214-15.

65 *The Morning Bulletin*, Rockhampton, 14 July 1975.

66 Letter from Hayden to Connor, 10 July 1975, Papers of Rex Connor, Series 1, D/61/2/8, University of Wollongong.

67 Quoted in Brian Galligan, *Utah and Queensland Coal: A Study in the Micro Political Economy of Modern Capitalism and the State*, University of Queensland Press, St Lucia, 1989, p. 128.

68 Statement by Rex Connor, Papers of Rex Connor, Series 1, D61/2/8, University of Wollongong.

69 Synopsis of Hail Creek Coking Coal project prepared for Rex Connor, 14 February 1973, Papers of Rex Connor, Series 1, D61/2/8, University of Wollongong.

70 "Whitlam Creates a Big Brother – The Catch in the New Foreign Investment Policy", *AFR*, 26 September 1975.

71 "Coal go ahead", *AFR*, 13 October 1975.

72 Galligan, op. cit., pp. 136–137.

73 See Alice Cawte, *Atomic Australia, 1994–1990* University of New South Wales Press, Kensington, 1992.

74 Wayne Reynolds, "Australia's Quest to Enrich Uranium and the Whitlam Government's Loans Affair", *Australian Journal of Politics and History* 54, no. 4 2008, p. 571.

75 Gough Whitlam, *The Whitlam Government* (Ringwood: Penguin, Ringwood, 1985, pp. 253, 261; minute paper for Executive Council, 13 December 1974, NAA: A571/144, 1974/96 part 1.

76 Frank Crowley, *Tough Times: Australia in the Seventies*, William Heinemann Australia, Richmond, 1986, p. 103.

77 Ibid.

78 Reynolds, op. cit., pp. 564–5.

79 Minute Paper for Executive Council, 13 December 1974, A571/144, 1974/96 part 1, NAA.

80 Reynolds, op. cit., p. 571.

81 Ibid., p. 572.

82 *National Times*, 4 July 1975.

83 *Review of Uranium Mining, Processing and Nuclear Energy in Australia*, June 2006 quoted in Reynolds op. cit., p. 562.

84 Wilson, op. cit., 94.

Chapter 7

1 Frank Crowley, *Tough Times: Australia in the Seventies*, William Heinemann Australia, Richmond, 1986, chapters 20 and 21.

2 Ian W. McLean, *Why Australia Prospered: The Shifting Sources of Economic Growth*, Princeton University Press, Princeton, 2013, p. 210.

3 Ibid., p. 214.

4 Brian Pinkstone, *Global Connections: A History of Exports and the Australian Economy*,

Australian Government Publishing Service, Canberra, 1992, p. 215.

5 Ibid., pp. 55, 388. In 1983 mining's share of goods and services exports was higher than that of the rural sector, manufactures and services. In 1903, influenced by the WA gold boom, minerals export had been 25.6 per cent of total exports.

6 Ben Smith, *The Role of Resource Development in Australia's Economic Growth*, Centre for Economic Policy Research, Occasional Paper no. 167, Australian National University, August 1987; Reserve Bank of Australia, *Occasional Paper No. 8, Australian Economic Statistics, 1949–1950 to 1996–1997.*

7 Chris Fisher, *Coal and the State*, Methuen, Melbourne, 1987, p. 196.

8 Bryan Frith, "High Cost of Cheap Coal", *Australian*, 21 April 1977.

9 Daniel Yergin, *The Quest: Energy, Security and the Remaking of the Modern World*, Penguin, New York, 2011, p. 400.

10 Australian Coal and Shale Employees' Federation, "The Commonwealth Government and Export Controls in the Australian Coal Industry", September 1986, NAA: A1209, 1987/951 part 5.

11 "Where Coal is King", *Age*, 6 November 1979.

12 Michael Byrnes, "Japanese hit big Aust coal takeovers", *Australian Financial Review*, 22 August 1977; Alan Jury, "Hopes High, But Profits Down", *Australian Business Coal Survey, Australian Business*, 24 September 1981.

13 Humphrey McQueen, *Gone Tomorrow: Australia in the 80s*, Angus and Robertson, Sydney, 1982, p. 44; Department of Trade and Resources, *Australian Resources in a World Context*, Australian Government Publishing Service, Canberra, 1982, p. 4.

14 Hugh Saddler, *Energy in Australia: Politics and Economics*, George Allen and Unwin, North Sydney, 1981, p. 105.

15 "The Great Australian Coal Rush", *Sun*, 20 October 1977.

16 John Byrne, "The Coal Rush", *Australian Financial Review* (hereafter *AFR*), 20 July 1977.

17 "Uranium takes back seat to King Coal", *Sun*, 23 August 1977.

18 Graham Larcombe, "New South Wales: The Political Economy of an Industrialised State" in Brian Head (ed.), *The Politics of Development in Australia*, Allen and Unwin, North Sydney, 1986, pp. 104–5.

19 Ibid., p. 107.

20 Peter Freeman, "Doubts over Warkworth Coal", *National Times*, 24–29 October 1977.

21 Christopher Jay, *The Coal Masters: The History of Coal & Allied 1844–1994*, Focus Publishing, Double Bay, 1994, pp. 208–10.

22 Ibid., pp. 218–19.

23 Larcombe, op. cit., pp. 98–99.

24 "A coal bonanza for NSW", *Courier Mail*, 19 July 1977; "Japan shares $40m project", *Sydney Morning Herald*, 31 August 1977.

25 Michael Byrnes, "$500m Japanese coal contract", *AFR*, 9 November 1977.

26 "State government changes its mind on coal", *Sydney Morning Herald* (hereafter *SMH*), 22 February 1978.

27 Jay, op. cit., p. 213.

28 Ibid., pp. 213-14.

29 John Byrne, "BHP cracks a winner", *AFR* 5 August 1977.

30 "New 200 mil. State coal scheme", *Courier Mail*, 5 August 1977.

31 "Bid to block BHP's huge coal buy", *AFR*, 2 March 1977.

32 Norman Hunter, "Japan to sign $2600m coal deal", *Courier Mail*, January 1981.

33 "State gives Oaky Creek go ahead to US firm", *Courier Mail*, 19 June 1977.

34 Minute from K.L. Mahar to Fraser, 'Oaky Creek', 1 October 1980, NAA: A1209,

1979/983 part 1.

35 Arthur Hadrich, "Canberra decision soon to Revive Coal Project", *Daily Telegraph*, 2 November 1980; minute from K.L. Mahar to Fraser, 30 October 1980, NAA: A1209, 1979/983 part 1. Tony Grant-Taylor, "Howard gives go-ahead to Oaky Creek coal project", *AFR*, 1 May 1981.

36 Arvi Parbo, *Down Under: Mineral Heritage in Australasia*, Australasian Institute of Mining and Metallurgy, Box Hill, 1992, p. 108.

37 "Two Major State Coal Projects to Cost $600 Mil", *Courier Mail*, 13 July 1978.

38 Max Griffiths, *Of Mines and Men: Australia's 20th Century Mining Miracle 1945-1985*, Kangaroo Press, East Roseville, 1998, p. 150.

39 Ibid.

40 Stuart Simson, "Conflicting policies hold up $500m Project", *National Times*, 17 April – 3 May 1980.

41 John Short, "Govt steps aside in coal takeover fight", *AFR*, 14 December 1978.

42 Letter from Howard to Fraser, 23 September 1980, NAA: A1209, 1979/983 part 1.

43 "Giant Coal deal Rejected", *Australian*, 16 April 1980.

44 Minute from Visbord to the Secretary, Prime Minister's Department, 9 September 1980, NAA: A1209, 1979/983 part 1.

45 Australian Business Coal Survey, *Australian Business*, 24 September 1981.

46 Joan Priest, *The Thiess Story*, Boolarong Publications, Ascot, 1981, p. 234.

47 "Sir Gordon Jackson": forthcoming article in *Australian Dictionary of Biography.*

48 John O'Hara, "Coal: NSW tortoise trails Queensland hare", *SMH*, 11 March 1981.

49 Department of the Prime Minister and Cabinet, "Coal Ports in New South Wales", 25 May 1982, NAA: A1209, 1987/951 part 1.

50 Jetse Kalma and Peter Laut, "The Hunter Region" in Robert Birrell, Doug Hill and John Stanley (eds), *Quarry Australia: Social and Environmental Perspectives on Managing the Nation's Resources*, Oxford University, Melbourne, 1982, p. 149.

51 Joanne Finley, "Coal Starts to Dominate Newcastle", *SMH*, 18 July 1978.

52 Errol Simper, "These Days it's not much use carrying coals to Newcastle", *Australian*, 15-16 December 1984.

53 Malcolm Knox, *Boom: The Underground History of Australia from Gold Rush to GFC*, Viking, Melbourne, 2013, pp. 269-70.

54 Ibid., p. 270.

55 Peter Robinson, "Coalminers increase pressure on Wran", *National Times*, 2 September 1978.

56 Milton Cockburn, "How NSW stumbled in the coal rush", *SMH*, 1 June 1982.

57 Larcombe, op. cit., pp. 108-9.

58 Department of the Prime Minister and Cabinet, Memorandum on Coal Ports in New South Wales, 25 May 1982, NAA: A1209, 1987/951 part 1.

59 Roger Stuart, "Resources Development Policy: The Case of Queensland Export Coal Industry" in Allan Patience (ed.), *The Bjelke-Petersen Premiership 1968–1983: Issues in Public Policy*, Longman Cheshire, Melbourne, 1985, p. 71.

60 "Slump dampens zeal for new coal projects", *SMH*, 1 December 1982.

61 Ibid.

62 Speech by Prime Minister Malcolm Fraser to the Institute of Australian Aluminium Producers, 14 April 1977, NAA: A1209, 1977/528 part 1.

63 Geoffrey Blainey, *White Gold: The Story of Alcoa of Australia*, Allen & Unwin, St Leonards, 1997, p. 191.

64 Ibid., pp. 143-4.

65 Susan Bambrick, *Australian Minerals and Energy Policy*, Australian National University Press, Canberra, 1979, p. 116.
66 Donald Lipscombe, "The World's Biggest Refinery to Rise in WA", *AFR*, 10 May 1977.
67 Arvi Parbo, *Down Under: Mineral Heritage in Australasia*, Australasian Institute of Mining and Metallurgy, Box Hill, 1992, p. 187.
68 McQueen, op. cit., p. 49.
69 Blainey, op. cit., p. 190.
70 John Byrne, "Comalco's Smelter on the Way", *AFR*, 21 April 1978.
71 McQueen, op. cit., p. 51.
72 Telex from Brian Loton to Sir Geoffrey Yeend, Secretary, Prime Minister's Department, 27 November 1979, NAA: A1209, 1977/528 part 3.
73 Kalma and Laut, op. cit., p. 153.
74 Larcombe, op. cit., p. 106.
75 "Smelter Viable without Alumax – Wran", *SMH*, 16 April 1981.
76 Letter from J.B. Connolly to Secretary, Department of the Prime Minister and Cabinet, 15 April 1980, NAA: A1209, 1977/528 part 3.
77 "BHP backs aluminium for 80's profit bonanza", *Australian*, 3 July 1980.
78 "Alcan Clinches $600m deal for Kurri Kurri", *SMH*, 10 July 1980. 6
79 J.M. Pierce, "Gove secures $600 million Tomago Deal", *SMH*, 23 January 1980.
80 Blainey, op. cit., p. 191.
81 Tim Colebatch, *Dick Hamer: The Liberal Premier*, Scribe, Brunswick, 2014, pp. 384–7.
82 Brian Galligan, "Victoria: the Political Economy of a Liberal State", in Brian Head (ed.), The *Politics of Development in Australia*, Allen & Unwin, North Sydney, 1986, p. 123.
83 Ibid., p. 124. Whereas Victoria had one-third of Australia's total manufacturing, it had 55 per cent of textile, clothing and footwear and 44 per cent of automobile manufacturing.
84 Colebatch, op. cit., p. 386.
85 Galligan, op. cit., p. 129.
86 Larcombe, op. cit., p. 111.
87 Crowley, op. cit., pp. 47–8.
88 Alice Cawte, *Atomic Australia 1944–1990*, University of New South Wales Press, Kensington, 1992, p. 153.
89 Quoted in Cawte, op. cit., p. 157.
90 Ibid., p. 159
91 Ibid., p. 160.
92 Ibid., p. 161.
93 Ibid., p. 162.
94 Ibid., pp. 163-6.
95 Pinkstone, op. cit., p. 216.
96 Knox, op. cit., p. 290.
97 Susan Bambrick, *Australian Minerals and Energy Policy*, Australian National University Press, Canberra, 1979, p. 115.
98 Ibid., pp. 124-5.
99 John Byrne, "Aust's iron ore miners wait for better times", *AFR*, 9 July 1979.
100 Jeffrey D. Wilson, *Governing Global Production: Resource Networks in the Asia Pacific Steel Industry*, Palgrave Macmillan, Basingstoke, 2013, pp. 105–6.
101 Pinkstone, op. cit., p. 214.

102 John Byrne, "Japan exploits Canberra's Development Mania", *AFR*, 30 August 1978.
103 Max Suich, *National Times*, 15-20 November 1976.
104 N.F. Dufty, *Industrial Relations in the Pilbara Iron Ore Industry*, Western Australian Institute of Technology, Perth, 1984, pp. 246-7.
105 Ibid., p. 257.
106 Peter Maher, "Iron Ore Industry Weathering Worst Setback since the 60's", *AFR*, 8 October 1979.
107 Letter from Court to Fraser, 25 September 1978, NAA: MA1209, 1978/745.
108 "Divided and Conquered – Again", *AFR*, 25 August 1978.
109 "Hamersley's record profit", *AFR*, 25 January 1978.
110 McQueen, op. cit., p. 82.
111 John Byrne, "Iron ore development struggle hots up", *AFR*, 25 June 1976.
112 John Byrne, "Goldsworthy coup may be heavy blow to Newman", *AFR*, 7 April 1977.
113 "Japan sets tough terms for $4.5m Iron Contract", *Financial Times*, 17 September 1976
114 Ibid.
115 John Byrne, "Robe River's $70m expansion plan", *AFR*, 14 May 1976.
116 John Byrne, "Japanese steel mills play off Aust mines in Iron Ore Race", *AFR*, 3 August 1976.
117 John Byrne, "Japan Exploits Canberra's Development Mania", *AFR*, 30 August 1978.
118 John Byrne, "Hamersley's $375m expansion", *AFR*, 7 January 1977.
119 John Byrne, 'Goldsworthy coup may be heavy blow to Newman', *AFR*, 7 April 1977.
120 'Mining still the future', *National Times*, 23-28 January 1978.
121 John McIlwraith, "Hancock's $22m Robe River deal", *AFR*, 11 February 1976.
122 John McIlwraith, "Hancock's $22m Robe River deal", *AFR*, 11 February 1976
123 Tery Ogg, "$22m Robe Sale", *AFR*, 1 April 1976.
124 John Byrne, "Robe River's $70m expansion plan", *AFR*, 14 May 1976.
125 John McIlwraith, *Mesa Harvest: Robe River's First 25 Years, Robe River Iron Associates*, Wickham, 1997, p. 20.
126 Michael Byrnes, "Go-ahead for Japanese equity in Robe River", *AFR*, 21 June 1977.
127 Michael Byrnes, "Japan's Iron Ore bombshell', *AFR*, 5 July 1978.
128 Max Suich and Bob Mills, "In the Pilbara a sudden scent of mothballs", *National Times*, 22 July 1978.
129 Wilson, op. cit., pp. 96–7.
130 "Divided and Conquered – Again", *AFR*, 25 August 1978.
131 Telex from Court to Fraser, 4 August 1978 and minute from K. Heydon to Fraser, 14 August 1978, NAA: A1209, 1978/745.
132 Letter from Anthony to Fraser, 21 September 1978, NAA: A1209, 1978/745.
133 Ibid.
134 Letter from Anthony to Fraser, 10 August 1978, NAA: A1209, 1978/745.
135 Letter from Peter Field to K. Heydon, Assistant Secretary, Trade and Industries Division, Department of the Prime Minister and Cabinet, 17 October 1978, NAA: A1209, 1978/745.
136 Letter from Court to Anthony, 20 September 1978, NAA: A1209, 1978/745.
137 Letter from Court to Fraser, 25 September 1978, and telex from Court to Fraser, 1 October 1978, NAA: A1209, 1978/745.
138 Stuart Simson and Andrew Clark, "Export Controls – Why Anthony Bucked", *National Times*, 4 November 1978.

[139] Telex from Court to Fraser, 25 October 1978,NAA: A1209, 1978/745.
[140] Wilson, op. cit., pp. 96-7.
[141] "New Iron Ore Projects – Discussion Paper", n.d. 1981, 1982/042, State Records Office Western Australia.
[142] McQueen, op. cit., p. 83.
[143] Graeme Atherton and Rick Wilkinson, *Beyond the Flame*: The Story of Australia's North West Shelf Natural Gas Project, Woodside Offshore Petroleum, Perth, 1989, p. 24.
[144] Ibid., p. 28.
[145] Letter from Rex Connor to R.T. Madigan, Chairman, Interstate Oil Limited, 4 October 1973; Meeting between Connor, Hewitt and representatives of Woodside Burmah, 18 October 1973, NAA: A1690, 1973/2299.
[146] Atherton and Wilkinson, op. cit., p. 32
[147] Ibid., p.33.
[148] Ronda Jamieson, *Charles Court: I Love this Place*, St George Books, Osborne Park, 2011, pp. 318-19.
[149] Atherton and Wilkinson, op. cit., pp. 38-9.
[150] Jamieson, op cit., p. 327.
[151] *Technology in Australia 1788–1988*, www.austehc.unimelb.edu.au/tia/430.htm
[152] Pinkstone, op. cit., p. 256.
[153] Statement by the Prime Minister, "Loan Council Financing of Development Projects", 6 November 1978, NAA: M1263 727.
[154] *Business Week*, 2 June 1980 quoted in Crowley, op. cit., pp. 413-14.
[155] John Welfield, "Australia's Relations with Japan and the Korean Peninsula" in P.J. Boyce and J.R. Angell (eds), *Diplomacy in the Marketplace: Australia in World Affairs 1981-90*, Longman Cheshire, Melbourne, 1992, p. 254.
[156] George Megalogenis, The *Australian Moment: How We Were Made for These Times*, Viking, Melbourne, 2012, p. 134.

Chapter 8

[1] David Humphreys, *The Remaking of the Mining Industry*, Palgrave Macmillan, Basingstoke, 2015, pp. 22-3.
[2] Ibid., pp. 14-15.
[3] Brian Pinkstone, *Global Connections: A History of Exports and the Australian Economy*, Australian Government Publishing Service, Canberra, 1992, p. 388.
[4] Geoffrey Blainey, *The Rush that Never Ended: A History of Australian Mining*, Melbourne University Press, Carlton, 2003, p. 381.
[5] Stephen G. Bunker and Paul S. Ciccantell, *East Asia and the Global Economy: Japan's Ascent with Implications for China's Future*, The Johns Hopkins University Press, Baltimore, 2007, p. 106.
[6] Michael Byrnes, "The Coal Conspiracy", *National Times*, 28 November to 4 December 1982.
[7] Jeffrey D. Wilson, *Governing Global Production: Resource Networks in the Asia Pacific Steel Industry*, Palgrave Macmillan, Basingstoke, 2013, p. 114.
[8] Simon Halberton, "Iron Ore and Coal Miss the 'Year of Recovery'", *Age*, 26 December 1983.
[9] Alan Goodall, "Coal miners facing losses after Tokyo slashes prices", *Australian*, 4 April 1983. "Coal orders lost and so are 184 mine jobs", *Sydney Morning Herald* (hereafter

SMH) 17 March 1983.

10 John Edwards, *Keating: The Inside Story*, Penguin Books, Ringwood, 1996, p. 145.

11 The author is persuaded by the arguments of Ian McLean, *Why Australia Prospered: The Shifting Sources of Economic Growth*, Princeton University Press, Princeton, 2013, p. 222–3. For an alternative view see *Paul Kelly, The End of Certainty: The Story of the 1980s*, Allen & Unwin, St Leonards, 1992.

12 Colin Brammall, "Concern on Cuts on Coal Prices", *Canberra Times*, 30 March 1983.

13 Ibid.

14 Bunker and Ciccantell, op. cit., p. 106.

15 Robert MacDonald, "The heat is rising aboard Queensland coal trains", *Australian Financial Review* (hereafter *AFR*), 23 May 1984.

16 Australian Coal Association, "Export of Coal from Australia", Submission to John Dawkins, 22 February 1985, NAA: A1209, 1987/951 part 5.

17 Pinkstone, op. cit., p. 258.

18 Ibid., p. 258.

19 "Make or break Year for Crucial Sector of the nation's economy", *Australian*, 31 January 1984.

20 Robert MacDonald, "Walsh discounts coal industry protection", *AFR*, 5 April 1984.

21 Mike Taylor, "Australia leads as coal shipper", *AFR*, 28 June 1984. Although Australia exported 65 million tonnes in 1983-84 it was still behind the United States which exported 77 million tonnes, including 19 million tonnes by road to Canada. Wray Vamplew (ed.), *Australians: Historical Statistics*, Fairfax, Syme and Weldson Associates, Sydney, 1987, p, 87.

22 Mike Taylor, "Australia's coal status at US expense", *AFR* 16 October 1984.

23 John Welfield, "Australia's Relations with Japan and the Korean Peninsula" in P.J. Boyce and J.R. Angel (eds), *Diplomacy in the Marketplace: Australia in World Affairs 1981-90*, Longman Cheshire, Melbourne, 1992, p. 255.

24 Informal meeting at Japanese Embassy between Ambassador Yanagiya, Dr Ross Garnaut and David Buckingham, 19 September 1984, NAA A1209, 1987/951 part 1.

25 Letter from Lionel Bowen to Bob Hawke, 12 September 1984, NAA A1209, 1987/951 part 1.

26 Welfield, op. cit., p. 255.

27 John Welfield, "Australia's Relations with Japan and the Korean Peninsula" in P.J. Boyce and J.R. Angell (eds), *Diplomacy in the Marketplace: Australia in World Affairs 1981–90*, Longman Cheshire, Melbourne, 1992, pp. 255–56.

28 Peter Thompson and Robert Macklin, *The Big Fella: The Rise and Rise of BHP Billiton*, William Heinemann Australia, Sydney, 2009, p. 153.

29 Ibid.

30 Utah Development Company, News Release, 21 July 1986, NAA: A1209, 1987/951 part 5.

31 John Turner, *200 Years Of Transport in the Hunter: A History of Transport's Contribution to the Hunter*, Charters Institute of Transport and Logistics Australia, Charlestown, 2005, p. 81.

32 Welfield, op. cit., p. 257.

33 Ibid.

34 Edwards, op. cit., pp. 296-7.

35 Jade Davenport, *Digging Deep: A History of Mining in South Africa*, Jonathan Ball Publishers, Jeppestown, 2013, p. 387.

36 Ibid.

37 Welfield, op. cit., pp. 261-2.

38 Ibid., pp. 262–3.

39 Wilson, op. cit., p. 117.
40 Ibid., pp. 117-19.
41 Bart Lucarelli, *Australia's Black Coal Industry: Past Achievements and Future Challenges*, PESD Stanford, Working Paper, March 2011.
42 Ibid.
43 Rikki Kersten, "Australia and Japan", in James Cotton and John Ravenhill (eds), *The National Interest in a Global Era: Australia in World Affairs 1996–2000*, Oxford University Press, South Melbourne, 2002, p.83.
44 Thomas W. Zeiler, "Opening Doors in the World Economy" in Akira Iriye (ed.), *Global Interdependence*, Harvard University Press, Cambridge Mass, 2013, pp. 318–19.
45 Robert C. Allen, *Global Economic History: A Very Short Introduction*, Oxford University Press, New York, 2011, p. 139.
46 Wilson, op. cit., pp. 121-2.
47 Pinkstone, op. cit., p. 259.
48 Productivity Commission, *The Australian Black Coal Industry: Inquiry Report: Volume 2: Appendices, Report No. 1*, 3 July 1998.
49 RBA Historical Statistics.
50 Wilson, op. cit., p. 124.
51 John Turner, *200 Years Of Transport in the Hunter: A History of Transport's Contribution to the Hunter*, Charters Institute of Transport and Logistics Australia, Charlestown, 2005, p. 81.
52 Ibid., p. 83.
53 Ibid., p.84.
54 John McIlwraith, *The First 500 Million: The Mt Newman Story*, Iron Ore BHP-Utah Minerals International, Perth, 1988, p. 92.
55 Booth Connell-Hatch, op. cit., p. 28.
56 Pinkstone, op. cit., p. 214.
57 Humphreys, op. cit., p. 18.
58 Pinkstone, op. cit., p. 260.
59 Ibid., p. 260.
60 McIlwraith, op. cit., p. 91.
61 Paul Cleary, *Mine-Field*, Black Inc, Collingwood, 2012, p. 145.
62 Malcolm Knox, *Boom: The Underground History of Australia from Gold Rush to GFC*, Viking, Melbourne, 2013, pp. 326-8.
63 Ibid., p. 329.
64 Erik Eklund, *Mining Towns: Making a Living, Making a Life*, University of New South Wales Press, Sydney, 2012, p. 237.
65 McIlwraith op. cit., p. 91.
66 David Lee, *Iron Country: Unlocking the Pilbara*, Minerals Council of Australia, Canberra, 2015, pp. 69-70; McIlwraith, op. cit., p. 91
67 Pinkstone, op. cit., p. 260.
68 Ibid., p. 260.
69 M.A.B. Siddique, "Western Australia–Japan Mining Co-operation: An Historical Overview", University of Western Australia Business School, 2009.
70 John McIlwraith, "Miners eagerly wait while Japan ponders iron ore prospects", *AFR*, 9 June 1977.
71 McIlwraith, op. cit., p. 92.
72 Booth, Connell-Hatch, op. cit. p. 113.
73 Bradon Ellem, "Robe River Revisited: Geohistory and Industrial Relations", *Labour*

History, 109, November 2015, pp. 111–31.

74 Frank Bongiorno, *The Eighties: The Decade that Transformed Australia*, Black Inc, Collinwood, 2015, pp. 173–5.

75 Ibid., p. 174

76 Ibid.

77 Ibid., p. 175.

78 Lee, op. cit., p. 74.

79 Stephen G. Bunker and Paul Cicantell, *East Asia and the Global Economy: Japan's Ascent with Implications for China's Future*, The Johns Hopkins University Press, Baltimore, 2007, p. 153.

80 Lee, op. cit., p.p. 74-5.

81 Knox, op. cit., p. 266.

82 Ibid.

83 John McIlwraith, *Mesa Harvest: Robe River's First 25 Years*, Robe River Iron Ore Associates, Wickham, 1997, p. 20.

84 Ibid., p. 20.

85 Terry Reynolds and Virginia P. Dawson, *Iron Will: Cleveland Cliffs and the Mining of Iron Ore, 1847-2006*, Wayne State University Press, Detroit, 2011, p. 229.

86 John McIlwraith, *Mesa Harvest: Robe River's First 25 Years*, Robe River Iron Ore Associates, Wickham, 1997, p. 21; For the context see Gideon Haigh, *The Battle for BHP*, Allen & Unwin, North Sydney, 1987 and Bongiorno, op. cit., pp. 1336.

87 Humphreys, op. cit., p. 32.

88 Thompson and Macklin, op. cit., p. 344.

89 Davenport, op. cit., pp. 384–5.

90 James Goodman, "Australia and Beyond: Targeting Rio Tinto", in Ronaldo Munck (ed.), *Labour and Globalisation*, Liverpool University Press, Liverpool, 2004, p. 112,

91 Lisa Warell, "A horizontal merger in the iron ore industry: An event study approach", *Resources Policy*, vol. 32, 2007, pp.191–204.

92 P.T. Breaden and J.B. Parker, "Financing the Channar Project", the AusIMM Annual Conference, Perth–Kalgoorlie, May 1989; Rowan Callick, *Channar: A Landmark Venture in Australian Iron Ore: Australia and China Forge a Bold Partnership*, HardieGrant, Richmond, 2012. 2012.

93 Geoffrey Blainey, *White Gold: The Story of Alcoa of Australia*, Allen & Unwin, St Leonards, 1997, 207–11.

94 Blainey, op. cit., p. 213.

95 Parbo, op. cit., p. 187.

96 Arvi Parbo, *Down Under: Mineral Heritage in Australasia*, Australasian Institute of Mining and Metallurgy, Box Hill, 1992, p. 163-4.

97 Humphreys, op. cit., p. 26.

98 Davenport, op. cit., pp. 344-45.

99 Humphreys, op. cit., pp. 26-7.

100 Ibid., p. 27.

101 Pinkstone, op. cit., p. 261.

102 Ibid., pp. 261-2.

103 Albert Smith, "Western Mining looms up as CRA counts albatrosses", *AFR*, 3 September 1987.

104 Davenport, op. cit., p. 345,

105 Letter from Parbo to John Horgan, 28 July 1986, Arvi Parbo Papers, Acc09/45, Box 20, NLA.

[106] Arvi Parbo, Address to Liberal Party Business Dinner, Bendigo, 25 March 1987, Arvi Parbo Papers, Acc09/45, Box 20, NLA.
[107] Pinkstone, op. cit., p. 262.
[108] Humphreys, op. cit., p. 28.
[109] David Upton, *The Olympic Dam Story: How Western Mining Defied the Odds to Discover and Develop the World's Largest Mineral Deposit*, Upton Financial PR, Armadale, 2010, p. 25.
[110] Ibid., p. 127.
[111] Ibid., p. 126.
[112] Ibid., p. 136.
[113] Ibid., pp. 137-8.
[114] Ibid., p. 140.
[115] Ibid., pp. 167-68.
[116] Ibid., pp. 168-69.
[117] Bongiorno, op. cit., pp. 91-3.
[118] Address by Arvi Parbo, "A look back on business experiences", 12 November 1986, Parbo Papers, Acc 09.45, Box 20, NLA.
[119] Ronald T. Libby, *Hawke's Law: The Politics of Mining and Aboriginal Land Rights in Australia*, University of Western Australia Press, 1989, p. 57.
[120] A.H. Parbo, Address to Hailebury College. Melbourne, 21 July 1986, Parbo Papers, Acc09/45 Box 20, NLA.
[121] Letter from Parbo to John Nightingale, Senior Lecturer in Economics, University of New England, 29 July 1986, Parbo Papers, Acc09/45 Box 20, NLA.
[122] Libby, op, cit., p. xxiii.
[123] Ibid., p. 59.
[124] Quoted in ibid., p. 60.
[125] Ibid.
[126] Chamber of Mines of Western Australia advertisement quoted in ibid. p. 71.
[127] Christine Jennett and Randal G. Stewart (eds), *Hawke and Australian Public Policy: Consensus and Restructuring*, Macmillan, South Melbourne, 1990, p. 257.
[128] David Day, *Keating*, HarperCollins, Sydney, 2015, p. 400.
[129] Ibid.
[130] Ibid., pp. 400-4.
[131] Leon Davis, Speech to the Australian Institute of Company Directors, Melbourne 3 October 1995.
[132] Paul Kauffman, *Wik, Mining and Aborigines*, Allen & Unwin, St Leonards, 1998, p. 102.
[133] Benedict Scambary, *My Country, Mine Country: Indigenous People, Mining and Development Contestation in Remote Australia*, Australian National University E press, Canberra, 2013, p. 141; Marcia Langton, *From Conflict to Cooperation: Transformations and Challenges in the Engagement between the Minerals Industry and Australian Indigenous Peoples*, Minerals Council of Australia, Canberra, 2015.
[134] Kauffman, op. cit., p. 156.
[135] Bongiorno, op. cit., pp. 294–7.
[136] Ibid., pp. 296–7.
[137] Peter Brain and John Stanley, "The Economics of the Resources Boom", in Robert Birrell and John Stanley (eds), *Quarry Australia? Social and Environmental Perspectives on Managing the Nation's Resources*, Oxford University Press, Melbourne, 1982, pp. 107–8.

Chapter 9

1 John Edwards, *Beyond the Boom*, Penguin, Melbourne, 2014, p. 34.
2 Ibid., pp. 29–33.
3 Some economists argue that, despite the Global Financial Crisis, there has been a super-cycle "driven by the explosive emergence of the Chinese and Indian economies and by complex new trading relationships". John L. Brooke, *Climate Change and the Course of Global History: A Rough History*, Cambridge University, Press, Cambridge, 2014, pp. 500-1.
4 Ian W. McLean, *Why Australia Prospered: The Shifting Sources of Economic Growth,* Princeton University Press, Princeton, 2013, p. 230.
5 David Humphreys, *The Remaking of the Mining Industry*, Palgrave Macmillan, Basingstoke, 2015, p. 38.
6 Edwards, op. cit., pp. 1-3.
7 Chi-Kwan Mark, *China and the World Since 1945: An International History*, Routledge, London and New York, 2012, pp. 96-7.
8 Ibid., p. 98.
9 Thomas W. Zeiler, "Opening Doors in the World Economy" in Akira Iriye (ed.), *Global Interdependence*, Harvard University Press, Cambridge Mass., 2014, p. 320.
10 Odd Arne Westad, *Restless Empire: China and the World Since 1750*, The Bodley Head, London, 2012, p. 385.
11 David Uren, *The Kingdom and the Quarry*, Black Inc, Collingwood, 2012, p. 50.
12 Humphreys, op. cit., p. 38.
13 Department of Foreign Affairs and Trade, Australia's Direction of merchandise trade.
14 Ibid., p. 46.
15 Mark, op. cit., p. 123; Westad, op. cit., p. 385.
16 Zeiler, op. cit., p. 323.
17 Humphreys, op. cit., p. 42.
18 Brooke, op. cit., p. 541.
19 Mark, op. cit., p. 124.
20 Elizabeth C. Economy and Michael Levi, *By All Means Necessary: How China's Resource Quest is Changing the World*, Oxford University Press, New York, 2014, p. 22.
21 Quoted in ibid., p. 23.
22 Humphreys, op. cit., pp. 47-8.
23 Ibid., p. 48.
24 Ibid., pp. 44-5.
25 Ibid., p. 46.
26 Uren, op. cit., p. 51.
27 Economy and Levi, op. cit., 2014, p. 24.
28 Humphreys, op. cit., p. 63.
29 Ibid., p. 64.
30 Economy and Levi, op. cit., p. 118.
31 Edwards, op. cit., p. 73.
32 Department of Foreign Affairs and Trade. Australia's Direction of Merchandise Trade.
33 Humphreys, op. cit., p. 69.
34 Economy and Levi, op. cit., p. 118.
35 Department of Foreign Affairs and Trade, "Australia's Liquefied Natural Gas (LNG) exports 2003-4 to 2013-14 and beyond", Department of Foreign Affairs and Trade, February 2015.

36 Uren, op. cit., p. 50.
37 Humphreys, op. cit., pp. 160-1.
38 Paul Cleary, *Too Much Luck: The Mining Boom and Australia's Future*, Black Inc, Collingwood, 2011, p. 5.
39 Ibid.
40 McLean, op. cit., p. 232.
41 Edwards, op. cit., pp. 73, 16.
42 Ibid., p. 43.
43 Ibid., p. 45.
44 Ibid., pp. 44-5.
45 Ibid., p. 44.
46 Ibid., p. 49.
47 Geoffrey Blainey, *The Rush that Never Ended: A History of Australian Mining*, Melbourne University Press, Carlton, 2003, p. 380.
48 Mimi Sheller, *Aluminum Dreams: The Making of Light Modernity*, Massachusetts Institute of Technology Press, Cambridge, 2014, p. 29.
49 Peter Thompson and Robert Macklin, *The Big Fella: The Rise and Rise of BHP Billiton*, William Heinemann Australia, Sydney, 2009, p. 313.
50 Ibid., p. 298.
51 Ibid., p. 334; Humphreys, op. cit, pp. 125-8.
52 Thompson and Macklin, op. cit., pp. 408-9.
53 David Upton, *The Olympic Dam Story: How Western Mining Defied the Odds to Discover and Develop the World's Largest Mineral Deposit*, Upton Financial PR, Armadale, 2010, p. 171.
54 Wilson, op. cit., p. 135.
55 Ibid.
56 Ibid., p. 141.
57 Macklin and Thompson, op. cit., p. 423.
58 Ibid., p. 424.
59 Ibid., p. 425.
60 Ibid., p. 426.
61 Wilson, op. cit., p.141.
62 Ibid., p. 142.
63 Humphreys, op. cit., p. 81.
64 Economy and Levy, op. cit., p. 38.
65 Ibid., p. 38.
66 Wilson, op. cit., p. 147.
67 Ibid., p. 166.
68 Ibid., p. 156
69 Uren, op. cit., p. 158.
70 Uren, op. cit., pp. 158–9.
71 Ibid., p. 158; Wilson, op. cit., pp. 166-7.
72 Uren, op. cit., pp. 159-60.
73 Uren, op. cit. pp. 164-70; Wilson, op. cit., pp. 168-9.
74 Humphreys, op. cit., pp. 110-3.
75 Ibid., p. 161.
76 Economy and Levi, op. cit., p. 40.
77 Wilson, op. cit., pp. 169-70.
78 Humphreys, op. cit., pp. 216-17.
79 Andrew Burrell, *Twiggy: The High-Stakes Life of Andrew Forrest*, Black Inc, Collingwood,

2013, pp. 6-7.
80 Uren, op. cit., p. 60.
81 Nicola Garvey, *A Sense of Purpose: Fortescue's 10-year Journey 2003-2013*, Fortescue Metals Group, Perth, 2013, p. 120.
82 Burrell, op. cit., p. 119.
83 Ibid., p. 113.
84 Garvey, op. cit., p. 134.
85 Ibid., p. 134.
86 Uren, op. cit., p. 60
87 Ibid.
88 Garvey, op. cit., p. 143.
89 Burrell, op. cit., pp. 114–15; Garvey, op. cit., p. 150.
90 Burrell, op. cit., p. 116.
91 Uren, op. cit., pp. 60–3.
92 Garvey, op. cit., p. 149;
93 Richard A. Posner, *A Failure of Capitalism: The Crisis of '08 and the Descent into Depression*, Harvard University Press, Cambridge MA, 2009.
94 Thompson and Macklin, op. cit., p. 346.
95 Uren, op. cit., p. 67.
96 Ibid., p. 68.
97 Ibid.,p. 68.
98 Quoted in Thompson and Macklin, op. cit., p. 435.
99 Humphreys, op. cit., p. 121.
100 Uren, op. cit., pp. 65–6.
101 Uren, op. cit., p. 70.
102 Thompson and Macklin, op. cit., p. 437.
103 Quoted in Uren, op. cit., p. 73.
104 Economy and Levi, op. cit., p. 123.
105 Quoted in Uren, op. cit., pp. 95–6.
106 Quoted in Economy and Levi, op. cit., pp. 123–4.
107 Quoted in Uren, op. cit., p. 98.
108 Ibid., p. 106.
109 Humphreys, op. cit., p. 53.
110 Ibid., p. 54.
111 Humphreys, op. cit., p. 55–6.
112 Ibid., p. 52.
113 Ibid.
114 Paul Kelly, *Triumph and Demise: The Broken Promise of a Labor Generation*, Melbourne University Publishing, Carlton, 2014, p. 297; David Lee, *Iron Country: Unlocking the Pilbara*, Minerals Council of Australia, Canberra, 2015, pp. 88–9.
115 Pearse, Guy, David McKnight and Bob Burton, *Big Coal: Australia's Dirtiest Habit*, University of New South Wales Press, Sydney, 2013, p. 30.
116 Pearse, McKnight and Burton, op. cit., p. 29.
117 Ibid., p. 5.
118 Paul Cleary, *Mine-Field: The Dark Side of Australia's Resources Rush*, Black Inc, Collinwood, 2012, p. 57.
119 Pearse, McKnight and Burton, op. cit., p. 36.
120 Ibid., p. 30.
121 Sinclair Davidson and Ashton de Silva, *The Australian Coal Industry – Adding Value to*

the Australian Economy, Paper commissioned by the Australian Coal Association, April 2013.

122 Pearse, McKnight and Burton, op. cit., p. 61.

123 Ibid., p. 37.

124 Jeff Goodell, *Big Coal: The Dirty Secret Behind America's Energy Future*, Houghton Mifflin, Boston, 2007, pp. 186–7.

125 Ibid., p. 181.

126 Jade Davenport, *Digging Deep: A History of Mining in South Africa*, Jonathan Ball Publishers, Jeppestown, 2013, pp. 384-5.

127 Thompson and Macklin, op. cit., p. 407.

128 Ibid., p. 407.

129 Ibid., p. 408. Humphreys, op. cit., p. 114.

130 BBC News, "Xstrata to buy Australian rival", 7 April 2003.

131 Paul Cleary, *Mine-Field*, Black Inc, Collingwood, 2012, p. 58; Pearse, McKnight and Burton, op. cit., p. 37.

132 Daniel Ammann, *The Secret Lives of Marc Rich: The King of Oil*, St Martin's Press, New York, 2009.

133 Pearse, McKnight and Burton, op. cit., p. 76.

134 Brooke, op. cit., pp. 560-4.

135 Goodell, op. cit., p. 177.

136 Yergin, op. cit., pp. 422-3.

137 Brooke, op. cit., p. 575.

138 Ibid., 462-3.

139 Yergin, op. cit., p. 464.

140 Barbara Freese, *Coal: A Human History*, William Heinemann, London, 2003, pp. 185-6.

141 Goodell, op. cit., p. 180.

142 Brooke, op. cit., p. 576.

143 Ibid., p. 184.

144 Pearse, McKnight and Burton, op cit., p. 4.

145 Ibid., p. 137.

146 Yergin, op. cit., p. 396.

147 Pearse, McKnight and Burton, op. cit., p. 4.

148 Ibid., p. 201.

149 Yergin, op. cit., p. 489.

150 Pearse, McKnight and Burton, op. cit., p. 6.

151 Goodell, op. cit., p. xv.

152 Pearse, McKnight and Burton, op. cit. Chapter 6.

153 Yergin, op. cit., p. 402.

154 Ibid., p. 139.

155 Brooke, op. cit., p. 579.

156 Pearse, McKnight and Burton, op. cit., p. 204.

157 Edwards, op. cit., p. 1.

158 Ibid., p. 2.

159 Ibid.

Bibliography

Archival Sources

National Archives of Australia

Cabinet Files, 'C' Series,
Menzies and Holt Ministries, A4940
Department of Trade, A1313
Department of the Treasury, A571
Department of External Affairs, A1838
Department of National Development,
Minerals and Energy, National Development [II],
Resources and Energy, A695, A1690
Department of Trade and Customs, A425
Department of the Prime Minister,
Department of the Prime Minister and Cabinet, A461, A462, A463, A1209
Papers of Sir John McEwen, 1915-1982, M58
Folders of Papers maintained by Robert Gordon Menzies as Prime Minister, M2576
Correspondence maintained by Harold Holt as Treasurer, 1950-1966, M2568
Correspondence files maintained by Harold Holt as Prime Minister, 1929-1968, M2684

Noel Butlin Archives Centre, Australian National University

New South Wales Combined Colliery Proprietors, Z224

State Records Office of Western Australia

Correspondence files of Department of Industrial Development

National Library of Australia

Papers of Wal Fife, 1942-2014, MS7626
Papers of Sir John Gorton, 1959-2003, MS7984
Papers of Sir Gordon Jackson, 1943-1990 MS8353
Papers of William McMahon, 1949-1987, MS3926
Papers of Billy Snedden, 1955-1987, MS7389
Papers of Sir Arvi Parbo, 1949-2008, MSAcc09.045
Papers of Sir James Vernon, 1936-2000, MSAcc01/08
Papers of Sir Frederick Wheeler, 191-1994, MS8096

University of Melbourne Archives

Papers of Sir Maurice Mawby, 1904-1977

University of Wollongong, University of Wollongong Library
Papers of Reginald (Rex) Francis Xavier Connor, 1955-1957

Newspapers and Journals
Australian Business
Business Week
Daily Telegraph
Financial Times
Nation
The *Age*
The *Australian*
The Canberra Times
The *Courier Mail*
The *Australian Financial Review*
The *Economist*
The Morning Bulletin (Rockhampton)
The *National Times*
The *Sun-Herald*
The *Sydney Morning Herald*

Secondary Sources

Books

Allen, Robert C., *Global Economic History: A Very Short Introduction*, Oxford University Press, New York, 2011.

Alcan Australia Limited, *40 Years on: A History of Alcan in Australia*, Alcan, Granville, 1981.

Altman, J.C., *Aborigines and Mining Royalties in the Northern Territory*, Australian Institute of Aboriginal Studies, Canberra, 1983.

Ammann, Daniel, *The King of Oil: The Secret Lives of Mark Rich*, St Martin's Press, New York, 2009.

Appleyeard, R.T. and Schedvin, C.B., *Australian Financiers: Biographical Essays*, Macmillan, South Melbourne, 1988.

Ashton, S.R., Bridge, Carl and Ward, Stuart (eds), *Documents on Australian Foreign Policy: Australia and the United Kingdom 1960–1975*, Department of Foreign Affairs and Trade, Canberra, 2010.

Atherton, Graeme and Wilkinson, Rick, *Beyond the Flame: The Story of Australia's North West Shelf Natural Gas Project*, Woodside Offshore Petroleum, Perth, 1989.

Attwood, Bain, *Rights for Aborigines*, Allen & Unwin, Crows Nest, 2003.

Avery, David, *Not on Queen Victoria's Birthday: The Story of the Rio Tinto Mines*, Collins, London, 1974.

Bambrick, Susan, *Australian Minerals and Energy Policy*, Australian National University Press, Canberra, 1979.

Barnett, Donald W., *Minerals and Energy in Australia*, Methuen, North Ryde, 1979.

Birrell, R, Jill, D. and Stanley, J., (eds), *Quarry Australia?* Oxford University Press, Melbourne, 1982.

Blainey, Geoffrey *The Rush that Never Ended: A History of Australian Mining*, Melbourne University Press, Carlton, 2003.

Blainey, Geoffrey, *The Steel Master: A Life of Essington Lewis*, Sun Books, Melbourne, 1981.

Blainey, Geoffrey, *Mines in the Spinifex: The Story of Mount Isa Mines*, Angus and Robertson, Sydney, 1960.

Blainey, Geoffrey, *White Gold: The Story of Alcoa of Australia*, Allen & Unwin, St Leonards, 1997.

Bolton, Geoffrey, *Land of Vision and Mirage: Western Australia Since 1826*, University of Western Australia Press, Crawley, 2008.

Bongiorno, Frank, *The Eighties: The Decade that Transformed Australia*, Black Inc, Collingwood, 2015.

Boyce, P.J. and Angel, J.R., *Diplomacy in the Marketplace: Australia in World Affairs 1981–90*, Longman Cheshire, Melbourne, 1992.

Brain, Peter and Stanley, John, "The Economics of the Resources Boom: in Birrell, Robert and Stanley, John (eds), *Quarry Australia? Social and Environmental Perspectives on Managing the Nation's Resources*, Oxford University Press, Melbourne, 1982.

British Aluminium Company, *The History of the British Aluminium Company Limited 1894–1955*, Curwen Press, London, c1956.

Brooke, John L., *Climate Change and the Course of Global History: A Rough History*, Cambridge University Press, Cambridge, 2014.

Brubaker, Sterling, *Trends in the World Aluminium Industry*, Johns Hopkins University Press for Resources for the Future, Baltimore, 1967.

Bunker, Stephen G. and Ciccantell, Paul S., *East Asia and the Global Economy: Japan's Ascent with Implications for China's Future*, The Johns Hopkins University Press, Baltimore, 2007.

Bunker, Stephen G. and Ciccantell, Paul S., *Globalization and the Race for Resources*, The Johns Hopkins University Press, Baltimore, 2005.

Burrell, Andrew, *Twiggy: The High-Stakes Life of Andrew Forrest*, Black Inc, Collingwood, 2013.

Butlin, N.G., A Barnard and Pincus, J.J., *Government and Capitalism: Public and Private Choice in Twentieth Century Australia*, George Allen and Unwin, Sydney, 1982.

Carr, Charles C., *Alcoa: An American Enterprise*, Rinehart & Co., New York, 1952.

Cawte, Alice, *Atomic Australia 1944-1990*, University of New South Wales Press,

Kensington, 1992
Clark, Lindesay, *Built on Gold: Recollections of Western Mining*, Hill of Content, Melbourne, 1983.
Cleary, Paul, *Mine-Field*, Black Inc, Collingwood, 2012.
Cleary, Paul, *Too Much Luck: The Mining Boom and Australia's Future*, Black Inc, Collingwood, 2011.
Colebatch, Tim, *Dick Hamer: The Liberal Liberal*, Scribe, Melbourne, 2014.
Comalco, *The Bluff Smelter*, Comalco, Auckland, 1971.
Cook L.H. and Porter, M.G. (eds), *The Minerals Sector and the Australian Economy*, George Allen and Unwin, North Sydney, 1984.
Crawford, J.G., *Australian Trade Policy, 1942–1966*, Australian National University Press, Canberra, 1968.
Crough, Greg and Wheelwright, Ted, *Australia: A Client State*, Penguin, Melbourne, 1982.
Crowley, F.K., *Australia's Western Third A History of Western Australia*, William Heinemann Australia, Melbourne, 1960.
Crowley, Frank, *Tough Times: Australia in the Seventies*, William Heinemann Australia, Richmond, 1986.
Davenport, Jade, *Digging Deep: A History of Mining in South Africa*, Jonathan Ball Publishers, Jeppestown, 2013.
Davidson, F.G. and Stewardson, B.R., *Economics and Australian Industry*, Second Edition, Longman Cheshire, Melbourne, 1979.
Davis, Mark, *The Land of Plenty: Australia in the 2000s*, Melbourne University Press, Carlton, 2008.
Day, David, *Keating*, HarperCollins, Sydney, 2015.
Department of Trade and Resources, *Australian Resources in a World Context*, Australian Government Publishing Service, Canberra, 1982.
Docherty, J.C., *Newcastle: The Making of an Australian City*, Hale & Iremonger, Sydney, 1983.
Donovan, P.F., *At the Other End of Australia: The Commonwealth and the Northern Territory*, University of Queensland Press, St Lucia, 1981.
Doran, D.R. ,"An Historical Perspective on Mining and Economic Change" in Cook, L.H. and Porter, M.G.(eds), *The Minerals Sector and the Australian Economy*, Allen & Unwin, Sydney, 1984, pp. 37–84.
Duffield, Robert, *Rogue Bull: The Story of Lang Hancock King of the Pilbara,* Collins, Sydney, 1979.
Dyster, Barrie and Meredith, David, *Australia in the Global Economy: Continuity and Change*, Second Edition, Cambridge University Press, Port Melbourne, 2012.
Economy, Elizabeth C. and Levi, Michael, *By All Means Necessary: How China's Resource Quest is Changing the World*, Oxford University Press, New York, 2014.
Edwards, John, *Keating: The Inside Story*, Penguin Books, Ringwood, 1996.

Edwards, John, *Beyond the Boom*, Penguin, Melbourne, 2014.

Eklund, Erik, *Mining Towns: Making a Living, Making a Life*, University of New South Wales Press, Kensington, 2012.

Eklund, Erik, *Steel Town: The Making and Breaking of Port Kembla*, Melbourne University Press, Carlton, 2002.

Ellis, M.H., *A Saga of Coal*, Angus & Robertson, Sydney, 1969.

Ferguson, Adele, *Gina Rinehart: The Untold Story of the Richest Woman in the World*, Macmillan, Sydney, 2012.

Fisher, Chris, *Coal and the State*, Methuen, Melbourne, 1987.

Fitzgerald, Anne, *Mining Agreements: Negotiated Frameworks in the Minerals Sector*, Prospects Books, Chatswood, 2010.

Fitzgerald, Ross, Megarrity, Lyndon and Symons, David, *Made in Queensland: A New History*, University of Queensland Press, St Lucia, 2009.

Fitzgerald, T.M., *The Contribution of the Mineral Industry to Australian Welfare*, Australian Government Publishing Service, Canberra, 1974.

Fleming, Grant, Merrett, David and Ville, Simon, *The Big End of Town: Big Business and Corporate Leadership in Twentieth Century Australia*, Cambridge University Press, Cambridge, 2004.

Fogarty, John and Wootton, Noel, *Breakthrough: The Establishment and Development of Queensland Alumina*, Hargreen Publishing Company in conjunction with Queensland Alumina, North Melbourne, 1992.

Freese, Barbara, *Coal: A Human History*, William Heinemann, London, 2003.

Furphy, Samuel (ed.), *The Seven Dwarfs and the Age of the Mandarins: Australian Government Administration in the Post-War Reconstruction Era*, ANU Press, Canberra, 2015.

Galligan, Brian, *Utah and Queensland Coal: A Study in the Micro Political Economy of Modern Capitalism and the State*, University of Queensland Press, St Lucia, 1989.

Galligan, Brian, "Victoria: The Political Economy of a Liberal State" in Head, Brian (ed.), *The Politics of Development in Australia*, Allen and Unwin, North Sydney, 1986.

Gao, Bai, *Japan's Economic Dilemma: The Institutional Origins of Prosperity and Stagnation*, Cambridge University Press, Cambridge, 2001.

Garnaut, Ross, *Australia and the Northeast Asian Ascendancy*, Australian Government Publishing Service, Canberra, 1989.

Garvey, Nicola, *A Sense of Purpose: Fortescue's 10-year Journey 2003–2013*, Fortescue Metals Group, Perth, 2013.

Glezer, Leon "Sir Ian Potter and his Generation" in Appleyard, R.T., and Schedvin, C.B. (eds), *Australian Financiers: Biographical Essays*, Macmillan, South Melbourne, 1988, pp. 401–27.

Golding, Peter, *Black Jack McEwen: Political Gladiator*, Melbourne University Press,

Carlton, 1996.

Goodell, Jeff, *Big Coal: The Dirty Secret Behind America's Energy Future*, Houghton Mifflin Company, Boston, 2007.

Gordon, Richard, *World Coal*, Cambridge University Press, Cambridge, 1987.

Griffiths, Max, *Of Mines and Men: Australia's 20th Century Mining Miracle 1945–1985*, Kangaroo Press, East Roseville, 1998.

Harding, R.H., *Wholeheartedly and at Once: A History of the First Operation of Mary Kathleen Ltd, 1954–1964*, CRA, Melbourne, 1994.

Harman, E.J. and Head, B.W., (eds) *State, Capital and Resources in the North and West of Australia*, University of Western Australia Press, Nedlands, 1982.

Hasegawa, H., *The Steel Industry in Japan: A Comparison with Britain*, Routledge, London and New York, 1996.

Hazlehurst, Cameron, *Gordon Chalk: A Political Life*, Darling Downs Institute Press, Toowoomba, 1987.

Head, Brian (ed.), *State and Economy in Australia*, Oxford University Press, Melbourne, 1983.

Head, Brian (ed.), *The Politics of Development in Australia*, Allen & Unwin, Sydney, 1986.

Hein, L., *Fueling Growth: The Energy Revolution and Economic Policy in Postwar Japan*, Harvard University Press, Cambridge Mass., 1990.

Hill, David, *The Gold Rush: The Fever that Changed Australia*, William Heinemann, Sydney, 2011.

Hocking, Jenny, *Gough Whitlam: His Time: The Biography Volume II*, Miegunyah Press, Carlton, 2012.

Hocking, Jenny, *Gough Whitlam: A Moment in History*, Miegunyah Press, Carlton, 2008.

Hocking, Jenny and Lewis, Colleen, *It's Time Again: Whitlam and Modern Labour*, Melbourne Publishing Group, Armadale, 2003.

Hogan, William, *The Steel Industry in China: Its Present Status and Future Potential*, Lexington Books, Lannham, Md, 1999.

Horne, Donald, *Time of Hope: Australia 1966–72*, Angus and Robertson, Sydney, 1980.

Horne, Donald, *The Next Australia*, Angus and Robertson, Sydney, 1970.

Horne, Donald, *The Lucky Country*, Penguin, Melbourne, 2009 (originally published 1964).

Hughes, Helen, *The Australian Iron and Steel Industry 1948–1962*, Melbourne University Press, Carlton, 1964.

Humphreys, David, *The Remaking of the Mining Industry*, Palgrave Macmillan, Basingstoke, 2015.

Jamieson, Ronda, *Charles Court: I Love This Place*, St George Books, Osborne Park, 2011.

Jay, Christopher, *The Coal Masters: The History of Coal & Allied 1844–1994*, Focus Publishing, Double Bay, 1994.

Jennett, Christine and Stewart, Randal G., *Hawke and Australian Public Policy: Consensus and Restructuring*, Macmillan, South Melbourne, 1990.

Johnson, Chalmers, *MITI and the Japanese Miracle: the Growth of Industrial Policy, 1972–1975*, Stanford University Press, Stanford, 1982.

Johnston-Liik, E.M., Liik, George and Ward, R.M., *A Measure of Greatness: The Origins of the Australian Iron and Steel Industry*, Melbourne University Press, Carlton, 1998.

Joyce, John and Tilley, Alan, *Railways in the Pilbara*, J & A Publications, Wembley, 1980.

Kalma, Jetse and Laut, Peter, "The Hunter Region" in Birrell, Robert, Hill, Doug and Stanley, John (eds), *Quarry Australia: Social and Environmental Perspectives on Managing the Nation's Resources*, Oxford University Press, Melbourne, 1982.

Katz, Richard, *Japan: The System that Soured*, M.E. Sharpe, Armonk, N.Y., 1998.

Kauffman, Paul, *Wik, Mining and Aborigines*, Allen & Unwin, St Leonards, 1998.

Kelly, Paul, *The End of Certainty: The Story of the 1980s*, Allen & Unwin, St Leonards, 1992.

Kelly, Paul, *Triumph and Demise: The Broken Promise of a Labor Generation*, Melbourne University Press, Carlton, 2014.

Kennedy, K.H., *Mining Tsar: The Life and Times of Leslie Urquhart*, Allen & Unwin, North Sydney, 1986.

Kersten, Rikki, "Australia and Japan" in Cotton, James and Ravenhill, John (eds), *The National Interest in a Global Era: Australia in World Affairs 1996–2000*, Oxford University Press, South Melbourne, 2002.

Knox, Malcolm, *Boom: The Underground History of Australia from Gold Rush to GFC*, Viking, Melbourne, 2013.

Kojima, Kiyoshi and Ozawa, Terutomo, *Japan's General Trading Companies: Merchants of Economic Development*, OECD, Paris, 1984.

LaFeber, Walter, *The Clash: US–Japanese Relations Throughout History*, W.W. Norton & Company, New York, 1998.

Langmore, Diane, (ed.), *Australian Dictionary of Biography, Vol. 17*, Melbourne University Press, Carlton, 2007.

Larcombe, Graham, "New South Wales: The Political Economy of an Industrialised State" in Head, Brian, *The Politics of Development in Australia*, Allen and Unwin, North Sydney, 1986.

Lewis, G., "Queensland Nationalism and Australian Capitalism" in Wheelwright, E.L. and Buckley, K. (eds), *Essays in the Political Economy of Australian Capitalism, Vol. 2*, Australia and New Zealand Book Co., Sydney, 1978.

Libby, Ronald, T. *Hawke's Law: The Politics of Mining and Aboriginal Land Rights in Australia*, University of Western Australia Press, Nedlands, 1989.

Lucarelli, B., *Australia's Black Coal Industry: Past Achievements and Future Challenges*, Program on Energy and Sustainable Development, Stanford University, 2011.

Manners, Gerald, *The Changing World Market for Iron Ore, 1950–1980: An Economic Geography*, The Johns Hopkins University Press, Baltimore, 1971.

Mark, Chi-Kwan, *China and the World since 1945: An International History*, Routledge, London and New York, 2012.

McDonald, T.K., *The Aluminium Industry: An Australian Perspective*, Comalco, Melbourne, 1984.

McKern, R.B., *Multinational Enterprise and Natural Resources Policy*, McGraw Hill, Sydney, 1976.

McEwen, John, *John McEwen: His Story*, Privately Published, Canberra, 1982.

McIlwraith, John, *Mesa Harvest: Robe River's First 25 Years*, Robe River Iron Associates, Perth, 1997.

McIllwraith, John and Woldendorp, Richard, *Hamersley Iron Twenty Five Years*, Hamersley Iron, Perth, 1997.

McLean, Ian W., *Why Australia Prospered: The Shifting Sources of Economic Growth*, Princeton University Press, Princeton, 2013.

McMinn, W.G., *A Constitutional History of Australia*, Oxford University Press, Melbourne, 1979.

McQueen, Humphrey, *Gone Tomorrow: Australia in the 80s*, Angus and Robertson, Sydney, 1982.

Megalogenis, George, *The Australian Moment: How We were Made for These Times*, Viking, Melbourne, 2012.

Mellor, D.P., *The Role of Science and Industry*, Australian War Memorial, Canberra, 1958.

Meredith, David and Dyster, Barrie, *Australia in the Global Economy: Continuity and Change*, Cambridge University Press, New York and Melbourne, 1999.

Millmow, Alex, "Australia and the Keynesian Revolution" in Samuel Furphy (ed.), *The Seven Dwarfs: the Age of the Mandarins*, ANU Press, Canberra, 2015.

Munck, Ronalso (ed.), *Labour and Globalisation: Results and Prospects*, Liverpool University Press, Liverpool, 2004.

Neale, R.G. (ed), *Documents on Australian Foreign Policy. Vol. I 1937–38*, Australian Government Publishing Service, Canberra, 1975.

Nolan, Melanie (ed.), *Australian Dictionary of Biography, Vol. 18*, Melbourne University Press, Carlton, 2012.

O'Faircheallaigh, Ciaran *Mining and Development: Foreign-Financed Mines in Australia, Ireland, Papua New Guinea and Zambia*, Croom Helm, Sydney, 1984.

O'Faircheallaigh, Ciaran, "Minerals and Energy Policy" in Jennett, Christine and Stewart, Randall G., (eds), *Hawke and Australian Public Policy: Consensus and Restructuring*, Macmillan, South Melbourne, 1990, pp. 137–55.

Österhammel, Jurgen, *The Transformation of the World: A Global History of the Nineteenth Century*, Princeton University Press, Princeton, 2014.

Parbo, Arvi, *Down Under: Mineral Heritage in Australasia*, Australian Institute of Mining and Metallurgy, Box Hill, 1992.

Patience, A. and Head, B.W., (eds), *From Whitlam to Fraser*, Oxford University Press, Melbourne, 1979.

Patience, Allan (ed.), *The Bjelke-Petersen Premiership 1968–1983: Issues in Public Policy*, Longman Cheshire, Melbourne, 1985.

Pearse, Guy, McKnight, David and Burton, Bob, *Big Coal: Australia's Dirtiest Habit*, University of New South Wales Press, Kensington, 2013.

Peck, Merton, J., *Competition in the Aluminium Industry, 1945–1958*, Harvard University Press, Cambridge MA., 1961.

Phillipson, Neil, *Hancock: Man of Iron*, Wren Publishing, Melbourne, 1974.

Pike, Francis, *Hirohito's War: The Pacific War 1941–1945*, Bloomsbury, London, 2015.

Pilling, David, *Bending Adversity: Japan and the Art of Survival*, Penguin, London, 2014.

Pinkstone, Brian, *Global Connections: A History of Exports and the Australian Economy*, Australian Government Publishing Service, Canberra, 1992.

Posner, Richard A., *A Failure of Capitalism: The Crisis of '08 and the Descent to Depression*, Harvard University Press, Cambridge MA, 2009.

Priest, Joan, *The Thiess Story*, Boolarong Publications, Ascot, Queensland, 1981.

Raggatt, Harold, *Mountains of Ore*, Landsdowne Press, Melbourne, 1968.

Reynolds, Terry S. and Dawson, Virginia P., *Iron Will: Cleveland-Cliffs and the Mining of Iron Ore, 1847–2006*, Wayne State University Press, Detroit, 2011.

Ritchie, John (ed.), *Australian Dictionary of Biography, Vol. 12*, Melbourne University Press, Carlton, 1990.

Ritchie, John (ed.), *Australian Dictionary of Biography, Vol. 13*, Melbourne University Press, Carlton, 1993.

Ritchie, John (ed.), *Australian Dictionary of Biography, Vol. 14*, Melbourne University Press, Carlton, 1996.

Ritchie, John (ed.), *Australian Dictionary of Biography, Vol. 15*, Melbourne University Press, Carlton, 2000.

Ritchie, John and Langmore, Diane (eds), *Australian Dictionary of Biography, Vol. 16*, Melbourne University Press, Carlton, 2002.

Rix, Alan, *Coming to Terms: The Politics of Australia's Trade with Japan 1945–57*, Allen & Unwin, North Sydney, 1986.

Saddler, Hugh, *Energy in Australia: Politics and Economics*, Allen & Unwin, Sydney, 1981.

Sexton, Michael, *Illusions of Power: the Fate of a Reform Government*, George Allen and Unwin, North Sydney, 1979.

Shaw, Murray, *Moving Mountains: The Evolution of Port Hedland Harbour*, Hesperian Press, Carlisle, 2006.

Sheller, Mimi, *Aluminum Dreams: The Making of Light Modernity*, Massachusetts Institute of Technology Press, Cambridge, 2014.

Shields, Jerry, *The Invisible Billionaire, Daniel Ludwig*, Houghton Mifflin, Boston, 1986.

Smith, George David, *From Monopoly to Competition: The Transformations of Alcoa, 1888–1986*, Cambridge University Press, Cambridge, 1986.

Smith, Gary, "Minerals and Energy" in Patience, Allan and Head, Brian (eds), *From Whitlam to Fraser: Reform and Reaction in Australian Politics*, Oxford University Press, Melbourne, 1979.

Smith, Keith R., *The Great Challenge: The Saga of Yampi*, Keith R. Smith, Perth, 1978.

Spillman, Ken, *A Rich Endowment: Government and Mining in Western Australia*, University of Western Australia Press, Nedlands, 1993.

Stockwin, J.A.A. (ed.), *Japan and Australia in the Seventies*, Angus and Robertson, Sydney, 1972.

Stuart, Roger, "Resources Development Policy: The Case of Queensland's Export Coal Industry" in Patience, Allan (ed.), *The Bjelke-Petersen Premiership: 1968–1983: Issues in Public Policy*, Longman Cheshire, Melbourne, 1985.

Sumiya, Mikio (ed.), *A History Japanese Trade and Industry Policy*, Oxford University Press, Oxford.

Sykes, Trevor, *The Money Miners: Australia's Mining Boom 1969–1970,* Wildcat Press, Sydney, 1978.

Thompson, Peter and Robert Macklin, *The Big Fella: The Rise and Rise of BHP Billiton*, William Heinemann Australia, Sydney, 2009.

Trengove, Alan, *Adventure in Iron: Hamersley's First Decade*, Stockwell Press, Clayton, 1976.

Trengove, Alan, *Discovery: Stories of Modern Mineral Exploration*, Stockwell Press, Clayton, 1979.

Tsokhas, Kosmas, *Beyond Dependence: Companies, Labour Process and Australian Mining*, Oxford University Press, Melbourne, 1986.

Tsokhas, Kosmas, *A Class Apart? Businessmen and Australian Politics 1960–1980*, Oxford University Press, Melbourne, 1984.

Turner, John, *200 Years of Transport in the Hunter: A History of Transport's Contribution to the Hunter*, Chartered Institute of Transport and Logistics, Australia, Charlestown, 2005.

Upton, David, *The Olympic Dam Story: How Western Mining Defied the Odds to Discover and Develop the World's Largest Mineral Deposit*, Upton Financial PR, Armadale, 2010.

Uren, David, *The Kingdom and the Quarry*, Black Inc, Collingwood, 2012.

Vamplew, Wray (ed.), *Australians: Historical Statistics*, Fairfax, Syme and Weldon Associates, Sydney, 1987.

Voynick, Stephen M., *Climax: the History of Colorado's Climax Molybdenum Mine*, Mountain Press Club Co., Missoula, 1996.

Welfield, John, "Australia's Relations with Japan and the Korean Peninsula" in Boyce, P.J., and Angel, J.R. (eds), *Diplomacy in the Marketplace: Australia in World Affairs 1981–90*, Longman Cheshire, Melbourne, 1992.

West, Richard, *River of Tears, The Rise of the Rio Tinto Zinc Corporation Ltd*, Earth Island Corporation, London, 1972.

Wheelwright, E.L. and Buckley, Ken (eds), *Essays in the Political Economy of Australian Capitalism vol. 4*, ANZ Book Company, Sydney, 1980.

Whitmore, R.L., *Coal in Queensland: the First Fifty Years, University of Queensland Press*, St Lucia, 1981.

Wilkinson, Rick, *Rocks to Riches: The Story of Australia's National Geological Survey*, Allen & Unwin, Sydney, 1986.

Wilson, Jeffrey D., *Governing Global Production: Resource Networks in the Asia-Pacific Steel Industry*, Palgrave Macmillan, Basingstoke, 2013.

Woodcock, J.T., (ed.), *Mining and Metallurgical Practices in Australasia: The Sir Maurice Mawby Volume*, Australian Institute of Mining and Metallurgy, Melbourne, 1980.

Yergin, Daniel, *The Quest: Energy, Security, and the Remaking of the Modern World*, The Penguin Press, New York, 2011.

Yule, Peter, *Ian Potter: Financier, Philanthropist and Patron of the Arts*, Miegunyah Press, Melbourne, 2006.

Yule, Peter, *William Laurence Baillieu: Founder of Australia's Greatest Business Empire*, HardieGrant Books, Richmond, 2012.

Zeiler, Thomas W., "Opening Doors in the World Economy" in Iriye, Akira (ed.), *Global Interdependence*, Harvard University Press, Cambridge MA., 2013.

Articles

Ellem, Bradon, "Robe River Revisited: Geohistory and Industrial Relations", *Labour History* 109, November 2015, pp. 111–31.

Frost, D., "The Revitalisation of Queensland Railways through Export Coal Shipments", *Journal of Transport History*, Vol. 5, no. 2, 1984, pp. 47–56.

Lee, David, "Comalco's Development of Queensland Bauxite, 1955-1975", *Australian Journal of Politics and History*, Vol 62, Num 1, 2016, pp. 44-58.

Lee, David, "The Development of Bauxite at Gove, 1955-1975", *Journal of Australasian Mining History*, Vol. 12, October 2014, pp. 131-47.

Lee, David, "The Emergence of Iron Ore Giants: Hamersley Iron and the Mount Newman Company", *Journal of Australasian Mining History*, Vol. 11, no. 1, 2013, pp. 61-77.

Lee, David, "Reluctant Relaxation: The End of the Iron Ore Export Embargo and the Origins of Australia's Mining Boom", *History Australia*, Vol. 10, no. 3, 2013, pp. 149-170.

Lee, David, "The 1949 Federal Election: A Reinterpretation", *Australian Journal of Political Science*, Vol. 29, no. 3, November 1994, pp. 501-30.

McDonald, K.D., "The Negotiation and Enforcement of Agreements with State Governments Relating to the Development of Mineral Ventures", *Australian Mining and Petroleum Law Journal*, Vol. 1 1977, pp. 29-47.

Reynolds, Wayne, "Australia's Quest to Enrich Uranium and the Whitlam Government's Loans Affair", *Australian Journal of Politics and History*, Vol. 54, no. 4, pp. 362–78.

Richardson, Peter, "The Origins and Development of the Collins House Group, 1915-1951", *Australian Economic History Review*, vol. 27, no. 1, March 1987, pp. 3-29.

Sukugawa, Paul, "Is Iron Ore Priced as a Commodity? Past and Current Pactice", *Resources Policy*, 35, 2010, pp. 54–63.

Walker, William, "The Genesis of Heavy Haul Railroads in the Pilbara", *Journal of Australian Mining History*, Vol. 13, October 2015, pp. 198-222.

Warell, Lisa, "A horizontal merger in the iron ore industry: an event study approach", Resource Policy, Vol. 32, 2007, pp. 191-204.

White, Jon, "The Port Kembla Pig Iron Strike of 1938", Labour History, Vol. 37, November 1979, pp. 63-77.

Monographs and Papers

Booth Connell-Hatch, *The Pilbara Iron Ore Industry: A Report for the Department of Resources Development Reviewing Issues Likely to Affect the Industry to the Year 2000*, Booth Connell Hatch, Western Australia, 1983.

Breaden, P.T. and Parker, J.B., "Financing the Channar Project" The AisIMM Annual Conference, Perth-Kalgoorlie, May 1989.

Comalco Limited Submission to the Senate Standing Committee on Natural Resources, *The Development of the Bauxite, Alumina and Aluminium Industries*, Comalco, Melbourne, 1981.

Davidson, Sinclair and de Silva, Ashton, *The Australian Coal Industry – Adding Value to the Australian Economy*, paper commissioned by the Australian Coal Association, April 2013.

Dufty, N.F., *Industrial Relations in the Pilbara Iron Ore Industry*, Western Australian Institute of Technology, Perth, 1984.

Lee, David, *Iron Country: Unlocking the Pilbara*, A Public Policy Analysis Produced for the Minerals Council of Australia, Minerals Council of Australia, Canberra, 2015.

Lucarelli, Bart, *Australia's Black Coal Industry: Past Achievements and Future Challenges*,

PESD Stanford, Working Paper, March 2011.
Productivity Commission, *The Australian Black Coal Industry: Inquiry Report: Volume 2: Appendices, Report No. 1*, 3 July 1998.
Reserve Bank of Australia, Occasional Paper No. 8, *Australian Economic Statistics, 1949-1950 to 1996-1997.*
Ryan, Janette, *The Development of the Australian Aluminium Industry*, Occasional Paper No 6, Transnational Corporations Research Project, Sydney, 1984.
Siddique, M.A.B., "Western Australia–Japan Mining Cooperation: An Historical Overview", University of Western Australia Business School, 2009.
Smith, Ben, *The Role of Resource Development in Australia's Economic Growth*, Centre for Economic Policy Research, Occasional Paper No. 167, Australian National University, August 1987.

Acknowledgments

This book arose from a research question: to explain how in the early 1960s the Australian government was so alarmed about the prospective entry of its major trading partner, Britain, into the European Economic Community but why its concerns had been substantially assuaged by the early 1970s. The answer appeared to lie in the extraordinary transformation wrought in Australia by the development of mining. From an effort to answer this question, the modern history of Australian mining became my major personal research project between 2012 and 2016 in addition to official historical work for the Department of Foreign Affairs and Trade. Throughout this period I incurred many debts to people and organisations. I gratefully acknowledge the National Library of Australia and the Minerals Council of Australia for supporting my research through a Minerals Council Fellowship, awarded by the National Library of Australia in 2013. This gave me an invaluable opportunity to conduct research in repositories in Canberra. A public lecture presented in association with the fellowship put me in touch with many people in the industry. I am also grateful to the staff at the National Archives of Australia, the University of Melbourne Archives, the Noel Butlin Archives, Australian National University, and the State Records Office of Western Australia for facilitating necessary archival research into the mining industries. The H.V. Evatt Library, Department of Foreign Affairs and Trade, was especially helpful. In 2015 the Department of Foreign Affairs and Trade supported a secondment to the Minerals Council of Australia to research the history of the iron ore industry, an essential part of this larger study of Australian mining.

I thank David Keating and Julie McDonald for their editorial work on the manuscript. Frank Bongiorno and David Moore were generous in commenting on drafts. I also thank Mike Adams, Nicholas Brown, Selwyn Cornish, James Cotton, Neal Davis, Kent Fedorowich, Jeremy Hearder,

John Kunkel, Christopher Lang, John and Sonia Lee, Terry Larkin, David Lowe, Chris McSweeney, Humphrey McQueen, Chad Mitcham, John Nethercote, Gregory Pemberton, Wayne Reynolds, Rob Schaap, Stuart Macintyre, John Rogis, Max Suich and Christopher Waters for general comments and observations. I thank Dr Brendan Whyte for drawing the maps and Sherry Quinn for compiling the index. From Connor Court I gratefully acknowledge Dr Anthony Cappello and his wife Julie Cappello for bringing the book into print. The responsibility for any errors, however, is mine only. The work was written in a private capacity and does not represent the views of the Department of Foreign Affairs and Trade.

Index

www.ingramcontent.com/pod-product-compliance
Lightning Source LLC
Chambersburg PA
CBHW060309100426
42812CB00003B/715

* 9 7 8 1 9 2 5 5 0 1 1 4 8 *